Nanotechnology in Modern Animal Biotechnology

Concepts and Applications

Nanotechnology in Modern Animal Biotechnology
Concepts and Applications

Edited by

PAWAN KUMAR MAURYA, PHD
Associate Professor
Department of Biochemistry
Central University of Haryana
Mahendergarh, India

SANJAY SINGH, PHD
Associate Professor
Division of Biological and Life Sciences
School of Arts and Sciences
Central Campus, Ahmedabad University
Ahmedabad, Gujarat, India

ELSEVIER

NANOTECHNOLOGY IN MODERN ANIMAL BIOTECHNOLOGY   ISBN: 978-0-12-818823-1

Copyright © 2019 Elsevier Inc. All rights reserved.

Publisher: Oliver Walter
Acquisition Editor: Sean Simms
Editorial Project Manager: Thomas Van der Ploeg
Production Project Manager: Sreejith Viswanathan
Cover Designer: Alan Studholme

3251 Riverport Lane
St. Louis, Missouri 63043

List of Contributors

Taru Aggarwal
Amity Institute of Biotechnology
Amity University Uttar Pradesh
Noida, Uttar Pradesh, India

Homica Arya
Biological and Life Sciences
School of Arts and Sciences
Ahmedabad University
Central Campus
Ahmedabad, Gujarat, India

Aditya Arya, PhD
Staff Scientist
Centre for Bioinformatics
Computational and Systems Biology
Pathfinder Research and Training Foundation
Greater Noida, Uttar Pradesh, India

Malvika Bakshi
Biological and Life Sciences
School of Arts and Sciences
Ahmedabad University
Central Campus
Ahmedabad, Gujarat, India

Ivneet Banga
Amity Institute of Biotechnology
Amity University
Noida, Uttar Pradesh, India

Department of Bioengineering
University of Texas
Dallas, TX, United States

Subhashini Bharathala
Centre for Biomedical Technology
Amity Institute of Biotechnology
Amity University
Noida, Uttar Pradesh, India

Sagarika Biswas, PhD
Senior Principal Scientist
Department of Genomics & Molecular Medicine
CSIR-Institute of Genomics & Integrative Biology
New Delhi, India

Pranjal Chandra, MSc, MTech, PhD
Laboratory of Bio-Physio Sensors and
    Nanobioengineering
Department of Biosciences and Bioengineering
Indian Institute of Technology Guwahati
Guwahati, Assam, India

Dinesh Kumar Chellappan
Department of Life Sciences
School of Pharmacy
International Medical University
Bukit Jalil, Malaysia

Trudi Collet, PhD
Indigenous Medicines Group
Institute of Health & Biomedical Innovation
Queensland University of Technology
Brisbane, QLD, Australia

Vikram Dalal
Department of Biotechnology
Indian Institute of Technology Roorkee
Roorkee, Uttarakhand, India

Shrusti Dave
Biological and Life Sciences
School of Arts and Sciences
Ahmedabad University
Central Campus
Ahmedabad, Gujarat, India

Kamal Dua, PhD
Discipline of Pharmacy
Graduate School of Health
University of Technology Sydney
Broadway, NSW, Australia
Priority Research Centre for Healthy Lungs
University of Newcastle & Hunter Medical
    Research Institute
Newcastle, NSW, Australia
Centre for Inflammation
Centenary Institute
Sydney, NSW, Australia

Akhil Gajipara
Biological and Life Sciences
School of Arts and Sciences
Ahmedabad University
Central Campus
Ahmedabad, Gujarat, India

Sonu Gandhi
DBT-National Institute of Animal Biotechnology
Hyderabad, Telangana, India

Risha Ganguly
Department of Biochemistry
University of Allahabad
Allahabad, Uttar Pradesh, India

Anamika Gangwar, MSc
Defence Institute of Physiology and Allied Sciences
Delhi, India

Monica Gulati, PhD
School of Pharmaceutical Sciences
Faculty of Applied Medical Sciences
Lovely Professional University
Phagwara, Punjab, India

Ashutosh Gupta
Department of Biochemistry
University of Allahabad
Allahabad, Uttar Pradesh, India

Gaurav Gupta
School of Pharmaceutical Sciences
Jaipur National University
Jagatpura, Rajasthan, India

Philip Michael Hansbro, PhD
School of Life Sciences
Faculty of Science
University of Technology Sydney (UTS)
Ultimo, NSW, Australia
Centre for Inflammation
Centenary Institute
Sydney, NSW, Australia
Priority Research Centre for Healthy Lungs
University of Newcastle & Hunter Medical Research
    Institute
Newcastle, NSW, Australia

Ashutosh Kumar
Laboratory of Bio-Physio Sensors and Nano
    bioengineering
Department of Biosciences and Bioengineering
Indian Institute of Technology Guwahati
Guwahati, Assam, India

Ramesh Kumar
Department of Biochemistry
University of Allahabad
Allahabad, Uttar Pradesh, India

Amit Kumar, PhD
International Research Centre
Sathyabama Institute of Science and Technology
Chennai, Tamil Nadu, India

Kuldeep Mahato
Laboratory of Bio-Physio Sensors and Nano
    bioengineering
Department of Biosciences and Bioengineering
Indian Institute of Technology Guwahati
Guwahati, Assam, India

Pawan Kumar Maurya, PhD
Associate Professor
Department of Biochemistry
Central University of Haryana
Haryana, India

V.G.M. Naidu
Department of Pharmacology and Toxicology
NIPER Guwahati
Guwahati, Assam, India

Mastan Mukram Naikwade
Department of Pharmacology and Toxicology
NIPER Guwahati
Guwahati, Assam, India

Rony Nunes, SPC
Laboratory of Extracellular Matrix and Gene Expression
  Regulation
Department of Structural and Functional Biology —
  Institute of Biology — UNICAMP, Brazil

Brian Oliver, PhD
School of Life Sciences
Faculty of Science
University of Technology Sydney (UTS)
Ultimo, NSW, Australia

Abhay K. Pandey
Department of Biochemistry
University of Allahabad
Allahabad, Uttar Pradesh, India

Akhilesh Kumar Pandey
Department of Surgery and Neurology
Texas Tech University Health Sciences Centre
Lubbock, TX, United States

Lilian Cristina Pereira, PhD
Department of Bioprocesses and Biotechnology
College of Agronomy Sciences (FCA)
São Paulo State University (UNESP) - Center for
  Evaluation of Environmental Impact on Human
  Health (TOXICAM)
Botucatu, São Paulo, Brazil

Buddhadev Purohit
Laboratory of Bio-Physio Sensors and Nano
  bioengineering
Department of Biosciences and Bioengineering
Indian Institute of Technology Guwahati
Guwahati, Assam, India

Thania Rios Rossi Lima, MSc
Department of Pathology
Botucatu Medical School
São Paulo State University (UNESP) - Center for
  Evaluation of Environmental Impact on Human
  Health (TOXICAM)
Botucatu, São Paulo, Brazil

Sharmili Roy
Laboratory of Bio-Physio Sensors and
  Nanobioengineering
Department of Biosciences and Bioengineering
Indian Institute of Technology Guwahati
Guwahati, Assam, India

Vijaylaxmi Saxena
Biological and Life Sciences
School of Arts and Sciences
Ahmedabad University
Central Campus
Ahmedabad, Gujarat, India

Juhi Shah
Biological and Life Sciences
School of Arts and Sciences
Ahmedabad University
Central Campus
Ahmedabad, Gujarat, India

Maitri Shah
Biological and Life Sciences
School of Arts and Sciences
Ahmedabad University
Central Campus
Ahmedabad, Gujarat, India

Rutvi Shah
Biological and Life Sciences
School of Arts and Sciences
Ahmedabad University
Central Campus
Ahmedabad, Gujarat, India

Deepshikha Shahdeo
DBT-National Institute of Animal Biotechnology
  (DBT-NIAB)
Hyderabad, Telangana, India

Anmol Shamal
Biological and Life Sciences
School of Arts and Sciences
Ahmedabad University
Central Campus
Ahmedabad, Gujarat, India

Pankaj Sharma, PhD
Professor
Centre for Biomedical Technology
Amity Institute of Biotechnology
Amity University
Noida, Uttar Pradesh, India

Amit Kumar Singh
Department of Biochemistry
University of Allahabad
Allahabad, Uttar Pradesh, India

Sanjay Singh, PhD
Associate Professor
Division of Biological and Life Sciences
School of Arts and Sciences
Central Campus, Ahmedabad University
Ahmedabad, Gujarat, India

Ananya Srivastava
Department of Pharmacology and Toxicology
NIPER Guwahati
Guwahati, Assam, India

Prachi Thakore
Biological and Life Sciences
School of Arts and Sciences
Ahmedabad University
Central Campus
Ahmedabad, Gujarat, India

Noopur Thapliyal
Amity Institute of Biotechnology
Amity University Uttar Pradesh
Noida, Uttar Pradesh, India

Roshika Tyagi
Amity Institute of Biotechnology
Amity University
Noida, Uttar Pradesh, India

Akdasbanu Vijapura
Biological and Life Sciences
School of Arts and Sciences
Ahmedabad University
Central Campus
Ahmedabad, Gujarat, India

Aashna Vyas
Biological and Life Sciences
School of Arts and Sciences
Ahmedabad University
Central Campus
Ahmedabad, Gujarat, India

Ridhima Wadhwa
Faculty of Life Sciences and Biotechnology
South Asian University
New Delhi, India

Kylie Williams, PhD
Discipline of Pharmacy
Graduate School of Health
University of Technology Sydney
Broadway, NSW, Australia

Contents

CHAPTER 1

Biosensors and Their Application for the Detection of Avian Influenza Virus

IVNEET BANGA • ROSHIKA TYAGI • DEEPSHIKHA SHAHDEO • SONU GANDHI

INTRODUCTION

Avian influenza virus (AIV), a type A influenza virus, belongs to the Orthomyxoviridae family. Its genetic makeup includes eight negative-sense, single-stranded RNAs. It is pleomorphic in nature with average particle size ranging between 80 and 120 nm. The viral genome is associated with many different proteins such as nucleoprotein (NP) and transcriptome complex (RNA polymerase proteins such as PB1, PB2, and PA), which together form the ribonucleoproteins (RNPs). The cell surface glycoproteins comprise hemagglutinin (HA) protein and neuraminidase (NA) protein majorly, and a small number of M2 proteins are also expressed, which form the ion channels.

AIVs infect a large volume of the poultry every year and also pose a severe threat to human health. Between 1997 and 2008, the H_5N_1 AIV is said to have led to an economic loss of 10 billion dollars in the poultry industry (Burns et al., 2006). According to reports, AIVs have affected approximately 62 countries worldwide, and the number is increasing at a very fast pace, wherein new cases are being reported in both animals and humans.

Conventional diagnostic techniques used for the detection of AIV include virus isolation and screening, polymerase chain reaction (PCR)—based assays, and enzyme-linked immunosorbent assays (ELISA). However, these techniques are labor-intensive, time-consuming, and less cost-effective and cannot be used for on-site monitoring. To reduce the yearly socioeconomic loss due to AIVs, there is a need for development of advanced techniques that can be used for early diagnostics. Advancements in the field of biosensors technology have paved way for the development of diagnostic platforms that have increased sensitivity and specificity. Biosensors can be used for on-site monitoring and as a rapid detection tool to control disease outbreaks. In this chapter, we have highlighted the molecular biology of AIVs and principle of different biosensors developed and applied for easy and quick detection of AIVs.

AVIAN INFLUENZA VIRUS: MOLECULAR BIOLOGY

Influenza is a contagious disease, which is caused by influenza viruses. Influenza viruses are members of the Orthomyxoviridae family and can be divided into three subtypes, A, B, and C depending on the composition of the nucleoprotein and matrix protein (Murphy and Webster, 1996). AIVs belong to type A influenza virus and are commonly isolated from as well as adapted to an avian host. The virions are quite unstable in the environment and can easily be degraded by action of heat, dryness, and detergents (Swayne and Halvorson, 2003).

The virions are generally 80—120 nm in diameter. The AIV genome comprises negative-sense, single-stranded RNA consisting of eight fragments that encode for 10—11 proteins within the viral envelope. The segments and the proteins that they encode have been described in Table 1.1. However, the arrangement of these viral segments is still under examination and not yet fully understood.

The outermost lipid layer of AIV arises from the plasma membrane of the host in which the virus replicates (Nayak et al., 2009). Thousands of projections are present on the surface of the virus particles in the form of spikes. Around 80% of these spikes are HA protein, whereas the rest are made up of NA proteins. Fig. 1.1 illustrates the structure of the AIV particle. Different strains of type A viruses are found depending on the antigenic differences of HA, cell surface glycoproteins, and NA proteins (Burns et al., 2006). Cell surface glycoproteins such as HA and NA are found in AIV in the ratio of 4:1. Some copies of the M2 proteins are also found in the outer lipid membrane of the virus particle,

Nanotechnology in Modern Animal Biotechnology. https://doi.org/10.1016/B978-0-12-818823-1.00001-6

TABLE 1.1
Single-Stranded RNA Segments Present in AIV and the Proteins They Encode.

Name of the Segment	Protein Encoded	Function
Segment 1	PB$_2$ protein	Forms polymerase subunit
Segment 2	PB$_1$ protein	Forms polymerase subunit
Segment 3	PA protein	Forms polymerase subunit and demonstrates protease activity
Segment 4	HA protein	Forms cells surface glycoproteins
Segment 5	NP protein	Forms RNA binding protein and demonstrates nuclear import regulation
Segment 6	NA protein	Helps in release of virus, and forms surface glycoproteins
Segment 7	M1 protein	Forms matrix protein
Segment 8	NS1 and NS2/NEP protein	Helps in nuclear export of RNA and regulates host gene expression

which play a significant role in maintaining the ion channel activity (Pielak and Chou, 2011). On the other hand, M1 protein, one of the most abundant proteins found inside the virus, aids in the attachment of the RNP. RNP consists of approximately 50 copies of RNA-dependent RNA polymerase and is made up of three proteins— PB1, PB2, and PA (Boivin et al., 2010; Noda et al., 2006).

The replication process of the virus particles mainly takes place inside the nucleus. The various stages involved in the replication of AIV (Fig. 1.2) are divided into five steps:
1. Adsorption of virus, entry inside the cell and uncoating
2. Synthesis of messenger RNA and RNA replication
3. Posttranscriptional processing
4. Translational and posttranslational modifications
5. Assembly and release from the cell

Replication process begins with the binding of AIV particles to the sialic acids present on the cell surface as receptors (Skehel and Wiley, 2000) and are internalized by the process of endocytosis, primarily clathrin-mediated endocytosis. In some cases, non—clathrin- and non—caveolae-mediated endocytosis pathways have also been observed during virus adsorption. During the process of entry of virion inside the cell, signaling process is activated, which further facilitates the process of endocytosis and release of viral RNP content into the cytoplasm (Elbahesh et al., 2014; Stegmann, 2000).

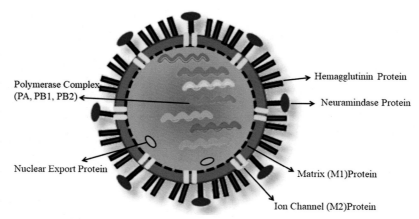

FIG. 1.1 Diagrammatic representation of the avian influenza virus particle. The viral projections are present in the form of spikes and arise from the lipid envelope. Most of these projections resemble rods and are made of hemagglutinin protein, whereas the remaining mushroom-like projections are made of neuraminidase protein. Some copies of M2 protein are also present on the membrane. Presence of matrix protein (M1) plays a crucial role in the attachment of RNP complex.

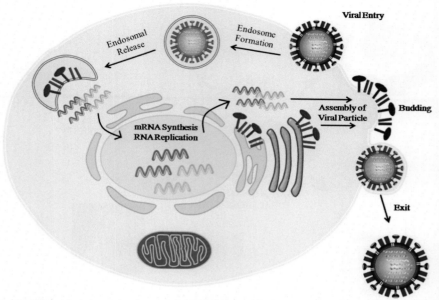

FIG. 1.2 Replication cycle of the viral particle inside the host begins with the endocytic uptake of the viral particle by the host followed by endosomal release of the viral ribonucleoprotein inside the cytoplasm. The RNP is transported to the nucleus where it is used as a template for the synthesis of mRNA and RNA replication. Progeny virions are transported from the nucleus onto the cell surface where they are reassembled, lyse the cell, and released.

Inside the nucleus, this viral RNP content is used as a template for the synthesis of cDNA (complementary DNA) and viral mRNA (messenger RNA). Viral RNA-dependent RNA polymerase, made up of PA, PB1, and PB2, controls the synthesis of mRNA. The nuclear export of viral mRNA utilizes the "machinery" of the host cell, but is selective; export is controlled by the viral nonstructural protein NS1 (Deng et al., 2005).

Viral proteins, mainly NS2, NP, M1, and polymerases, are transported to the nucleus to complete the replication process and aid in assembly of the viral particle. Progeny virions are assembled at the apical surface of the plasma membrane followed by export of newly synthesized RNPs from the nucleus to the plasma membrane to allow their incorporation into budding virions.

Talking about the virulence factors of AIV, there is little information known in hosts other than gallinaceous birds (such as chickens, turkeys, and quail). Recently, evidence has emerged about the importance of nonstructural (NS1) gene in the virulence of AIV (Basler et al., 2001; Li et al., 2006; Quinlivan et al., 2005). Moreover, it has been observed that HA glycoprotein is an important molecular determinant for pathogenicity of AIV. HA, which is synthesized as a precursor molecule, is activated by posttranslational cleavage by host proteases to obtain its full biological properties. What is unique is that the sequence of the proteolytic site is a determining factor of the level of pathogenicity, that is, will the infection be systemic of restricted to systems such as respiratory or enteric (Taubenberger and Morens, 2008).

Conventional techniques used for the detection of AIV are immunohistochemical assays, virus isolation, immunofluorescence antibody (FA) test, ELISA, gene isolation, PCR-based analysis, sequencing, and microarray. However, these molecular diagnostic techniques lack sensitivity and specificity, require trained personnel, cannot be used for on-site monitoring, and are not cost-effective. Therefore, advancements are required in the development of rapid detection methods to reduce socioeconomic burden and provide cost-effective, sensitive, specific, and quick detection technique. Advancements in the field of nanotechnology have led to the development of biosensors that can be used for disease detection or detect the present of target analyte/biomolecule present in biological samples. It is imperative to incorporate the following characteristics (Fig. 1.3) into biosensors so that they can be used as a diagnostic tool to increase its applicability.

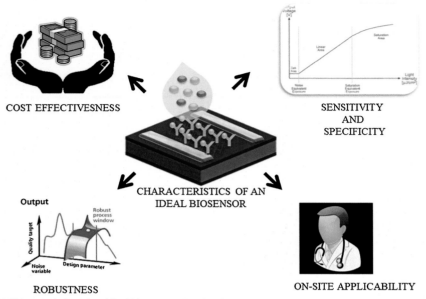

FIG. 1.3 Characteristics of an ideal biosensor. An ideal biosensor should be highly specific in detection of the target analyte. Sensitivity of the biosensor defines its limit of detection and has a great effect on its usability. Development and manufacturing of an ideal biosensor should be cost-effective in nature to increase its onsite applicability.

BASICS OF BIOSENSORS TECHNOLOGY

The term biosensor was pioneered by Clark and Lyons with the development of their oxygen detecting biosensors for blood. It can be defined as diagnostic tool that generates signal by interpreting the biophysical or biochemical interactions with biological components. Constructing a biosensor requires some basic knowledge from various fields of science such as physics, chemistry, biology, microbiology, bioengineering, and nanoscience for the kind of materials, devices, and techniques involved. A biosensor is based on the principle of signal transduction and primarily contains a biosensing layer, an organic element responsible for the selective and specific nature of sensor, present to discern any particular analyte from the interested medium, and a sensor element that represents the signal transducing segment of the biosensors is responsible for the generation of signal and an electronic system for displaying, processing and amplification. The sensing element or transducer can make use of many transducing mechanisms such as optical, magnetic, piezoelectric, and electrochemical. On the basis of their transducer, there are different types of biosensors such as piezoelectric, optical, electrochemical, and thermistor with its own subtype.

The basic principle governing the working of biosensor (Fig. 1.4) is the recognition of a specific analyte (biomolecule) by the receptor and then converting this biorecognition (bioaffinity or biometabolic) into quantifiable signal by transducer. Concentration of the analyte and intensity of generated signals are directly proportional to each other. These signals are then processed by electronic system. For the coupling of bioelement and sensor element, mainly four mechanisms have been used, i.e., physical adsorption, membrane immobilization, covalent amalgamation, and matrix entrapment (Ali et al., 2017). A productive and efficient construction of a high-performance biosensors take into account very crucial factors such as immobilization of bioanalyte in its native form, receptor sites availability for the species of interest, and a productive adsorption of the analyte on the supportive medium.

Biosensors can be classified on the basis of its transducing element (electrochemical, mass dependent, optical, radiation sensitive, and so on), bioanalyte (classes of DNA, glucose, toxins, mycotoxins, drugs, or enzymes), and sensing element (enzyme, nucleic acid, proteins, saccharides, oligonucleotides, ligands, etc.) (Monošík et al., 2012; Turner, 2000). Biosensors have been used in various disciplines such as medicine,

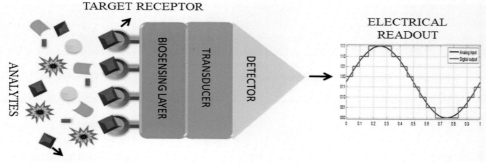

FIG. 1.4 Basic working principle of biosensor technology. Biosensors are based on specific interaction between the target analyte and the recognition element immobilized onto the sensing layer, which leads to the generation of an electrical response that can be correlated with the analyte-recognition molecule interaction and thus can aid in detection.

food, environment, and military (Edelstein et al., 2000; Situ et al., 2010; Tamayo et al., 2001). They provide ease of portability, fast response time, high specificity, less response time, high productivity, and long stability over various conventional methods for analysis.

BIOSENSORS IN DETECTION OF AVIAN INFLUENZA

On the basis of signal transduction principle, three types of biosensors are currently being used for the detection of AIV, viz., electrochemical biosensors, piezoelectric biosensors, and optical biosensors. On the basis of biological component used in the detection of AIV, biosensors can be aptasensors, immunosensors, DNA or RNA probe sensors, or more. This chapter will traverse different types of biosensors used for the detection of AIV, describing their principle with suitable examples that are currently being used.

Piezoelectric Biosensors

Piezoelectric sensors exploit the principle of piezoelectricity, to discern mass changes in the sensing environment. This principle states that an electric dipole is generated known as piezoelectricity, if a mechanical pressure is exerted on an anisotropic material/crystal, i.e., a crystal without symmetry and vice versa (Wang, 2012). Piezoelectricity can be executed by various organic (e.g., polyamides, Rochelle salts) (Pukada et al., 2000; Sawyer and Tower, 1930), inorganic (e.g., zinc oxide, SiO_2) (Lazovski et al., 2012; Meyers et al., 2013), and even biomolecules (e.g., collagen, cellulose) (Rajala et al., 2016; Ravi et al., 2012).

These biosensors are constructed by placing electrodes on either sides of a crystal, and an electric field is applied causing changes in the dimensions or oscillation of crystal, at its natural resonant frequency (Fig. 1.5). A decrease in the resonant frequency is observed with every increase in the mass on the surface

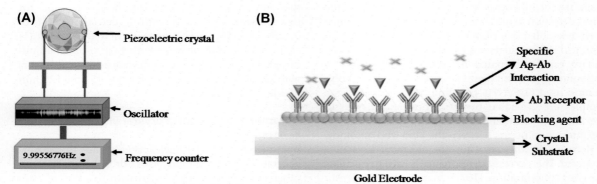

FIG. 1.5 **(A)** Schematic representation of piezoelectric biosensors, **(B)** piezoelectric immunosensor illustrating specific interaction between the antigen and antibody.

TABLE 1.2
Piezoelectric QCM-Based Biosensors for AIV Detection, Along With Their Detection Limit, Detection Time, and Bioreactor-Target Combination.

Biosensor Type	Influenza Subtype	Target	Bioreceptor	Detection Time	Detection Limit	References
Immunosensor	H_3N_2	Virus	Antibody	—	4 virus particles/mL	Owen et al. (2007)
Immunosensor	H_5N_3	Virus	Antibody	—	2.03×10^{-10} M	Wangchareansak et al. (2013)
Immunosensor	H_5N_1	Virus	Antibody	2 h	0.128–12.8 HAU	Li et al. (2011)
Immunosensor	H_1N_1	Virus	Antibody	45 min	10^4 pfu/mL	Peduru Hewa et al. (2009)
Aptasensors	H_5N_1	Virus	Aptamer	1 h	1 HAU	Brockman et al. (2013)
Aptasensors	H_5N_1	Virus	Aptamer	30 min	0.0128 HAU	Wang and Li (2013)

of the crystal because of presence of antibody or binding of antigen.

Among various types of piezoelectric biosensors, quartz crystal imbalance (QCM) is well learned and thus has been intensively used in the detection of AIV. QCM measures mass variation per unit area with even a slight change in the resonant frequency of the crystal. Theses QCM biosensors are further divided on the basis of the bioreceptor used for the detection of AIV (Table 1.2).

Antibody-based QCM sensors used antibody as the bioreceptor and specific immunological interaction occurred between antigen and antibody as shown in Fig. 1.1B. The biorecognition signal produced because of the interaction between antigen and antibody, immobilized on the quartz crystal, is then converted into a quantifiable signal by the transducer.

First report on the detection of AIV using QCM biosensor was given by Owen et al. (2007). They made use of microgravimetric immunosensor for the direct detection of aerolized influenza A virus, in which a self-assembled monolayer (SAM) of mercaptoundecanoic acid is fabricated on a QCM gold electrode so that surface can be prepared for the immobilization of antibodies against influenza virus. The surface immobilization used EDC/NHS coupling. The aerosols of the samples are generated by the nebulizer, attached to the chamber with antibody-immobilized crystal. As the interaction occurs between the virus and antibody, a decrease in the resonant frequency of crystal is observed. This decrease in resonance is used to calculate the virus concentration. Detection range with biosensor for the detection of AIV is 0.02–3.0 HAU with limitation of 4 virus/mL (Owen et al., 2007). A similar type

of QCM-based immunosensor was developed by Wangchareansak et al. (2013), making use of self-assembled glucosamine monolayer for the detection three strains of AIV A, viz., H_5N_3, H_5N_1, H_1N_3 with a sensitivity of Ka 2.03×10^{10} M, Ka 4.35×10^{10} M, and Ka 2.56×10^{10} M, respectively. In 2017, Wang et al. developed another aptasensor for H_5N_1 AIV detection with detection time of 10 min, detection limit of 2^{-4} HAU/50 μL, and detection range of 2^{-4} to 2^4 hemagglutination units (HAUs)/50 μL. They prepared nanoporous gold film of pore size 20 nm and thickness of 120 nm for immobilization on gold electrode. Fabrication of the developed aptasensor was achieved by attachment of H_5N_1-specific single-stranded DNA (ssDNA) aptamer with an NH_2 conjugated 5′-terminal (Wangchareansak et al., 2013).

Li et al. modeled a similar sensor as developed by Owen et al. for AIV detection, but the sensitivity was enhanced by using magnetic nanobeads coated with the anti-H_5N_1 antibodies. The detection time for this biosensor is 1.5 h with a detection limit of 0.128–12.8 HAU and sensitivity of 0.0128 HAU (Li et al., 2011). There are several other QCM-based immunosensors for AIV detection developed over the years with detection time ranging between 45 and 120 min (Table 1.2).

Aptamer-based quartz crystal imbalance sensors

Aptamers are artificial nucleic acid ligands that can be generated against amino acids, drugs, proteins, and other molecules, thereby often used in place of antibodies as a sensing element. There are several reasons for them to be advantageous overusing antibodies such as no use of animals to obtain these aptamers, as they are synthesized in vitro; in nonphysiologic condition, antibodies get irreversibly denatured, whereas denaturation of aptamers is reversible; low immunogenicity of aptamers provides an edge over antibodies, their smaller size makes it easy to access protein epitopes that cannot be accessed by antibodies and producing these aptamers is cheaper and easier than producing a batch of homogeneous antibodies. Therefore, an aptamer offers high affinity for its target molecule, reusability, stability, specificity, and nonimmunogenicity over antibodies. Biosensors using these aptamers as their receptor are known as aptasensors. These aptasensors are developed in vitro using SELEX (systematic evolution of ligands by exponential enrichment) system.

The most important reasons to use the aptamer for constructing AIV-detecting biosensors are (1) the availability of a number of aptamers such as DNA, RNA, and others, developed to prevent or treat influenza (most of these aptamers are against HA protein, restricting virus entry and fusion) and (2) aptasensors give more consistent outcomes because of their temperature stability and boundless shelf life. This section will discuss some examples of aptasensors constructed for AIV detection (Table 1.2).

Brockman et al. (2013) developed a QCM-based aptasensors for the detection of H_5N_1 influenza virus. They immobilized aptamer specific against H_5N_1 AIV on the surface of QCM gold electrode by using biotic and streptavidin conjugation to seize the target virus. They intensified the signal by using magnetic nanobeads as they have large surface to volume ratio and hence can increase the binding rate of target. A significant change in the resonant frequency was observed as AIV concentration increases. Amplification via nanobead was effective till the concentration of the target virus is lower because when the concentration reached to 1 HAU or higher, nanobeads did not prove to be effective in causing change in the frequency. This sensor gives result in a time period of 1 h with a detection limit of 1 HAU.

Wang et al. (2013) developed a hydrogel-based aptasensor for the detection AIV H_5N_1. They used ssDNA cross-linked polymeric hydrogel and hybridized it with high-affinity AIV H_5N_1-specific aptamer to cross-link with the hydrogel. This hybridized gel is then used as a coating material for the QCM sensor. When no virus is present, the gel remains in the shrinking form because of the cross-linker formed owing to hybridization of ssDNA and aptamer, whereas in the presence of virus bound to the antibody, the hydrogel

swells owing to the cessation of the linkage between ssDNA and aptamer. This swelling of gel is detected by the sensor owing to decrease in the resonant frequency. The developed aptasensor obtained highest sensitivity with detection limit of 0.0128 HAU and a detection time of 30 min, which is less than reported immunosensors.

QCM-based piezoelectric biosensors have several advantages over the conventional methods such as the simplicity of protocol, inexpensiveness, sensitivity, low detection level, durability, real time output, label free, and low detection time. But there exist some disadvantages also such as long incubation time, lack of stability, immobilized ligands loss due to multiple washing and regeneration, nonspecific binding of some other targeted biomolecules, and difficulty in sensor surface regeneration. Therefore, it is pertinent that further development should be done to reduce the incubation time and other disadvantages.

Optical Biosensors

Optical biosensors comprise a biorecognition sensing element unified with an optical transducer system. Optical biosensors performed detection via two ways—first, when optical properties of the sensing environment are influenced directly by the analytes, such as in surface plasmon resonance (SPR) or absorption methods, and second, when optical episode produced due to the analyte being tagged, as in fluorescence methods. Various types of optical computation prevail, such as absorption, reflection, fluorescence, chemiluminescence, and phosphorescence, but fluorescence- and SPR-based biosensors are most commonly used techniques for AIVs detection (Table 1.3).

Fluorescence-based optical biosensors

Fluorescence is one of the most widely studied optical phenomena, which includes labeling for the detection of analytes or molecules and provides high sensitivity. Because of this, this phenomenon has gained a lot of attraction for the construction of biosensor system. Various types of fluorescent dyes are used in such biosensor system such as dyes, quantum dots (QDs), and fluorescent proteins.

Pang et al. (2015) constructed a fluorescent aptasensor for the sensitive detection of pathogenic AIV H_5N_1. They layered Ag@SiO2 core-shell nanoparticles with the anti-HA (antihemagglutinin) aptamers. A G-quadruplex complex was formed on attachment of aptamer-HA protein, which captured thiazole orange (TO) and reported the fluorescent signal of TO. It also instigated augmentation in a SPR and acted as a

TABLE 1.3
Types of Optical Biosensors for AIV Detection, With Detection Limit, Detection Time, and Bioreceptor-Target Combination.

Biosensor Type	Influenza Subtype	Target	Bioreceptor	Detection Time	Detection Limit	References
Fluorescence-based aptasensor	H_5N_1	HA	Aptamer	30 min —	2–3.5 ng/mL 0.5 nM–1.0 µM	Pang et al. (2015) Chou and Huang (2012)
Fluorescence-based immunosensor	H_1N_1 H_1N_1 H_3N_3	Virus	Antibody	— — —	10^{-13} g/mL 0.1 pg/mL 50 PFU/mL	Li et al. (2014) Lee et al. (2014)
SPR-based immunosensor	H_5N_1	Virus	Antibody	—	193.3 nM/mL	Wong et al. (2017)
SPR-based aptasensor	H_5N_1 H_1N_1	Virus HA	Aptamer	90 min· —	0.128 HAU 67 fM	Bai et al. (2012) Gopinath et al. (2013)
LRET-based aptasensor	H_7N_9	HA gene	Aptamer	2 h	7 pM	Ye et al. (2014)
Waveguide-mode immunosensor	H_3N_2	Virus HA	Antibody	— —	— —	Gopinath et al. (2010)

metal-enhanced fluorescence sensing. The detection of HA protein can be operated both in aqueous buffer and in human serum with the detection limit of 2 and 3.5 ng/mL, respectively. The total detection time was 30 min. Li et al. (2014) designed a highly sensitive fluorescent immunosensor by using antibodies as a bioreceptor for detection of H_1N_1, with sensitivity up to 1.0×10^{-12} to 1.0×10^{-8} g/mL with a detection limit of 10^{-13} g/mL.

Many optical biosensors use fluorescence resonance energy transfer (FRET) phenomenon, which involves transfer of energy between two light-sensitive molecules, i.e., an acceptor chromophore and a donor. In case of DNA/RNA-based FRET, a nucleic acid probe has a reporter and a quencher coupled at its terminal. A fluorescence quenching occurs when there is no target as these two molecules coupled together to form a duplex, whereas in presence of a target, a conformational change occurs in the probe, which breaks this complex resulting in the emission of fluorescence. A QDs-induced FRET system was designed using two oligonucleotides, which identified two regions of the AIV H5 sequence, as the capture and reporter probes (Chou and Huang, 2012). They were conjugated to QD_{655} (donor) and Alexa Fluor 660 dye (acceptor), respectively. When the concentration of target ranged from 0.5 nM to 1 µM, the QD emission decreased at 653 nm and dye emission increased at 690 nm. Lee et al. (2014) developed a plasmon-assisted fluoroimmunosensor by the linking antibody onto the surface of cadmium telluride QDs and the Au nanoparticle-immobilized on carbon nanotubes (AuCNTs) for the detection of H_1N_1 and H_3N_2, acquiring a detection limit of 0.1 pg/mL and 50 pfu/mL for H_1N_1 and H_3N_2, respectively.

Even with simplicity, high sensitivity, and reduced detection time, fluorescence-based biosensors possess several disadvantages such as quenching of probe. For the application of this fluorescent biosensor as a detector of AIVs from real samples with complex background, such as tracheal and cloaca swabs from birds, further work would be required to combine an effective sample pretreatment method.

Surface plasmon resonance–based optical biosensors

These biosensors use the principle of SPR, which occurs on the surface of conducting material. When a light is illuminated at the interface of two media, it results in the formation of surface plasmon and eventually leads to the decline in the intensity of the reflected light at a resonance angle. This effect is proportionate to the

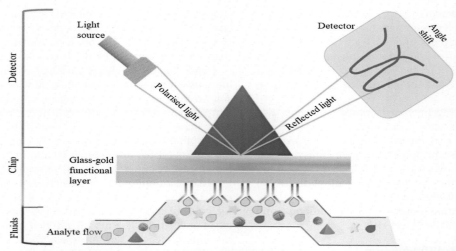

FIG. 1.6 A surface plasmon-based optical biosensors showing a gold plate immobilized with antibodies that when binds to the specific antigen cause change in the angle of reflection of the polarized light incident at the other side of the gold film.

mass of the surface. SPR comprises an optical detector, a gold surface with a sensor chip, and a layer for ligand immobilization unified with a fluidic system so that the sample to be detected can flow through the immobilized layer (Fig. 1.6).

The most favored optical setup for SPR applications is Kretschmann configuration. Any slight variation in the refractive index as a result of attachment of biomolecule in close distance of the sensor is measured by SPR biosensors (Kretschmann, 1971).

Nilsson et al. (2010) used SPR to determine influenza virus for vaccine production via an antibody inhibition assay using HA proteins immobilized on the sensor surface. In 2013, Gopinath et al. designed an SPR-based aptasensor as an applicable tool for detection of AIVs using RNA aptamers targeting HA protein or HA of H_1N_1 AIV. It has a detection limit of K_d^- 67 fM.

Biacore, an automated SPR technique for AIVs detection, provides an edge over other technology as it has the ability to detect even a weak macromolecular interactions. SPR was used to assess the binding of influenza A virus H_1N_1 directly to a neomembrane of bovine brain lipid or an egg yolk lecithin fraction along with the examining of the interactions between influenza HA and glycan or aptamers. Biacore instruments are an advantage for the detection of AIVs, but they are quite expensive and need proper maintenance. In 2012, a portable handheld SPR-based biosensor, namely Spreeta, USA, were designed by Bai et al. (2012) that can be commissioned for the diagnosis of AIV H_5N_1. This sensor used streptavidin-biotin binding

for the fabrication. On the gold surface, first streptavidin was immobilized directly, followed by the immobilization of biotinylated ssDNA. On the binding of targeted analyte or AIVs, increase in the refractive index occurred. This gave result in 90 min with a lower detection limit of 0.128 HAU, when used to detect AIV H_5N_1 in poultry swab.

In 2017, Wong et al. developed a phase-intensity SPR immunosensor and applied it for the detection of H_5N_1 AIV. This sensor has a detection limit of 193.3 ng/mL, which was way better than the Biocare one as described earlier. A pair of polarizer sandwiched the prism sensor head, with a forbidden transmission path and a perpendicular angle of orientation. A sharp phase variation occurs, at SPR, between the p- and s-polarized light causing the rotation of the polarization ellipse of the transmission beam. This leads to the transmission of the light at resonance, and an analogous intensity variation can be measured.

SPR has several advantages over other techniques such as real-time monitoring, ability to replicate measurements, ability to handle complex samples, label-free detection, and small sample size.

Other optical biosensors for AIV detection

A luminescence resonance energy transfer (LRET)—based optical biosensor was also designed by Ye et al. (2014) for H_7N_9 AIV detection, with the use of upconversion nanoparticles (UCNPs) and gold nanoparticles (AuNPs). They functionalized AuNPs surface with H7 HA gene, and amino acids were used as probe or modified complementary sequence. Quenching of fluorescence of UCNPs occurs as the energy donor and acceptor come in propinquity of each other because of the hybridization of H7 gene complementary strands and their respective probe. This sensor has a detection limit of around 7 pM and a long detection time up to 2 h.

There are waveguide-based optical biosensors also that work on the same principle as that of SPR-based sensors with the only difference in their measurement. In case of SPR, conduction of measurement is done using surface mode, whereas waveguide-based sensors use wave mode for measurement. Gopinath et al. (2010) developed two waveguide-based sensors for H_3N_2AIV detection, i.e., immunosensors using antibody and aptasensor specific for HA of H_3N_2.

Electrochemical Biosensors

Electrochemical biosensors are the most widely developed sensors that work on the biochemical reaction because of change in electric current between the biorecognition element and target analyte. Electrochemical can be of four types based on the electric signal generated or on its operating principle:

1. Potentiometric biosensors: These sensors make use of ion-selective electrodes and ion-sensitive field effect transistors to convert biochemical reaction into potential signal.
2. Amperometric biosensors: They worked in two or three electrode configuration. When a constant potential is applied across an electrode, electroactive species are generated due to the interaction between biorecognition element and the target analyte, and the resulting current is then measured. These sensors are more sensitive, selective, and more suitable for mass production as compared with potentiometric.
3. Conductometric biosensors: They measure any variation in the conductivity of the medium.
4. Impedance biosensors: These devices follow either change in impedance or resistance and capacitance or inductance of the material in response to the excitation signal. It uses a low amplitude AC voltage to the sensor electrode and then measuring the in/out-of-phase current response as a function of frequency

In this chapter, we will discuss electrochemical biosensors for AIV detection on the basis of bioelement used, i.e., antibody or DNA or protein (Table 1.4).

Electrochemical immunosensors

These sensors recruit capture antibody, which, on binding to the target analyte and the antibody, generates an electric signal owing to the catalytic reaction of enzyme labeled with the detection antibody, which then can be quantified (Fig. 1.7). The products containing electric charges can be detected by the electrode.

Depending on the signal generated, electrochemical immunosensors can have operating principle of impedance, conductometry, potentiometry, or amperometry. Out of these developed electrochemical immunosensors, impedance-based immunosensor provides the most optimistic approach to detect AIV. Impedance biosensors are further divided into faradic and nonfaradic biosensors. Wang et al. (2009) developed a faradic impedance immunosensors to detect H_5N_1 surface antigen HA. They immobilized polyclonal antibodies on the gold microelectrode via protein A. Binding of the target antigen to the antibody resulted in a change in the impedance of the microelectrode surface. To amplify the signal generated due to the binding of antigen to the antibody, they used red blood cells. This immunosensor had a lower detection limit of 10^3EID/50 mL with detection time of 2 h. In 2011, Wang et al. developed another nonfaradic portable impedance immunosensor for AIV detection, by using magnetic nanobeads along with interdigitated gold microelectrode. This sensor had a detection time of less than 1 h.

An electrochemical impedance spectroscopy—based immunosensor developed by Hassen et al. (2011). This sensor quantified AIVs in the medium containing other viruses and proteins by making use of gold microelectrode, which is immobilized with an antibody-neutravidin-thiol architecture. This had a detection limit of 8 mg/mL. In the same year, an electrochemical impedance spectroscopy—based universal immunosensor was developed for the detection of AIV as well as human influenza virus, i.e., H_5N_1, H_7N_1, H_1N_1, using antibodies again M1 protein of the virus with a detection limit of 20 pg/mL (80—100 virus/μL) (Nidzworski et al., 2014). Electrochemical impedance spectroscopy—based immunosensors attracted more glances among all types of electrochemical biosensors as they are a powerful and nondestructive tool analyzing interfacial properties and biological interaction and their drawbacks such as lower sensitivity and lower detection limit can be overcome by use of microelectrodes.

In 2014, an impedance immunosensor was developed against a novel antigen, i.e., PB1-F2 instead of HA or whole virus with a detection limit of 0.42 nm. They immobilized anit-PB1-F2 antibodies on the copolypyrrole layer via biotin-streptavidin system (Miodek et al., 2014). A low-cost electrode was developed using impedance immunosensor for the detection AIV H5,

TABLE 1.4
Electrochemical Biosensors for AIV Detection, Along With Their Detection Limit, Detection Time, and Bioreactor-Target Combination.

Biosensor Type	Influenza Subtype	Target	Bioreceptor	Detection Time	Detection Limit	References
Faradic impedance immunosensors	H_5N_1	HA	Antibody	2 h	10^3 EID/50 mL	Wang et al. (2009)
Nonfaradic impedance immunosensor	H_5N_2	HA	Antibody	1 h	10^3 EID/50 mL	Wang et al. (2011)
Electrochemical impedance spectroscopy—based immunosensor	H_3N_2	Virus	Antibody	—	8 ng/mL	Hassen et al. (2011)
	H_5N_2, H_7N_1, H_1N_1	M_1 protein		—	20 pg/mL	Nidzworski et al. (2014)
Impedance immunosensor	H_1N_1	PB1-F2	Antibody	—	0.42 nM	Miodek et al. (2014)
	H_5N_1	Virus		1 h	2^{-1}HAU/50 µL	Lin et al. (2010)
	H_7N_1	Virus		—	1.6 pg/mL	Huang et al. (2016)
Electrochemical impedance genosensor	H_7N_9	A fragment of the HA gene sequence	DNA probe		100 fM	Dong et al. (2015)
	H_5N_1	HA	Glycan	—	Attomolar level (aM)	Hushegyi et al. (2015)
	H_7N_9	A fragment of the HA gene	Primer and DNAzyme	—	9.4 fM	Yu et al. (2015)
	H_5N_1	Full lengthH_5 gene	DNA probe	—	0.03 fM (20-mer ssDNA); 0.08 fM (ds DNA)	Grabowska et al. (2014)
Electrochemical amperometric genosensors	H_5N_1	Viral proteins	Aptamer-antibody pair	—	100 fM	Diba et al. (2015)

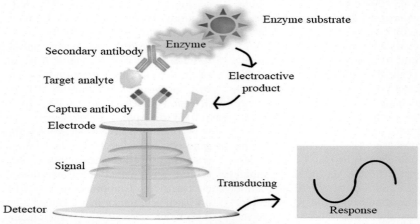

FIG. 1.7 An electrochemical biosensor—a sensor that makes use of antibody or DNA or aptamers to capture target analyte, a detection antibody conjugated with enzyme bind with antibody/DNA/aptamer and addition of the substrate generates electric signal which is detected by the electrode surface.

with a detection time of less than an hour and detection limit of 2^{-1}HAU/50 μL (Lin et al., 2015).

A highly sensitive sandwich immunosensor was developed by Huang et al. (2016), for detection of AIV H7. This sensor recruited a gold microelectrode coated with gold nanoparticles-graphene nanocomposites (AuNPs-g) on which monoclonal antibodies against H7 (capture antibodies) were immobilized, and to amplify the signal, they used polyclonal antibodies against H7 attached on silver nanoparticles-graphene nanocomposite (AgNPs-g) as the detection antibodies. This sensor is highly sensitive, with a lower detection limit of 1.6 pg/mL.

Electrochemical sensors have lots of advantages over conventional methods for AIV detection such as low cost, simplicity, high sensitivity, ease of miniaturization, and lower power requirements, but they also have drawbacks of long-term stability and sensitivity for the sample with intricate backcloth.

Electrochemical genosensors

Genosensors can be defined as a class of electrochemical immunosensors wherein the biologically active recognition element is usually represented by an ssDNA. Efficiency of the developed immunosensors is mainly dependent on the method of immobilization used for the fabrication of the transducer. Various immobilization methods have been used till date, linking of ssDNA probes to carboxy-terminated thiols by using carbodiimide chemistry, attachment of self-assembling sulfur containing ssDNA probes on the surface of gold electrodes, using specific biotin-avidin interactions by immobilizing biotinylated DNA probe onto streptavidin-modified electrode (Bonanni and Del Valle, 2010; Davis et al., 2003; Chenet al, 2009; Chung et al., 2011; Grabowska et al., 2013; Grabowska et al, 2014a,b; Liu et al., 2005; Loget et al., 2013; Malecka et al., 2012; Reddy et al., 2012; Stobiecka et al., 2007).

The basic principle of genosensor was originally described in 2012 (Paleček and Bartošík, 2012). The presence of nucleobases (such as guanine, cytosine, or adenine) leads to the changes in electrochemical activity onto the surface of carbon or mercury electrode. Applying this concept for the detection of AIV, an electrochemical immunosensor was developed for the detection of avian influenza using glassy carbon electrode surface as a platform (Zhu et al., 2009). The multiwalled carbon nanotubes-cobalt phthalocyanine (MWNTs—CoPc) nanocomposite and poly(amidoamine) (PAMAM) dendrimer were fabricated onto the electrode surface, and immobilization was done. DNA probes were able to bind to the electrode surface, and

the interactions were measured by differential pulse voltammetry. Readings were taken before and after hybridization of the probe onto the electrode. The electrical readout was based on the oxidation signal of guanine without any external label. The limit of detection for the developed genosensor was observed to be as 1.0 pg/mL.

An electrochemical genosensor using gold electrode as the sensor platform was developed (Dong et al., 2015). The working principle of the sensor was based on the recognition of portion of HA gene sequence. To accomplish this, a DNA tetrahedral probe having long nucleotide sequence made up of complementary DNA that could specifically recognize and bind to the target ssDNA was prepared and immobilized onto the gold electrode. After binding of the target, biotinylated-ssDNA oligonucleotide was added as a detection probe that hybridized with the target, followed by addition of avidin-horseradish peroxidase. Interaction between avidin-horseradish peroxidase with 3,3′,5,5′-tetramethylbenzidine substrate leads to the generation of amperometric signal that can be easily detected and recorded. The developed genosensor is highly specific and sensitive, having a limit of detection as 100 fm.

On the other hand, electrochemical amperometric-based genosensors are based on signal generation because of current generation when potential is applied across the electrodes. This principle has been successfully used to develop gold nanoparticle modified screen printed carbon electrode involving detection by sandwich assay (Diba et al., 2015). Firstly, interaction of DNA-aptamer immobilized onto electrode and H_5N_1 protein was achieved, followed by addition of alkaline phosphatase—linked monoclonal antibody. This resulted in the formation of sandwich complex, which was detected by using 4-amino phenyl phosphate (APP). The enzyme substrate interaction resulted in an electrocatalytic reaction that could be easily detected. The magnitude of current increased with increasing concentration of H_5N_1 with limit of detection observed as 100 fM with linear range of 100 fM to 10 pM.

A common problem associated with electrochemical genosensors is relatively low signal strength. Most often, scientists use redox labels on the surface of electrode to achieve better signal as a consequence of nucleobases electroactivity. Grabowska et al. (2014a,b) developed one such genosensor that included 3-iron bis(dicarbollide) redox active complex. The oligonucleotide probe bearing the redox label was fabricated onto the gold electrode. The oligonucleotide probe was successfully deposited onto the electrode surface using carbodiimide chemistry using EDC/NHS coupled with 3-mercaptopropionic acid SAM. 3-Mercaptopropionic acid SAM was already deposited onto the surface, and the changes in redox activity of Fe (3+) before and after the process of hybridization were recorded and used as signal for detection. The limit of detection for PCR products derived from DNA obtained from H_5N_1 AIV type and 20-mer complementary ssDNA as 0.08 and 0.03 fM, respectively. The developed diagnostic platform was highly sensitive and specific and was able to differentiate between variations in complementary sequences present at different regions in the PCR-amplified products.

To conclude, electrochemical genosensors have been successfully used as a diagnostic tool for the detection of AIV with increased sensitivity and specificity. They also display good limit of detection in the range of sub-femtomolar to nanomolar, which eliminates room for error and aid in early diagnosis of the disease. They can easily replace conventional diagnostic techniques and can lead to early diagnosis and treatment, thereby reducing socioeconomic burden on the society.

CONCLUSION AND FUTURE PROSPECTS

Biosensors, with all their advantages (high sensitivity, simplicity, reasonable cost, portability, selectivity, low volume of sample) and availability of variety of transducers, provide us a *carte blanche* to design sophisticated sensor for the detection of AIV. Conventional methods always lack at one or the other stage (such as detection time, detection limit, portability, price, etc.) and hence always put us few steps back in early detection of virus or the target analyte, whereas biosensors provide comparatively low detection time and limit proving to be more agile.

With all the positives, biosensor still has some setbacks stopping it to be an advanced option to be used for the AIV detection. A biosensor specifically designed as a diagnostic tool needs to have long-term stability and to detect the target antigen or analyte in real sample even with a complex background. However, with recent advances in nanotechnology, we can design more efficient and sensitive biosensors with advanced features and properties such as using nanoparticles (gold or silver), graphene, carbon nanotubes, and sheets will help in elevating the signals leading to increase in the sensitivity and lowering of detection time. Novel conducting materials and nanoparticles belong to one of the most swiftly growing research areas. These new advanced materials hold a very high potential to refine biosensor properties, specifically their sensitivity, hence promising a future with sophisticated biosensors as diagnostic tools for AIV detection.

ACKNOWLEDGMENTS
Sonu Gandhi gratefully acknowledges the financial assistance from the Department of Science and Technology (DST-SERB), New Delhi, India (Grant number-DST/ECR/2016/000075), and Department of Biotechnology (DBT-BIOCARE), New Delhi, India (Grant number-BT/PR18069/BIC/101/574/2016).

REFERENCES
Ali, J., Najeeb, J., Asim Ali, M., Farhan Aslam, M., Raza, A., 2017. Biosensors: their fundamentals, designs, types and most recent impactful applications: a review. Journal of Biosensors & Bioelectronics 08, 1–9. https://doi.org/10.4172/2155-6210.1000235.

Bai, H., Wang, R., Hargis, B., Lu, H., Li, Y., 2012. A SPR aptasensor for detection of avian influenza virus H5N1. Sensors (Switzerland) 12, 12506–12518. https://doi.org/10.3390/s120912506.

Basler, C.F., Reid, A.H., Dybing, J.K., Janczewski, T.A., Fanning, T.G., Zheng, H., Taubenberger, J.K., 2001. Sequence of the 1918 pandemic influenza virus nonstructural gene (NS) segment and characterization of recombinant viruses bearing the 1918 NS genes. Proceedings of the National Academy of Sciences 98, 2746–2751. https://doi.org/10.1073/pnas.031575198.

Boivin, S., Cusack, S., Ruigrok, R.W.H., Hart, D.J., 2010. Influenza A virus polymerase: structural insights into replication and host adaptation mechanisms. Journal of Biological Chemistry 285, 28411–28417. https://doi.org/10.1074/jbc.R110.117531.

Bonanni, A., Del Valle, M., 2010. Use of nanomaterials for impedimetric DNA sensors: a review. Analytica Chimica Acta 678, 7–17. https://doi.org/10.1016/j.aca.2010.08.022.

Brockman, L., Wang, R., Lum, J., Li, Y., 2013. QCM aptasensor for rapid and specific detection of avian influenza virus. Open Journal of Applied Biosensor 2, 97–103. https://doi.org/10.4236/ojab.2013.24013.

Burns, A., Mensbrugghe, D. van der, Timmer, H., 2006. Evaluating the Economic Consequences of Avian Influenza, vol. 1. http://documents.worldbank.org/curated/en/977141468158986545/Evaluating-the-economic-consequences-of-avian-influenza.

Chen, Y., Nguyen, A., Niu, L., Corn, R.M., 2009. Fabrication of DNA microarrays with poly(L-glutamic acid) monolayers on gold substrates for SPR imaging measurements. Langmuir 24, 5054–5060. https://doi.org/10.1021/la804021t.

Chou, C.C., Huang, Y.H., 2012. Nucleic acid sandwich hybridization assay with quantum dot-induced fluorescence resonance energy transfer for pathogen detection. Sensors

(Switzerland) 12, 16660–16672. https://doi.org/10.3390/s121216660.

Chung, D.J., Kim, K.C., Choi, S.H., 2011. Electrochemical DNA biosensor based on avidin-biotin conjugation for influenza virus (type A) detection. Applied Surface Science 257, 9390–9396. https://doi.org/10.1016/j.apsusc.2011.06.015.

Davis, B.C., Shamansky, L.M., Rosenwald, S., Stuart, J.K., Kuhr, W.G., Brazill, S.A., 2003. Independently-addressable micron-sized biosensor elements. Biosensors and Bioelectronics 18, 1299–1307. https://doi.org/10.1016/S0956-5663(03)00071-X.

Deng, T., Sharps, J., Fodor, E., Brownlee, G.G., 2005. In vitro assembly of PB2 with a PB1-PA dimer supports a new model of assembly of influenza A virus polymerase subunits into a functional trimeric complex. Journal of Virology 79, 8669–8674. https://doi.org/10.1128/JVI.79.13.8669-8674.2005.

Diba, F.S., Kim, S., Lee, H.J., 2015. Amperometric bioaffinity sensing platform for avian influenza virus proteins with aptamer modified gold nanoparticles on carbon chips. Biosensors and Bioelectronics 72, 355–367. https://doi.org/10.1016/j.bios.2015.05.020.

Dong, S., Zhao, R., Zhu, J., Lu, X., Li, Y., Qiu, S., Song, H., 2015. Electrochemical DNA biosensor based on a tetrahedral nanostructure probe for the detection of avian influenza a (H7N9) virus. ACS Applied Materials & Interfaces 7, 8834–8842. https://doi.org/10.1021/acsami.5b01438.

Edelstein, R.L., Tamanaha, C.R., Sheehan, P.E., Miller, M.M., Baselt, D.R., Whitman, L.J., Colton, R.J., 2000. The BARC biosensor applied to the detection of biological warfare agents. Biosensors and Bioelectronics 14, 805–813. https://doi.org/10.1016/S0956-5663(99)00054-8.

Elbahesh, H., Cline, T., Baranovich, T., Govorkova, E.A., Schultz-Cherry, S., Russell, C.J., 2014. Novel roles of focal adhesion kinase in cytoplasmic entry and replication of influenza a viruses. Journal of Virology 88, 6714–6728. https://doi.org/10.1128/JVI.00530-14.

Gopinath, S.C.B., Awazu, K., Fujimaki, M., 2010. Detection of influenza viruses by a waveguide-mode sensor. Analytical Methods 2, 1880–1884. https://doi.org/10.1039/c0ay00491j.

Gopinath, S.C.B., Kumar, P.K.R., 2013. Aptamers that bind to the hemagglutinin of the recent pandemic influenza virus H_1N_1 and efficiently inhibit agglutination. Acta Biomaterialia 9, 8932–8941. https://doi.org/10.1016/j.actbio.2013.06.016.

Grabowska, I., Malecka, K., Stachyra, A., Góra-Sochacka, A., Sirko, A., Zagórski-Ostoja, W., Radecki, J., 2013. Single electrode genosensor for simultaneous determination of sequences encoding hemagglutinin and neuraminidase of avian influenza virus type H_5N_1. Analytical Chemistry 85, 10167–10173. https://doi.org/10.1021/ac401547h.

Grabowska, I., Singleton, D.G., Stachyra, A., Góra-Sochacka, A., Sirko, A., Zagórski-Ostoja, W., Radecki, J., 2014a. A highly sensitive electrochemical genosensor based on Co-porphyrin-labelled DNA. Chemical Communications 50, 4196–4199. https://doi.org/10.1039/C4CC00172A.

Grabowska, I., Stachyra, A., Góra-Sochacka, A., Sirko, A., Olejniczak, A.B., Leśnikowski, Z.J., Radecka, H., 2014b. DNA probe modified with 3-iron bis(dicarbollide) for electrochemical determination of DNA sequence of Avian Influenza Virus H_5N_1. Biosensors and Bioelectronics 51, 170–176. https://doi.org/10.1016/j.bios.2013.07.026.

Hassen, W.M., Duplan, V., Frost, E., Dubowski, J.J., 2011. Quantitation of influenza A virus in the presence of extraneous protein using electrochemical impedance spectroscopy. Electrochimica Acta 56, 8325–8328. https://doi.org/10.1016/j.electacta.2011.07.009.

Huang, J., Xie, Z., Xie, Z., Luo, S., Xie, L., Huang, L., Zeng, T., 2016. Silver nanoparticles coated graphene electrochemical sensor for the ultrasensitive analysis of avian influenza virus H7. Analytica Chimica Acta 913, 121–127. https://doi.org/10.1016/j.aca.2016.01.050.

Hushegyi, A., Bertok, T., Damborsky, P., Katrlik, J., Tkac, J., 2015. An ultrasensitive impedimetric glycan biosensor with controlled glycan density for detection of lectins and influenza hemagglutinins. Chemical Communications 51, 7474–7477. https://doi.org/10.1039/c5cc00922g.

Kretschmann, E., 1971. The determination of the optical constants of metals by excitation of surface plasmons. Die Bestimmung Optischer Konstanten von Metallen Durch Anregung von Oberflächenplasmaschwingungen 241, 313–324. https://doi.org/10.1007/BF01395428.

Lazovski, G., Wachtel, E., Lubomirsky, I., 2012. Detection of a piezoelectric effect in thin films of thermally grown SiO_2 via lock-in ellipsometry. Applied Physics Letters 100, 262905. https://doi.org/10.1063/1.4731287.

Lee, J., Ahmed, S.R., Oh, S., Kim, J., Suzuki, T., Parmar, K., Park, E.Y., 2014. A plasmon-assisted fluoro-immunoassay using gold nanoparticle-decorated carbon nanotubes for monitoring the influenza virus. Biosensors and Bioelectronics 64, 311–317. https://doi.org/10.1016/j.bios.2014.09.021.

Li, D., Wang, J., Wang, R., Li, Y., Abi-Ghanem, D., Berghman, L., Lu, H., 2011. A nanobeads amplified QCM immunosensor for the detection of avian influenza virus H5N1. Biosensors and Bioelectronics 26, 4146–4154. https://doi.org/10.1016/j.bios.2011.04.010.

Li, Y., Hong, M., Qiu, B., Lin, Z., Chen, Y., Cai, Z., Chen, G., 2014. Highly sensitive fluorescent immunosensor for detection of influenza virus based on Ag autocatalysis. Biosensors and Bioelectronics 54, 358–364. https://doi.org/10.1016/j.bios.2013.10.045.

Li, Z., Jiang, Y., Jiao, P., Wang, A., Zhao, F., Tian, G., Chen, H., 2006. The NS1 gene contributes to the virulence of H5N1 avian influenza viruses. Journal of Virology 80, 11115–11123. https://doi.org/10.1128/JVI.00993-06.

Lin, J., Lum, J., Wang, R., Tung, S., Hargis, B., Li, Y., Berghman, L., 2010. A portable impedance biosensor instrument for rapid detection of avian influenza virus. Proceedings of IEEE Sensors 1558–1563. https://doi.org/10.1109/ICSENS.2010.5690302.

Lin, J., Wang, R., Jiao, P., Li, Y., Li, Y., Liao, M., Wang, M., 2015. An impedance immunosensor based on low-cost microelectrodes and specific monoclonal antibodies for rapid detection of avian influenza virus H5N1 in chicken swabs. Biosensors and Bioelectronics 67, 546–552. https://doi.org/10.1016/j.bios.2014.09.037.

Liu, S.F., Li, Y.F., Li, J.R., Jiang, L., 2005. Enhancement of DNA immobilization and hybridization on gold electrode modified by nanogold aggregates. Biosensors and Bioelectronics 21, 789–795. https://doi.org/10.1016/j.bios.2005.02.001.

Loget, G., Wood, J.B., Cho, K., Halpern, A.R., Corn, R.M., 2013. Electrodeposition of polydopamine thin films for DNA patterning and microarrays. Analytical Chemistry 85, 9991–9995. https://doi.org/10.1021/ac4022743.

Malecka, K., Grabowska, I., Radecki, J., Stachyra, A., Góra-Sochacka, A., Sirko, A., Radecka, H., 2012. Voltammetric detection of a specific DNA sequence of avian influenza virus H5N1 using HS-ssDNA probe deposited onto gold electrode. Electroanalysis 24, 439–446. https://doi.org/10.1002/elan.201100566.

Meyers, F.N., Loh, K.J., Dodds, J.S., Baltazar, A., 2013. Active sensing and damage detection using piezoelectric zinc oxide-based nanocomposites. Nanotechnology 24. https://doi.org/10.1088/0957-4484/24/18/185501.

Miodek, A., Sauriat-Dorizon, H., Chevalier, C., Delmas, B., Vidic, J., Korri-Youssoufi, H., 2014. Direct electrochemical detection of PB1-F2 protein of influenza A virus in infected cells. Biosensors and Bioelectronics 59, 6–13. https://doi.org/10.1016/j.bios.2014.02.037.

Monošík, R., Streďanský, M., Šturdík, E., 2012. Biosensors – classification, characterization and new trends. Acta Chimica Slovaca 5, 109–120. https://doi.org/10.2478/v10188-012-0017-z.

Murphy, B.R., Webster, R.G., 1996. Orthomyxoviruses. In: Fields, B.N., Knipe, D.M., Howley, P.M. (Eds.), Fields Virology, third ed. Lippincott-Raven, Philadelphia, pp. 1397–1445.

Nayak, D.P., Balogun, R.A., Yamada, H., Zhou, Z.H., Barman, S., 2009. Influenza virus morphogenesis and budding. Virus Research 143, 147–161. https://doi.org/10.1016/j.virusres.2009.05.010.

Nidzworski, D., Pranszke, P., Grudniewska, M., Król, E., Gromadzka, B., 2014. Universal biosensor for detection of influenza virus. Biosensors and Bioelectronics 59, 239–242. https://doi.org/10.1016/j.bios.2014.03.050.

Nilsson, C.E., Abbas, S., Bennemo, M., Larsson, A., Hämäläinen, M.D., Frostell-Karlsson, Å., 2010. A novel assay for influenza virus quantification using surface plasmon resonance. Vaccine 28, 759–766. https://doi.org/10.1016/j.vaccine.2009.10.070.

Noda, T., Sagara, H., Yen, A., Takada, A., Kida, H., Cheng, R.H., Kawaoka, Y., 2006. Architecture of ribonucleoprotein complexes in influenza A virus particles. Nature 439, 490–492. https://doi.org/10.1038/nature04378.

Owen, T.W., Al-Kaysi, R.O., Bardeen, C.J., Cheng, Q., 2007. Microgravimetric immunosensor for direct detection of aerosolized influenza A virus particles. Sensors and Actuators B: Chemical 126, 691–699. https://doi.org/10.1016/j.snb.2007.04.028.

Paleček, E., Bartošík, M., 2012. Electrochemistry of nucleic acids. Chemical Reviews 122, 3427–3481. https://doi.org/10.1021/cr200303p.

Pang, Y., Rong, Z., Wang, J., Xiao, R., Wang, S., 2015. A fluorescent aptasensor for H_5N_1 influenza virus detection based-on the core-shell nanoparticles metal-enhanced fluorescence (MEF). Biosensors and Bioelectronics 46, 527–532. https://doi.org/10.1016/j.bios.2014.10.052.

Peduru Hewa, T.M., Tannock, G.A., Mainwaring, D.E., Harrison, S., Fecondo, J.V., 2009. The detection of influenza A and B viruses in clinical specimens using a quartz crystal microbalance. Journal of Virological Methods 162, 14–21. https://doi.org/10.1016/j.jviromet.2009.07.001.

Pielak, R.M., Chou, J.J., 2011. Influenza M2 proton channels. Biochimica et Biophysica Acta (BBA) – Biomembranes 1808, 522–529. https://doi.org/10.1016/j.bbamem.2010.04.015.

Pukada, E., 2000. History and recent progress in piezoelectric polymers. IEEE Transactions on Ultrasonics, Ferroelectrics, and Frequency Control 47, 1277–1290. https://doi.org/10.1109/58.883516.

Quinlivan, M., Zamarin, D., García-Sastre, A., Cullinane, A., Chambers, T., Palese, P., 2005. Attenuation of equine influenza viruses through truncations of the NS1 protein. Journal of Virology 79, 8431–8439. https://doi.org/10.1128/JVI.79.13.8431-8439.2005.

Rajala, S., Siponkoski, T., Sarlin, E., Mettänen, M., Vuoriluoto, M., Pammo, A., Tuukkanen, S., 2016. Cellulose nanofibril film as a piezoelectric sensor material. ACS Applied Materials and Interfaces 8, 15607–15614. https://doi.org/10.1021/acsami.6b03597.

Ravi, H.K., Simona, F., Hulliger, J., Cascella, M., 2012. Molecular origin of piezo- and pyroelectric properties in collagen investigated by molecular dynamics simulations. Journal of Physical Chemistry 116, 1901–1907. https://doi.org/10.1021/jp208436j.

Reddy, V., Ramulu, T.S., Sinha, B., Lim, J., Hoque, R., Lee, J.H., Kim, C.G., 2012. Electrochemical detection of single nucleotide polymorphism in short DNA sequences related to cattle Fatty acid binding protein 4 gene. International Journal of Electrochemical Science 7, 11058–11067.

Sawyer, C.B., Tower, C.H., 1930. Rochelle salt as a dielectric. Physical Review 35, 269. https://doi.org/10.1103/PhysRev.35.269.

Situ, C., Mooney, M.H., Elliott, C.T., Buijs, J., 2010. Advances in surface plasmon resonance biosensor technology towards high-throughput, food-safety analysis. TRAC Trends in Analytical Chemistry 29, 1305–1315. https://doi.org/10.1016/j.trac.2010.09.003.

Skehel, J.J., Wiley, D.C., 2000. Receptor binding and membrane fusion in virus entry: the influenza hemagglutinin. Annual Review of Biochemistry 69, 531–569. https://doi.org/10.1146/annurev.biochem.69.1.531.

Stegmann, T., 2000. Membrane fusion mechanisms: the influenza hemagglutinin paradigm and its implications for intracellular fusion. Traffic 1, 498–604. https://doi.org/10.1034/j.1600-0854.2000.010803.x.

Stobiecka, M., Ciela, J.M., Janowska, B., Tudek, B., Radecka, H., 2007. Piezoelectric sensor for determination of genetically modified soybean roundup ready in samples not amplified by PCR. Sensors 7, 1462–1479. https://doi.org/10.3390/s7081462.

Swayne, D.E., Halvorson, D.A., 2003. Influenza. In: Saif, Y.M., Fadly, A.M., Glisson, J.R., McDougald, L.R., Nolan, L.K., Swayne, D.E. (Eds.), Diseases of Poultry, twelfth ed. Iowa State University Press, Ames, pp. 135–160 https://himakahaunhas.files.wordpress.com/2013/03/disease-of-poultry.pdf.

Tamayo, J., Humphris, A.D.L., Malloy, A.M., Miles, M.J., 2001. Chemical sensors and biosensors in liquid environment based on microcantilevers with amplified quality factor. Ultramicroscopy 86, 167–173. https://doi.org/10.1016/S0304-3991(00)00082-6.

Taubenberger, J.K., Morens, D.M., 2008. The pathology of influenza virus infections. Annual Review of Pathology 3, 499–522. https://doi.org/10.1146/annurev.pathmechdis.3.121806.154316.

Turner, A.P.F., 2000. Biosensors–sense and sensitivity. Science 290, 1315–1317. https://doi.org/10.1126/science.290.5495.1315.

Wang, R., Li, Y., 2013. Hydrogel based QCM aptasensor for detection of avian influenza virus. Biosensors and Bioelectronics 42, 148–155. https://doi.org/10.1016/j.bios.2012.10.038.

Wang, R., Lin, J., Lassiter, K., Srinivasan, B., Lin, L., Lu, H., Li, Y., 2011. Evaluation study of a portable impedance biosensor for detection of avian influenza virus. Journal of Virological Methods 178, 52–58. https://doi.org/10.1016/j.jviromet.2011.08.011.

Wang, R., Wang, Y., Lassiter, K., Li, Y., Hargis, B., Tung, S., Bottje, W., 2009. Interdigitated array microelectrode based impedance immunosensor for detection of avian influenza virus H5N1. Talanta 79, 159–164. https://doi.org/10.1016/j.talanta.2009.03.017.

Wang, Z.L., 2012. Progress in piezotronics and piezo-phototronics. Advanced Materials 24, 4632–4646. https://doi.org/10.1002/adma.201104365.

Wangchareansak, T., Sangma, C., Ngernmeesri, P., Thitithanyanont, A., Lieberzeit, P.A., 2013. Self-assembled glucosamine monolayers as biomimetic receptors for detecting WGA lectin and influenza virus with a quartz crystal microbalance. Analytical and Bioanalytical Chemistry 405, 6471–6478. https://doi.org/10.1007/s00216-013-7057-0.

Wong, C.L., Chua, M., Mittman, H., Choo, L.X., Lim, H.Q., Olivo, M., 2017. A phase-intensity surface plasmon resonance biosensor for avian influenza a (H5N1) detection. Sensors 17, 2363. https://doi.org/10.3390/s17102363.

Ye, W.W., Tsang, M.K., Liu, X., Yang, M., Hao, J., 2014. Upconversion luminescence resonance energy transfer (LRET)-based biosensor for rapid and ultrasensitive detection of avian influenza virus H7 subtype. Small 10, 2390–2397. https://doi.org/10.1002/smll.201303766.

Yu, Y., Chen, Z., Jian, W., Sun, D., Zhang, B., Li, X., Yao, M., 2015. Ultrasensitive electrochemical detection of avian influenza A (H7N9) virus DNA based on isothermal exponential amplification coupled with hybridization chain reaction of DNAzyme nanowires. Biosensors and Bioelectronics 64, 566–571. https://doi.org/10.1016/j.bios.2014.09.080.

Zhu, X., Ai, S., Chen, Q., Yin, H., Xu, J., 2009. Label-free electrochemical detection of Avian Influenza Virus genotype utilizing multi-walled carbon nanotubes-cobalt phthalocyanine-PAMAM nanocomposite modified glassy carbon electrode. Electrochemistry Communications 11, 1543–1546. https://doi.org/10.1016/j.elecom.2009.05.055.

CHAPTER 2

Nanoparticle-Mediated Oxidative Stress Monitoring

THANIA RIOS ROSSI LIMA, MSC • RONY NUNES, SPC •
LILIAN CRISTINA PEREIRA, PHD

OXIDATIVE STRESS MECHANISMS AND CELLULAR EFFECTS

Free Radicals and Reactive Oxygen and Nitrogen Species

Free radicals are any species that can exist independently and that contains one or more unpaired electrons occupying atomic or molecular orbitals (see Fig. 2.1). In general, free radicals are unstable, short-lived molecules that rapidly react with various compounds and cellular targets. Free radicals can originate through single electron loss or addition to a nonradical compound (Krinsky, 1992). Higher organisms cannot live without oxygen, but even oxygen can be dangerous for their existence, a fact known as "The Oxygen Paradox." The harmful aspect of oxygen is directly related to the fact that each oxygen atom (O) has an unpaired electron in its valence layer, whereas each oxygen molecule (O_2) has two unpaired electrons. In other words, atomic oxygen is a free radical, and the oxygen molecule is a biradical (Krinsky, 1992; Davies, 1995).

Exploration of the science behind atoms and molecules began centuries ago. Studies on organic free radicals started in the 1900s, when the Russian researcher Moses Gomberg (1866 to 1947) discovered triphenylmethyl, a highly reactive free radical that is stable in the absence of air (see Fig. 2.1). Currently, we know that free radicals and other reactive oxygen and nitrogen species (RONS) originate in the cytoplasm, mitochondrion, endoplasmic reticulum, cellular membrane, and nucleus of all aerobic cells. In humans, oxidation-reduction reactions occurring mainly via oxidative phosphorylation generate free radicals. Oxidative phosphorylation is the last step of energy-producing metabolism in nonphotosynthetic aerobic organisms. It takes place in mitochondrion, a double membrane organelle that functions as the "cell powerhouse" because it is the main site producing energy from adenosine triphosphate (ATP). This cellular respiration step consists of two interrelated processes: electron transport chain (ETC) and chemiosmosis.

The superoxide anion radical (O_2^-), the hydroxyl radical ($^\cdot OH^-$), and nitric oxide ($^\cdot NO$) are the main free radicals arising in biological systems. Hydrogen peroxide (H_2O_2) and peroxynitrite (^-ONOO) are not radical species, but they are extremely reactive and have high oxidizing power, too. In addition, in the presence of transition metals such as iron and copper, hydrogen peroxide can form the hydroxyl radical, which is the most reactive radical (see Fig. 2.2).

Electron transport chain

ETC consists in electron transport via reduction and oxidation (redox) reactions happening simultaneously through a series of carrier proteins inserted into the inner mitochondrial membrane. The electrons participating in ETC emerge during the early cellular respiration stages, and the acceptor coenzymes nicotinamide adenine dinucleotide (NADH) and flavin adenine dinucleotide ($FADH_2$) collect these electrons. Next, coenzyme Q (or ubiquinone), cytochromes, and iron-sulfur proteins act as electron carriers along the ETC in four different multienzymatic complexes (Augusto, 2006; Nelson et al., 2013). Each carrier accepts an electron or electron pair from a carrier with lower reduction potential and transfers the electron to a carrier with higher reduction potential, favoring a unidirectional electron flow (Lodish et al., 2000). Complex I (NADH dehydrogenase) catalyzes electron transport from NADH to coenzyme Q (CoQ); after that, electron flows to flavin mononucleotide and proceeds to an iron-sulfur protein. Succinate conversion to fumarate releases two electrons in

Nanotechnology in Modern Animal Biotechnology. https://doi.org/10.1016/B978-0-12-818823-1.00002-8

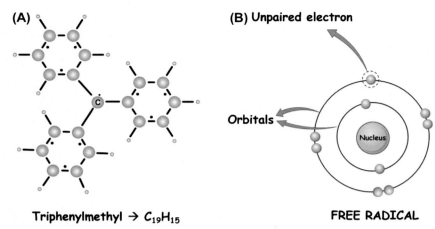

(A)

Triphenylmethyl → $C_{19}H_{15}$

(B) Unpaired electron

Orbitals

Nucleus

FREE RADICAL

FIG. 2.1 **(A)** Structural and molecular formula of the free radical discovered by Moses Gomberg; **(B)** schematic representation of a free radical.

$$M^{(x+)} + H_2O_2 \longrightarrow M^{(x+)} + OH^{\bullet} + OH^{-}$$

FIG. 2.2 Fenton reaction.

complex II (succinate dehydrogenase), and these electrons are transferred to flavin adenine dinucleotide in its quinone form (FAD) and later to an iron-sulfur protein and to CoQ, to yield the reduced $CoQH_2$. Reduced $CoQH_2$ releases two electrons in complex III (ubiquinol-cytochrome c reductase) and becomes oxidized CoQ. These electrons are transferred to an iron-sulfur protein and then follow to two oxidized cytochrome c molecules, to form cytochrome c. Cytochrome c carries electrons to complex IV (cytochrome c oxidase). These electrons are transferred to both copper ions and cytochrome. Finally, they are transferred to molecular oxygen, which is reduced to water (Lodish et al., 2000; Hüttemann et al., 2007) (see Fig. 2.3).

Chemiosmosis

Although electrons are transported through the respiratory chain, the released energy is used to pump H^+ ions out of the mitochondrial matrix and into the intermembrane space, which establishes an electrochemical gradient, also known as proton-motive force. The H^+ ions move across the inner membrane in favor of the electrochemical gradient through the ATP synthase proton channel. Thus, the energy emerging from H^+ ion pumping boosts ATP synthesis from ADP and phosphate (Augusto, 2006; Nelson et al., 2013). The amount of energy released by oxidative phosphorylation is high as compared with other cell metabolism phases; at the end of the process, a total of between 32 and 34 ATP molecules are obtained (see Fig. 2.3).

Along this sequence of events, especially during the passage of electrons by complexes I and III, an electron leak can occur, to afford RONS (Muller et al., 2004; Brand, 2016). The superoxide anion is the primary ROS arising from molecular oxygen monoelectronic reduction. Thereafter, in the presence of Fe^{2+}/Cu, the Fenton reaction generates the highly reactive hydroxyl radical (see Fig. 2.4) (Nelson et al., 2013). Cellular factors including the electron carrier reduction state and posttranslational modification or damage to the respiratory chain alter superoxide anion production, so the latter species can be directly associated with functional mitochondria status (Collins et al., 2012). Although blockage of the electron flow through the chain seems to amplify superoxide anion production, accelerated electron flow promoted by uncoupling reduces RONS (Blajszczak and Bonini, 2017). On the other hand, some studies have shown that the uncoupling mechanism can increase mitochondrial RONS production.

The reactive nitrogen species, derived from ˙NO, originates mainly from L-arginine amino acid oxidation catalyzed by constitutive enzymes called nitric oxide synthase (NOS). Nitric oxide reacts with molecular oxygen to give rise to nitrogen dioxide (˙NO_2) and dinitrogen trioxide (N_2O_3). Alternatively, nitric oxide reaction with the superoxide anion can generate peroxynitrite ($ONOO^-$). High nitric oxide concentrations

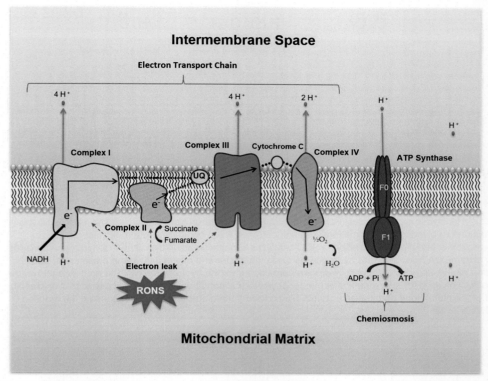

FIG. 2.3 The electron transport chain is a series of electron carriers in the inner mitochondrial membrane. Electrons are passed from NADH to oxygen, moving protons (H$^+$) from the matrix to the intermembrane space and creating a concentration gradient. ATP synthase uses the H$^+$ flow kinetic energy to catalyze ATP synthesis from ADP and phosphate (chemiosmosis).

favor such reactions because the enzyme superoxide dismutase (SOD) competes for the superoxide anion (Augusto, 2006; Na et al., 2008).

To reestablish the redox equilibrium, SOD, which is widely distributed in mitochondria, catalyzes superoxide anion dismutation: this anion is oxidized and reduced to molecular oxygen and hydrogen peroxide. On the other hand, the enzyme catalase (CAT) decomposes hydrogen peroxide into water and molecular oxygen. The redox reaction catalyzed by the enzymes glutathione peroxidase (GPx) and glutathione reductase (GSR) inhibits the action of the generated hydrogen peroxide; reduced glutathione (GSH) is oxidized; and to maintain the GSH cellular levels, NADPH electrons coming mainly from the pentoses-phosphate cycle recycle oxidized glutathione (GSSG) to its reduced form (see Fig. 2.4) (Lu, 2013).

Other endogenous compounds, such as uric acid, modulate RONS concentrations. However, the enzymes SOD, glutathione, and catalase are abundant in mammals, so they are considered the main antioxidant defenses. Antioxidant defenses are generally effective, but imbalance between prooxidants and antioxidants in favor of oxidants can occur, leading to systemic intracellular oxidative stress (see Fig. 2.5) (Sies, 2015).

Oxidative Stress Induced by Exogenous Agents and Its Consequence

As a result of both endogenous and exogenous factors, oxidative stress commonly occurs in organs with high energy demand, such as the brain, the heart, and the liver. Exogenous sources such as cigarette smoke, ionizing radiation, and pesticides can induce oxidative stress by multiple triggers, as described below (Birben et al., 2012; Puppel et al., 2015).

Exposure to ionizing radiation can overproduce ROS directly by water oxidation or indirectly via secondary reactions. The antioxidant status and the amount and availability of activating mechanisms modulate the cellular injury. Alterations in factors affecting these

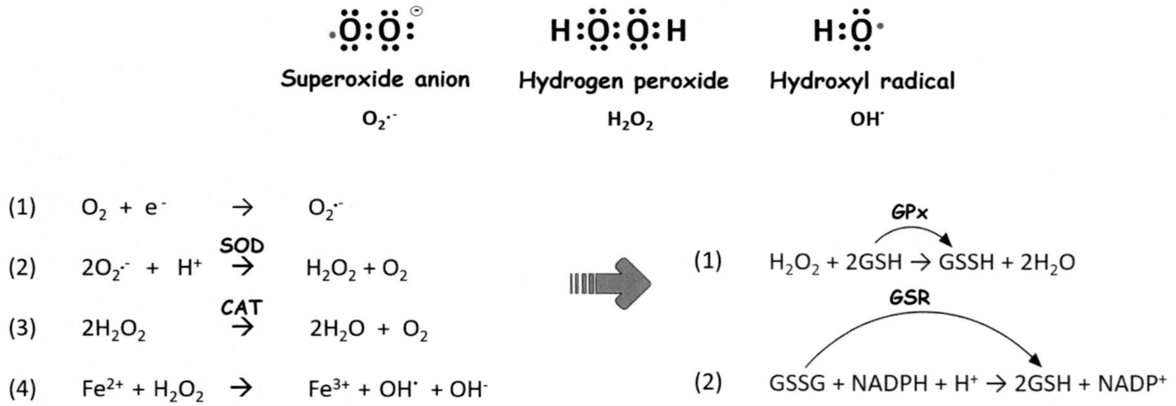

$$\cdot \ddot{O} : \ddot{O} : {}^{\ominus}$$

Superoxide anion

$O_2 {}^{\cdot -}$

$$H : \ddot{O} : \ddot{O} : H$$

Hydrogen peroxide

H_2O_2

$$H : \ddot{O} \cdot$$

Hydroxyl radical

$OH^{\cdot}$

(1) $O_2 + e^- \rightarrow O_2{}^{\cdot -}$

(2) $2O_2{}^{\cdot -} + H^+ \overset{SOD}{\rightarrow} H_2O_2 + O_2$

(3) $2H_2O_2 \overset{CAT}{\rightarrow} 2H_2O + O_2$

(4) $Fe^{2+} + H_2O_2 \rightarrow Fe^{3+} + OH^{\cdot} + OH^-$

(1) $H_2O_2 + 2GSH \overset{GPx}{\rightarrow} GSSH + 2H_2O$

(2) $GSSG + NADPH + H^+ \overset{GSR}{\rightarrow} 2GSH + NADP^+$

FIG. 2.4 Reactive oxygen species (ROS) formed during the electron transport chain. The left side illustrates the reactions involved in ROS formation; (1) the superoxide anion originates from molecular oxygen monoelectronic reduction, (2) the enzyme SOD catalyzes superoxide anion dismutation to hydrogen peroxide, (3) the enzyme CAT decomposes hydrogen peroxide into water and oxygen, (4) in the presence of some transition metals (e.g., iron, copper), hydrogen peroxide forms the hydroxyl radical. The right side shows the reactions involved in the antioxidant effect under physiological conditions; (1) hydrogen peroxide is oxidized by action of the glutathione peroxidase enzyme, (2) oxidized glutathione is recycled to maintain the reduced glutathione levels.

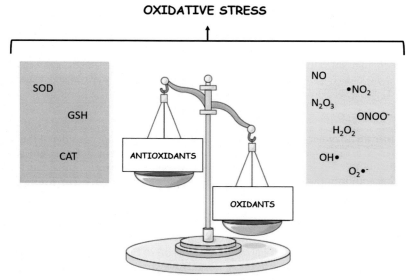

FIG. 2.5 Oxidative stress mechanism. Imbalance between oxidant generation and antioxidant action in favor of oxidants characterizes the oxidative stress condition.

mechanisms of cellular reestablishment may influence the biological response to radiation (Riley, 1994; Bhattacharyya et al., 2014).

Cigarette smoke consists of thousands of chemical substances and oxidizing agents that exist in two forms: gaseous and particulate. The gas phase comprises aldehydes, benzo(a)pyrene, nitric oxide, peroxyl radicals, and carbon-centered radicals; the particulate phase contains relatively stable semiquinones, nicotine, anabasine, phenol, catechol, and hydroquinone, among other components. In the presence of iron, the particulate phase semiquinones can generate hydroxyl radicals and hydrogen peroxide (Bhattacharyya et al., 2014).

Pesticide-induced oxidative stress can result from free radical overproduction, antioxidant defense mechanism changes, or increased lipid peroxidation because of interaction between free radicals and lipids in cellular membranes (Abdollahi et al., 2004; Hernández et al., 2013).

OS is formed in the body owing to external factors and as a result of metabolic processes in the body (endogenous pathway).

Mitochondrial energy metabolism is an important RONS source in most eukaryotic cell types. Although RONS play an essential role in several intra- and extracellular processes and participate in many signaling networks, extreme RONS generation impairs cellular components (i.e., DNA, proteins, carbohydrates, and lipids) (Boelsterli, 2002; Kowaltowski et al., 2009) and has been implicated in countless pathological conditions, including the well-known neurodegenerative Parkinson's and Alzheimer's diseases. Parkinson's disease is associated with dopaminergic (DA) neuron loss in the substantia nigra pars compacta of the brain. Even though the molecular mechanisms underlying DA neurotoxicity remain unclear, oxidative stress is believed to have a crucial part because DA neurons undergo selective degeneration involving ROS autooxidation and microglia activation during neuroinflammatory responses, to produce nitric oxide and superoxide. Molecules released from damaged DA neurons aggravate this condition (Hwang et al., 2012; Blesa et al., 2015).

Age is the major risk factor for Alzheimer's disease. In 1956, Denham Harman described a possible correlation between this disease and oxidative stress and hypothesized that aging results from deleterious cellular effects caused by reactive oxygen species (ROS). Increased ROS production associated with age- and disease-dependent mitochondrial function loss directly affects synaptic activity and neurotransmission in neurons, leading to cognitive dysfunction (Tönnies and Trushina, 2017; Zolkipli-Cunningham et al., 2017). Both Parkinson's and Alzheimer's disease pathogenesis are associated with environmental exposure inducing mitochondrial oxidative stress.

NANOPARTICLE APPLICATIONS IN EXPERIMENTAL BIOMEDICAL RESEARCH

Nanotechnology is a scientific area that develops different types of materials at the nanometric scale. Nanoparticles (NPs) are a broad class of materials that include particulate matter with dimension ranging from 1 to 100 nm and that are classified according to size, morphology, and physicochemical properties (see Fig. 2.6). The best known NP categories are gold (AuNPs) and silver (AgNPs) nanoparticles, which can be biologically synthesized or engineered in the laboratory (Hasan and Hasan, 2015; Khan et al., 2017).

Several natural phenomena can generate particulate matter. NPs originate from chemical, mechanical, or thermal processes in the terrestrial sphere; for example, volcanic eruptions and air currents arising from storms or strong winds. NPs can also be biologically generated through the metabolic activities of microorganisms. In addition, phenomena occurring in the extraterrestrial sphere such as disintegration of meteorites that enter the atmosphere or accumulation of cosmic dust generate NPs (Sharma et al., 2015).

In the laboratory, NPs are engineered by two basic approaches: bottom-up or top-down. The bottom-up strategy involves chemical reactions in the liquid or vapor phase for particles to nucleate and to grow. This strategy includes functionalization by conjugation with bioactive molecules. The top-down approach involves mechanical crushing of the source materials into nanocrystallites by a milling process. The nanoparticle chemical composition and desired features will define which approach will be used (Raab et al., 2011).

NPs can be applied in numerous fields, including the following.

Manufacture

The microelectronics, aerospace, and pharmaceutical industries use NPs in many marketable products because of their unique electrical, mechanical, optical, and image properties. In chemical sensor and biosensor manufacturing, noble metal NPs such as AuNPs and AgNPs are well explored. Moreover, metallic NPs, organic electronic molecules, and ceramics are used in functional link for printers (Khan et al., 2017).

Sustainable Energy

Limitations in fossil fuel sources have called for new strategies to produce renewable energy. In this context, NPs have been used to generate energy from photoelectrochemistry (PEC) and electrochemical water separation, and they have also been used in new nanogenerators that convert mechanical energy into electricity by using a piezoelectric sensor. To maintain product quality, to reduce wear, and to save energy,

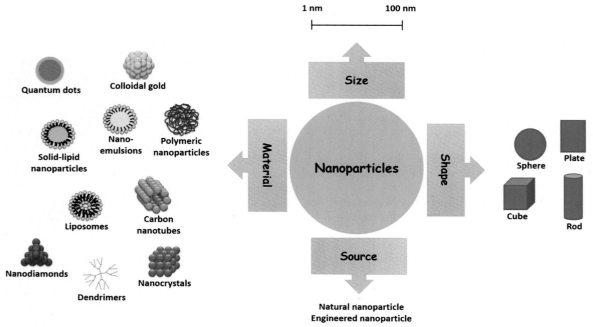

FIG. 2.6 Nanoparticle characteristics. (Adapted from Qin, W., Huang, G., Chen, Z., Zhang, Y., 2017. Nanomaterials in targeting cancer stem cells for cancer therapy. Frontiers in Pharmacology 8, 1–15; Fornaguera, C., Solans, C., 2017. Methods for the in vitro characterization of nanomedicines -biological component interaction. Journal of Personalized Medicine 7 (1).)

mechanical industries have also used NPs in coatings and lubricants (Khan et al., 2017).

Medicine

Some NP applications in medicine include tissue engineering, cancer therapy, and drug and gene delivery. NPs exist in the same size as biomolecules (see Fig. 2.7), which is a characteristic that allows NPs to pass through biological membranes. Consequently, NPs have been widely applied in the health area (Khan et al., 2017; Salata, 2004).

Nanotechnology has been rapidly and increasingly diffused in different sectors, which has stimulated studies on how nanomaterials affect human and animal health as well as the environment. Despite the various beneficial NP applications, researchers have reported on in vitro and in vivo NP cellular toxicity, immunotoxicity, and genotoxicity (Bahadar et al., 2016). Wen et al. (2017) described that intravenous administration of an acute AgNP dose in 16-week-old Sprague-Dawley rats induced extensive histopathologic damage in the liver, kidneys, thymus, and spleen; increased the levels of enzymes related to hepatic and renal functions, such as ALT, BUN, TBil,

and Cre; and significantly augmented chromosome breakage and polyploidy cell rates, pointing out that NPs are potentially genotoxic.

Another study showed that the mesenchymal stem cell profile changed after exposure to different AuNP concentrations. The CD 44 antigen level decreased in the presence of AuNPs at 6 μg/mL, and the CD 44 and CD 105 antigens levels dropped for AuNPs at 9 μg/mL. Furthermore, the latter AuNP concentration also reduced collagen type I synthesis and abated adipogenic differentiation, evidencing the toxic effect of these NPs (Volkova et al., 2017).

Zinc oxide nanoparticles (ZnONPs) are used as catalysts in the cosmetic industry, posing a great toxicity risk to whoever manipulates these NPs. Analyses conducted in rats demonstrated that ZnONPs induced inflammation and pulmonary oxidative stress as a result of Zn^{2+} release. In this sense, the antioxidant L-ascorbic acid (AA) could act as a potential preventive agent; AA is known to protect against membrane oxidation and injury and to prevent inflammatory processes by diminishing proinflammatory cytokine (IL-1 and IL-6) secretion in human keratinocytes stimulated with UVA irradiation. Likewise, A549 cells

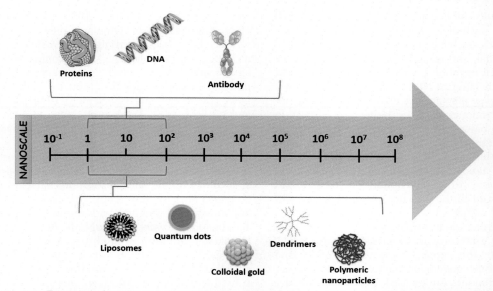

DNA

Proteins

Antibody

NANOSCALE

10^{-1} 1   10 10^2 10^3 10^4 10^5 10^6 10^7 10^8

Liposomes   **Quantum dots**   **Dendrimers**

Colloidal gold   **Polymeric nanoparticles**

FIG. 2.7 Examples of nanoparticles and biomolecules with the same size range in the nanometric scale. (Adapted from Fornaguera, C., Solans, C., 2017. Methods for the in vitro characterization of nanomedicines-biological component interaction. Journal of Personalized Medicine 7 (1).)

(human lung carcinoma cells) exposed to ZnONPs and treated with AA inhibited oxidative stress and inflammation. There are two explanations for this outcome: (1) AA removed the ROS generated by ZnONPs and (2) Zn^{2+} formed a complex with AA, an anionic antioxidant that can chelate metals, which limited oxidative stress and inflammation prevention. Therefore, techniques to study how to prevent nanotoxicity are welcome (Iwahashi et al., 2017). Compared with normal cells, tumor cells are more sensitive to oxidative stress because of their deficient active oxygen species (ROS) system. Cancer oxidation therapy consists in killing cancer cells by inducing oxidative stress. Thus, techniques in the field of nanomedicine have helped to integrate nanopathic therapies that increase oxidative stress in the tumor tissue while suppressing the cancer cell antioxidant capacity. To this end, poly(ethylene glycol)-b-poly[2-(((4-(4,4,5,5-tetramethyl-1,3,2-dioxaborolan-2-yl) benzyl) oxy) carbonyl) oxy) ethyl methacrylate (PEG-b-PBEMA) was integrated with palmitoyl ascorbate (PA) to form hybrid micelles (PA-Micelle). These micelles were exposed to tumor cells such as HeLa, A549, and 4T1 and to normal cells such as NIH3T3. The therapeutic efficiency was high for 4T1 cells; the PA-Micelle was not cytotoxic to normal cells in vitro or in vivo. This hybrid micelle apparently has potential

application in cancer therapy, especially breast cancer treatment (Yin et al., 2017).

The AgNP biological properties have allowed for a major breakthrough in medicine. AgNPs can induce programmed cell death mediated by oxidative stress through intracellular ROS accumulation in pathogens such as *Candida albicans* (Hwang et al., 2012). However, Radhakrishnan et al. (2018) observed that fungi treated with AA did not have the intracellular ROS level completely reversed, which altered membrane fluidity, the membrane microenvironment, the ergosterol content, and cellular morphology and ultrastructure. This "nano-based drug therapy" is likely to favor broad-spectrum activity, multiple cellular targets, and minimal host toxicity. In other words, AgNPs appear to have the potential to meet the challenges of multiple drug resistance and fungal therapy.

Oxidative stress induction has been speculated as one of the mechanisms through which NPs exert their toxic action. A study that investigated the molecular mechanisms underlying AgNP cytotoxicity in human liver cells verified higher ROS generation and reduced glutathione (GSH) suppression, which culminated in DNA cleavage, lipid membrane peroxidation, protein carbonylation, and lower cell viability due to apoptosis (Piao et al., 2011).

Because NPs can be easily transported through cell membranes, including mitochondria, they can act directly on the organelle, reduce mitochondrial membrane potential, and result in damage to the internal mitochondrial membrane and cell death. Of particular concern is the effect of NPs on mitochondria because these organelles play an essential role in cellular homeostasis (Pereira et al., 2012; Heller et al., 2012). In this scenario, bioenergetics and oxidative state of the organelles after subchronic exposure to AgNPs and TiNPs maintained increased ROS levels and depleted the endogenous antioxidant system, and the intensity of the toxic effect on the mitochondrial redox state was more significant in the presence of both types of NPs (Pereira et al., 2018).

Oxidative damage has been etiologically associated with the appearance of many disorders (Sies, 2015), so monitoring this damage is essential to understand the mechanisms underlying pathological conditions. Different techniques can be used to monitor oxidative stress, as discussed in the next topic.

NANOPARTICLE-MEDIATED OXIDATIVE STRESS MONITORING

Although nanomedicinal products have potential advantages such as higher effectiveness and lower toxicity and although diagnostic interventions are highly desirable, these products also raise questions concerning the currently used safety assessment procedures: do these procedures adequately evaluate the quality, safety, and efficacy of these products with regard to any possible nanospecific aspects? Recent advances in nanoscience and nanotechnology have elicited a growing demand for the assessment of the toxicological effects of nanomaterials on humans and other biological systems.

Isolated mitochondria could be exploited to develop in vitro model systems to investigate how nanomaterials interact with biological systems and constitute a potential tool to predict toxicity (e.g., oxidative stress) in humans (Heller et al., 2012; Pereira et al., 2018). To monitor the NP ability to induce oxidative stress, we can use mitochondrial redox state parameters because this organelle is the major source of reactive species in living organisms. One of the initial parameters is to verify whether lipoperoxidation is induced. The products resulting from lipid peroxidation help to monitor the damage that free radicals cause. The biophysical consequences of membrane lipid peroxidation ultimately lead to irreversible structural alterations in cells and their organelles, to culminate in cell death (Richter,

1987; Halliwell and Chirico, 1993; De Zwart et al., 1999; Gutteridge and Quinlan, 1983). Many biomarkers have been used to assess oxidative stress, including malondialdehyde (MDA), conjugated dienes, ethane and pentane gases, isoprostanes, 4-hydroxynonenal (4-HNE), modified proteins, and modified DNA (Esterbauer and Cheeseman, 1990; Dormandy and Wickens, 1987; Frank et al., 1980; Morrow et al., 1995; Esterbauer et al., 1991; Levine et al., 1994; Marnett, 1999).

In this scenario, the TBA reaction is the most often used. MDA is a peroxidized lipid degradation product, so if color with the same absorption characteristics (at 532–535 nm) or fluorimetry (excitation at 532 nm and emission at 553 nm) with the same features of a TBA-malonaldehyde chromophore develops, it can be taken as an index of lipid peroxidation in a given biological sample (Grotto et al., 2007). This is the so-called TBARS (thiobarbituric acid reactive substances) method. Although this method is easy and inexpensive, it has received criticism for its lack of specificity.

Numerous high-performance liquid chromatography methods for MDA determination with colorimetric and fluorimetric detection have been reported (Grotto et al., 2007; Bird et al., 1983; Tsikas, 2017). Indeed, chromatographic methods are the gold standard for MDA analysis.

RONS generation can be spectrofluorimetrically monitored with the fluorescence probe H_2DCFDA. DCFDA is a molecule that can easily cross membranes. It is hydrolyzed by esterases to release the nonfluorescent compound DCFH. In the presence of RONS, DCFH is rapidly oxidized and converted to DCF, which displays fluorescence with high quantum yield when excited at 503 nm. Therefore, an increase in DCF fluorescence is directly proportional to RONS production by mitochondria and cells (Cathcart et al., 1983).

Another way to evaluate whether NPs induce oxidative stress is to monitor the mitochondrial antioxidant systems. As described above, a balance between antioxidant molecules and antioxidants is essential to maintain homeostasis, and endogenous antioxidant system depletion favors the emergence of oxidative stress. Reduced pyridine nucleotides (NAD(P)H) are fluorescent, but the oxidized pyridine nucleotides $(NAD(P)^+)$ are not. Thus, fluorescence intensity changes reflect alterations in the pyridine nucleotide redox state (Oliveira et al., 2005). This is an important parameter to monitor the oxidative stress induced by xenobiotics, such as NPs, because NAD(P)H oxidation can initiate or amplify pathological and physiological events (Di Lisa and Ziegler, 2001; Takeyama et al., 1993).

To protect the cell against NP-induced oxidative stress, mitochondrial and cytosolic enzymes as well as nonenzymatic antioxidants, including glutathione (GSH), scavenge ROS (Piao et al., 2011; Chernyak et al., 2005). Glutathione is distributed in cells. In its oxidized form (GSSG), glutathione underlies several functions. The enzyme glutathione reductase rapidly converts GSSG to the reduced form (GSH). In this context, another method to monitor oxidative stress comprises the use of a fluorescent agent with affinity for GSH or GSSG depending on pH; NEM is used to avoid GSSG oxidation (Hissin and Hilf, 1976).

Nanotoxicology, or the study of the potential adverse effects that nanomaterials exert on the human body, especially oxidative stress, has become increasingly important and will continue to attract researchers' attention for many years to come because an ever-growing number of consumer products based on nanomaterials are continuously entering the market. As recently shown by our research team, oxidative stress is commonly the mechanism through which NPs exert their toxicity, and the aforementioned methods may contribute to better NP toxicity evaluation (Pereira et al., 2018).

REFERENCES

Abdollahi, M., Ranjbar, A., Shadnia, S., Nikfar, S., Rezaie, A., 2004. Pesticides and oxidative stress: a review. Medical Science Monitor 10 (6), RA141–RA147.

Augusto, O., 2006. Radicais Livres: Bons, maus e naturais. Oficina de Textos, São Paulo.

Bahadar, H., Maqbool, F., Niaz, K., Abdollahi, M., 2016. Toxicity of nanoparticles and an overview of current experimental models. Iranian Biomedical Journal 20 (1), 1–11.

Bhattacharyya, A., Chattopadhyay, R., Mitra, S., Crowe, S.E., April 2014. Oxidative stress: an essential factor in the pathogenesis of gastrointestinal mucosal diseases. Physiological Reviews 94 (2), 329–354.

Birben, E., Sahiner, U.M., Sackesen, C., Erzurum, S., Kalayci, O., 2012. Oxidative stress and antioxidant defense. World Allergy Organ J (5), 9–19.

Bird, R.P., Hung, S.S.O., Hadley, M., Draper, H.H., 1983. Determination of malonaldehyde in biological materials by high pressure liquid chromatography. Analytical Biochemistry 128, 24–244.

Blajszczak, C., Bonini, M.G., November 1, 2017. Mitochondria targeting by environmental stressors: implications for redox cellular signaling. Toxicology 391, 84–89.

Blesa, J., Trigo-Damas, I., Quiroga-Varela, A., Jackson-Lewis, V.R., 2015. Oxidative stress and Parkinson's disease. Frontiers in Neuroanatomy 9 (July), 1–9.

Boelsterli, U.A., 2002. Mechanistic Toxicology: The Molecular Basis of How Chemicals Disrupt Biological Targets. Taylor and Francis, London, p. 312.

Brand, M.D., 2016. Mitochondrial generation of superoxide and hydrogen peroxide as the source of mitochondrial redox signaling. Free Radical Biology and Medicine 100, 14–31.

Cathcart, R., Schwiers, E., Ames, B.N., 1983. Detection of picomole levels of hydroperoxides using a fluorescent dichlorofluorescein assay. Analytical Biochemistry 134 (1), 111–116.

Chernyak, B.V., Pletjushkina, O.Y., Izyumov, D.S., Lyamzaev, K.G., Avetisyan, A.V., February 2005. Biochemistry 70 (2), 240–245.

Collins, Y., Chouchani, E.T., James, A.M., Menger, K.E., Cocheme, H.M., Murphy, M.P., 2012. Mitochondrial redox signalling at a glance. Journal of Cell Science 125 (7), 1837–1837.

Davies, K.J., 1995. Oxidative stress: the paradox of aerobic life. Biochemical Society Symposia 61, 1–31.

De Zwart, L.L., Meerman, J.H., Commandeur, J.N., Vermeulen, N.P., January 1999. Biomarkers of free radical damage applications in experimental animals and in humans. Free Radical Biology and Medicine 26 (1–2), 202–226.

Di Lisa, F., Ziegler, M., March 9, 2001. Pathophysiological relevance of mitochondria in NAD$^+$ metabolism. FEBS Letters 492 (1–2), 4–8.

Dormandy, T.L., Wickens, D.G., 1987. Chemistry and Physics of Lipids 45, 353–364, 1987;45:353–64.

Esterbauer, H., Cheeseman, K.H., 1990. Determination of aldehydic lipid peroxidation products: malonaldehyde and 4-hydroxynonenal. Methods in Enzymology 186 (C), 407–421, 48. 1.

Esterbauer, H., Schaur, R.J., Zollner, H., 1991. Chemistry and biochemistry of 4-hydroxynonenal, malonaldehyde and related aldehydes. Free Radical Biology & Medicine 11, 81–128.

Fornaguera, C., Solans, C., 2017. Methods for the in vitro characterization of nanomedicines -biological component interaction. Journal of Personalized Medicine 7 (1).

Frank, H., Hintze, T., Bimboes, D., Remmer, H., 1980. Monitoring lipid peroxidation by breath analysis: endogenous hydrocarbons and their metabolic elimination. Toxicology and Applied Pharmacology 56 (3), 337–344.

Grotto, D., Santa Maria, L.D., Boeira, S., Valentini, J., Charão, M.F., Moro, A.M., et al., 2007. Rapid quantification of malondialdehyde in plasma by high performance liquid chromatography-visible detection. Journal of Pharmaceutical and Biomedical Analysis 43 (2), 619–624.

Gutteridge, J.M., Quinlan, G.J., August-October. Malondialdehyde formation from lipid peroxides in thethiobarbituric acid test: the role of lipid radicals, iron salts, and metal chelators. Journal of Applied Biochemistry 5 (4–5), 293–299.

Halliwell, B.E., Chirico, S., May 1993. Lipid peroxidation: its mechanism, measurement andsignificance. American

Journal of Clinical Nutrition 57 (5 Suppl. l), 715S–724S discussion 724S-725S.

Hasan, S., Hasan, S., 2015. A Review on Nanoparticles: Their Synthesis and Types, pp. 7–10.

Heller, A., Brockhoff, G., Goepferich, A., September 2012. Targeting drugs to mitochondria. European Journal of Pharmaceutics and Biopharmaceutics 82 (1), 1–18.

Hernández, A.F., Lacasaña, M., Gil, F., Rodríguez-Barranco, M., Pla, A., López-Guarnido, O., 2013. Evaluation of pesticide-induced oxidative stress from a gene-environment interaction perspective. Toxicology 307, 95–102.

Hissin, P.J., Hilf, R., 1976. A fluorimetric method for determination of oxidized and reduced glutathione in tissues. Analalytical Biochemistry 74, 214–226.

Hüttemann, M., Lee, I., Samavati, L., Yu, H., Doan, J.W., 2007. Regulation of mitochondrial oxidative phosphorylation through cell signaling. Biochimica et Biophysica Acta (BBA) — Molecular Cell Research 1773 (12), 1701–1720.

Hwang, I.S., Lee, J., Hwang, J.H., Kim, K.J., Lee, D.G., 2012. Silver nanoparticles induce apoptotic cell death in Candida albicans through the increase of hydroxyl radicals. FEBS Journal 279, 1327–1338.

Iwahashi, H., Nishio, K., Hagihara, Y., Yoshida, Y., Horie, M., September 2017. Ascorbic acid prevents zinc oxide nanoparticle-induced intracellular oxidative stress and inflammatory responses. Toxicology and Industrial Health 33 (9), 687–695.

Khan, I., Saeed, K., Khan, I., 2017. Nanoparticles: properties, applications and toxicities. Arabian Journal of Chemistry (In Press), https://doi.org/10.1016/j.arabjc.2017.05.011.

Kowaltowski, A.J., de Souza-Pinto, N.C., Castilho, R.F., Vercesi, A.E., 2009. Mitochondria and reactive oxygen species. Free Radical Biology and Medicine 47 (4), 333–343.

Krinsky, N.I., June 1992. Mechanism of action of biological antioxidants. Proceedings of the Society for Experimental Biology and Medicine 2000 (2), 248–254.

Levine, A., Tenhaken, R., Dixon, R.A., Lamb, C., 1994. H_2O_2 from the oxidative brust orchestrates the plant hypersensitive response. Cell 79, 583–593.

Lodish, H., Berk, A., Zipursky, S.L., et al., 2000. Molecular Cell Biology, fourth ed. W. H. Freeman, New York. Section 16.2, Electron Transport and Oxidative Phosphorylation. Available from: https://www.ncbi.nlm.nih.gov/books/NBK21528/.

Lu, S.C., May 2013. Glutathione synthesis. Biochimica et Biophysica Acta 1830 (5), 3143–3153.

Marnett, L.J., 1999. Lipid peroxidation—DNA damage by malondialdehyde. Mutation Research 424 (1–2), 83–95.

Morrow, J.D., Frei, B., Longmire, A.W., Gaziano, J.M., Lynch, S.M., Shyr, Y., et al., 1995. Smoking as a cause of oxidative damage. New England Journal of Medicine 332, 1198–1203.

Muller, F.L., Liu, Y., Van Remmen, H., 2004. Complex III releases superoxide to both sides of the inner mitochondrial membrane. Journal of Biological Chemistry 279 (47), 49064–49073.

Na, H.J., Chung, H.T., Ha, K.S., Lee, H., Kwon, Y.G., Billiar, T.R., Kim, Y.M., 2008. Detection and measurement for the modification and inactivation of caspase by nitrosative stress in vitro and in vivo. Methods in Enzymology 441, 317–327.

Nelson, D.L., Lehninger, A.L., Cox, M.M., 2013. Lehninger Principles of Biochemistry, sixth ed. W.H. Freeman, New York.

Oliveira, H.C., et al., February 2005. Oxidative stress in atherosclerosis-prone mouse is due to low antioxidant capacity of mitochondria. The FASEB Journal 19 (2), 278–280.

Pereira, L.C., de Souza, A.O., Dorta, D.J., 2012. Mitocôndria como alvo para Avaliação de Toxicidade de Xenobiótico. Revista Brasileira de Toxicologia 25 (1–2), 1–14.

Pereira, L.C., Pazin, M., Franco-Bernardes, M.F., Martins Jr., A.C., Barcelos, G.F.M., Pereira, M.C., Mesquita, J.P., Rodrigues, J.L., Barbosa Jr., F., Dorta, D.J., May 2018. A perspective of mitochondrial dysfunction in rats treated with silver and titanium nanoparticles (AgNPs and TiNPs). Journal of Trace Elements in Medicine & Biology 47, 63–69.

Piao, M.J., Kang, K.A., Lee, I.K., Kim, H.S., Kim, S., Choi, J.Y., et al., 2011. Silver nanoparticles induce oxidative cell damage in human liver cells through inhibition of reduced glutathione and induction of mitochondria-involved apoptosis. Toxicology Letters 201 (1), 92–100.

Puppel, K., Kapusta, A., Uczyńska, B., 2015. The etiology of oxidative stress in the various species of animals, a review. Journal of the Science of Food and Agriculture 95 (11), 2179–2184.

Qin, W., Huang, G., Chen, Z., Zhang, Y., 2017. Nanomaterials in targeting cancer stem cells for cancer therapy. Frontiers in Pharmacology 8, 1–15.

Raab, C., Simkó, M., Fiedeler, U., Nentwich, M., Gazsó, A., 2011. Production of nanoparticles and nanomaterials. Nano Trust Dossiers 6, 1–4.

Radhakrishnan, V.S., Dwivedi, S.P., Siddiqui, M.H., Prasad, T., March 15, 2018. In vitro studies on oxidative stress-independent, Ag nanoparticles-induced cell toxicity of Candida albicans, an opportunistic pathogen. International Journal of Nanomedicine 13, 91–96.

Richter, C., 1987. Biophysical consequences of lipid peroxidation in membranes. Chemistry and Physics of Lipids 44 (2–4), 175–189.

Riley, P.A., 1994. Free radicals in biology: oxidative stress and the effects of ionizing radiation. International Journal of Radiation Biology 65, 27–33.

Salata, O.V., 2004. Applications of nanoparticles in biology and medicine. Journal of Nanobiotechnology 2, 1–6.

Sharma, V.K., Filip, J., Zboril, R., Varma, R.S., 2015. Natural inorganic nanoparticles — formation, fate, and toxicity in the environment. Chemical Society Reviews 44 (23), 8410–8423.

Sies, H., 2015. Oxidative stress: a concept in redox biology and medicine. Redox Biology 4, 180–183.

Takeyama, N., Matsuo, N., Tanaka, T., September 15, 1993. Oxidative damage to mitochondria is mediated by the

Ca^{+2} dependent inner-membrane permeability transition. Biochemical Journal 294 (Pt 3), 719—725.

Tönnies, E., Trushina, E., 2017. Oxidative stress, synaptic dysfunction, and Alzheimer's disease. Journal of Alzheimer's Disease 57 (4), 1105—1121.

Tsikas, D., 2017. Assessment of lipid peroxidation by measuring malondialdehyde (MDA) and relatives in biological samples: analytical and biological challenges. Analytical Biochemistry 524, 13—30.

Volkova, N., Pavlovich, O., Fesenko, O., Budnyk, O., Kovalchuk, S., Goltsev, A., October 2017. Studies of the influence of gold nanoparticles on characteristics of mesenchymal stem cells. Journal of Nanomaterials 2017, 1—9.

Wen, H., Dan, M., Yang, Y., Lyu, J., Shao, A., Cheng, X., et al., 2017. Acute toxicity and genotoxicity of silver nanoparticle in rats. PLoS One 12 (9), 1—16.

Yin, W., Li, J., Ke, W., Zha, Z., Ge, Z., 2017. Integrated nanoparticles to synergistically elevate tumor oxidative stress and suppress antioxidative capability for amplified oxidation therapy. ACS Applied Materials & Interfaces 9 (35), 29538—29546.

Zolkipli-Cunningham, Z., Falk, M.J., Cunningham, Z.Z., 2017. Clinical effects of chemical exposures on mitochondrial function Clinical effects of chemical exposures on mitochondrial function. Toxicology 391 (April), 90—99.

CHAPTER 3

Nanoparticles as Modulators of Oxidative Stress

RISHA GANGULY • AMIT KUMAR SINGH • RAMESH KUMAR • ASHUTOSH GUPTA • AKHILESH KUMAR PANDEY • ABHAY K. PANDEY

INTRODUCTION

Nanoparticles (NPs) are particulate substances, which range between 1 and 100 nm (nm) in size. Because of the extremely small size, NPs demonstrate typical structural and biological properties, which include a greater reactive area and the potential to pass through cell and tissue barriers that render them useful for humans in several industrial applications, including agriculture, cosmetics, food and medicine, and particularly in drug delivery. With these applications in effect and increased exposure, humans are more vulnerable to the harmful effects that arise due to NPs, namely, oxidative stress, genetic damage, inflammation, and inhibition of cell division and apoptosis (Fu et al., 2014). The most frequently reported toxic effects occur due to changes in physicochemical properties of NPs at the structural level. This can subsequently lead to the production of reactive oxygen species (ROS) and reactive nitrogen species (RNS) causing oxidative stress. Oxidative stress results from engineered NPs because of noncellular factors such as composition, particle size, surface area, and the presence of metal ions. Cellular responses in the biological system such as mitochondrial respiration, changes in cell signaling pathways, cellular interaction with NPs, and activation of immune cells are also responsible for ROS/RNS-mediated cellular damage. Oxidative stress induced by NPs further initiates pathophysiological effects such as DNA damage, inflammation, and cytotoxicity (Manke et al., 2013).

NANOPARTICLES: PHYSICOCHEMICAL PROPERTIES AND NANOTOXICITY

NPs, owing to their small size, have a large surface area to volume ratio (Nel et al., 2006; Chiang et al., 2012). They are usually made up of three layers, that is, (1) the inner core, which is the central part of the NP,

(2) the middle layer, which is entirely made up of a different substance, and (3) the outer surface layer is composed of a number of metal ions, small molecules, surfactants, and polymers (Khan et al., 2017). The small size of the NPs can also influence their optical properties. For instance, gold (Au), silver (Ag), platinum (Pt), and palladium (Pd) NPs of size 20 nm have distinct wine red, black, yellowish gray, and dark black color, respectively. These unique physical characteristics of NPs attribute to different chemical and biological properties in comparison with the same compound in its bulk form (Dreaden et al., 2012). The small size of NPs allows them to cross cell membranes with ease. Hence, NPs can be conveniently taken up inside living organisms and can result in loss of cellular function. Their exceptional physicochemical, thermal, and electrical properties lead to their wide applications in medicine, cosmetics, and textiles (Maurya et al., 2012). This has led to increase in probability of human exposure to nanomaterials (Xia et al., 2008). In addition, because of these diverse structural properties, NPs are reactive and catalytic in function, which adds to their potential toxicity (Gonzalez et al., 2008). Of all the nanomaterials, metal-based NPs and carbon nanotubes (CNTs) have generated significant commercial interest because of their extraordinary intrinsic properties such as high conductivity and tensile strength. For example, zinc oxide NPs are most commonly used with a wide range of applications in medicine and cosmetic industries. Similarly, titanium dioxide NPs are used as food additives and in drug delivery, in personal care products, paints, plastics, and cosmetics. The major application of NPs is in drug release and drug targeting. NPs are also used as tools in diagnostic applications for the detection of biomarkers (Khanna et al., 2015). It has been established that particle size affects the drug release potential of

Nanotechnology in Modern Animal Biotechnology. https://doi.org/10.1016/B978-0-12-818823-1.00003-X

NPs. Because of small particles have large surface area, most of the drugs they carry are exposed to the particle surface, which leads to rapid drug release. In contrast, for large-sized NPs, the process of drug release and diffusion is slow (Oberdorster, 2004; Oberdorster et al., 2005; Karlsson et al., 2008).

OXIDATIVE STRESS AND CELLULAR DAMAGE

Oxidative stress is defined as a state in which the generation of free radicals (ROS and RNS) exceeds the ability of antioxidant systems to detoxify the formed reactive intermediates or repair the cellular damage (Kumar et al., 2015; Sharma et al., 2017). ROS produced during oxidative stress play an important role in the regulation of signal transduction and homeostasis (Yoshikawa and Naito, 2002). In general, production of ROS is a normal cellular process, which is correlated with several signaling pathways along with the defense machinery of the immune system. However, when produced in excess, it is known to damage the cellular membranes and biomolecules such as lipids, proteins, and DNA, resulting in deleterious effects (Mishra et al., 2013; Kumar and Pandey, 2015). ROS include a number of species such as superoxide anion ($O_2^{\cdot-}$), hydroxyl radical ($OH^{\cdot}$), hydrogen peroxide (H_2O_2), singlet oxygen ($^1O_2$), and hypochlorous acid (HOCl). Molecular oxygen (O_2) generates $O_2^{\cdot-}$, which is produced upon the reduction of a single electron catalyzed by the enzyme NADPH oxidase. Subsequent reduction of oxygen results in the formation of either H_2O_2 via dismutation or $OH^{\cdot}$ by Fenton reaction (Vallyathan and Shi, 1997; Thannickal and Fanburg, 2000). ROS are generated intrinsically through mitochondrial respiration, inflammatory responses, in microsomes and peroxisomes, whereas environmental pollutants and engineered NPs produce ROS extrinsically (Lin et al., 2008; Fahmy and Cormier, 2009; Sarkar et al., 2014). Inflammatory phagocytic cells such as macrophages and neutrophils release free radicals in response to microbes, tumor cells, and environmental pollutants as defense mechanism (Risom et al., 2005). A variety of NPs including metal-based particles cause cytotoxicity via ROS generation. NPs alter intracellular Ca^{2+} concentrations, activate several transcription factors, and modulate inflammatory cytokines via free radical generation (Li et al., 2010; Huang et al., 2010a,b). To counter the excess of ROS produced upon exposure to NPs, cells can activate the in-built antioxidant enzymes and nonenzymatic antioxidant systems in the body (Sies, 1991). At low levels of oxidative stress, phase II

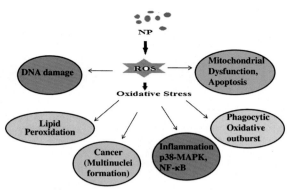

FIG. 3.1 Nanoparticle (NP)-induced oxidative stress pathway.

antioxidant enzymes are activated transcriptionally via nuclear factor (erythroid-derived 2)-like 2 (Nrf2) induction. At moderate levels of oxidative stress, cascades of redox-sensitive MAP kinase and nuclear factor NF-κB induce a proinflammatory response. However, severe oxidative stress results in mitochondrial membrane damage and subsequent impairment of the electron transport chain causing cell death. The prooxidant effects of engineered NPs include either the increased production of free radicals or the elimination of antioxidants. The production of free radicals and peroxide occurs due to interruption in the normal redox state of the system that in turn effect cellular components such as lipids, proteins, and DNA. This activates the signaling pathways that are associated with loss of cell growth, apoptosis, and initiation of tumorigenesis (Valco et al., 2006; Buzeaet al., 2007) (Fig. 3.1). Apart from cellular damage, ROS are also produced when NPs interact with biological targets via cellular respiration, metabolism, inflammation, ischemia/reperfusion (inadequate supply of blood to the heart), and metabolism of several nanomaterials. Most importantly, oxidative stress that occurs due to occupational NP exposures leads to airway inflammation and fibrosis (Donaldson et al., 2004; Nel, 2005).

GENERATION OF ROS IN NANOPARTICLE-CELL INTERACTIONS

Humans are exposed to NPs via several routes such as inhalation, injection, ingestion, and the skin. Specifically, the respiratory tract, gastrointestinal tract, the circulatory system, and the nervous system are most adversely affected by NPs (Medina et al., 2007). It has been reported that CNTs can cause granulomatous lesions in the lung epithelial tissue and continual interstitial inflammation upon dose-dependent chronic exposure (Lam et al., 2004; Morimoto et al., 2013). Furthermore, ceramic NPs that are made of calcium, silicon, and titanium oxides are also used in drug delivery and show evidence of oxidative stress and cytotoxic activity in the brain, heart, liver, and lungs (Singh et al., 2014). NP-induced oxidative stress is the principal cause of nanotoxicity and can be attributed to NP-cellular interaction or due to the presence of ROS generating functional groups on the particle surface. Studies with different sizes of silver NPs have revealed that they induce ROS, decrease levels of glutathione, and inhibit superoxide dismutase (SOD) (Avalos et al., 2014). Similarly, Au NPs of 5–250 nm also establish the inverse relationship between the size of NPs and higher production of ROS (Misawa and Takahashi, 2011). In vitro studies with cerium oxide NPs of different sizes ranging from 15 to 45 nm reveal that their toxicity occurs due to ROS-mediated oxidative stress, which further activates the enzyme heme oxygenase-1 (HO-1) mediated by Nrf2 (Eom and Choi, 2009). Studies have shown that single-dose exposure to silica NPs results in generation of ROS, which subsequently initiates inflammatory responses (Park and Park, 2009). Furthermore, single- and multiwalled CNTs are known to cause ROS induction, increased lipid peroxidation, decrease in mitochondrial membrane potential, and decrease in enzymatic antioxidants such as catalase, SOD, glutathione peroxidase, and glutathione reductase (Wang et al., 2011; Zhornik et al., 2012).

MECHANISM OF NANOPARTICLE-INDUCED OXIDATIVE STRESS

The key factors, which influence NP-induced ROS generation, are the presence of prooxidant functional groups on reactive surface of NPs and NP-cellular interactions (Knaapen et al., 2004). The reactive surface of NPs in particular plays an important role in causing oxidative stress. Surface-bound free radicals present on silica NPs such as $SiO^{\cdot}$ and $SiO_2^{\cdot}$ result in the formation of oxidative species such as $OH^{\cdot}$ and $O_2^{\cdot-}$ (Fubini and Hubbard, 2003). Molecules such as ozone and nitrogen dioxide (NO_2), which are adsorbed on the surface of NPs, also induce oxidative damage. Decrease in particle size along with changes in electronic properties leads to structural defects giving rise to reactive groups on the particle surface. These reactive groups interact with molecular oxygen (O_2) to form $O_2^{\cdot-}$, which is followed by further generation of ROS via Fenton reaction. For instance, Zn and Si NPs with similar shape and size lead to varied toxicity responses, which can be ascribed

$$O_2^{\cdot-} + H_2O_2 \longrightarrow O_2 + OH^{\cdot} + OH^{-} \quad \text{(Haber-Weiss Reaction)}$$

$$\left. \begin{array}{l} O_2^{\cdot-} + Fe^{3+} \longrightarrow Fe^{2+} + O_2 \\ Fe^{2+} + H_2O_2 \longrightarrow Fe^{3+} + OH^{\cdot} + OH \end{array} \right\} \quad \text{(Fenton Reaction)}$$

FIG. 3.2 Reactions involved in the generation of ROS.

to their reactive surface properties (Donaldson and Tran, 2002; Oberdorster et al., 2005). Dissolution of NPs, which have surface-bound free radicals, lead to their subsequent release and further enhance the ROS response and cytotoxicity in the cell. For example, quartz particles in aqueous suspensions generate $OH^{\cdot}$, $^1O_2$, and H_2O_2. Besides the surface-dependent factors, metals ions and other chemical compounds on the surface of NPs also influence ROS response (Schins, 2002; Wilson et al., 2002). Transition metals such as copper (Cu), iron (Fe), chromium (Cr), silica (Si), and vanadium (V) generate ROS response through Haber-Weiss and Fenton reactions (Fig. 3.2). Fenton reaction typically involves interaction of a transition metal ion with H_2O_2 to yield hydroxyl radical ($OH^{\cdot}$) and oxidized form of the metal ion. For instance, reduction of H_2O_2 with ferrous ion (Fe^{2+}) gives rise to $OH^{\cdot}$ that is extremely reactive and toxic to biological molecules. Fe- and Cu-based NPs are known to cause oxidative stress ($O_2^{\cdot-}$ and $OH^{\cdot}$) via Fenton reaction, whereas in the Haber-Weiss reaction, the oxidized metal ion reacts with H_2O_2 to form $OH^{\cdot}$. NPs with transition metals Cr, Co, and V can catalyze both Fenton and Haber-Weiss-type reactions (Aust et al., 1993). Glutathione reductase, an antioxidant enzyme, has the ability to reduce metal NPs into active intermediates that can promote ROS response. Besides, some metal-based NPs such as argon, beryllium, cobalt, and nickel can also accelerate the intercellular free radical–generating pathways involving MAPK and NF-κB (Smith et al., 2001). When ROS are produced endogenously, the mitochondrion plays an important role in NP-induced oxidative stress. The NPs are easily accessible into the mitochondria and stimulate ROS generation through impaired electron chain, structural damage, depolarization of the mitochondrial membrane, and activation of NADPH-like enzymes (Sioutas et al., 2005; Xia et al., 2006). Cellular internalization of NPs activates immune cells such as neutrophils and macrophages and contributes to the production of ROS and RNS. This process usually involves the activation of NADPH oxidase enzymes. In vivo NP exposures activate the large number of inflammatory phagocytes within the lungs causing them to stimulate an oxidative burst

(Coccini et al., 2013). The respiratory responses, which are induced by small NPs upon inhalation, are much greater than those produced by larger particles because of the increased surface area to particle mass ratio. Larger surface area ensures that the majority of the molecules are exposed to the surface than the interior of the NPs. Similarly, SiO_2 and TiO_2 NPs and CNT induce greater ROS production as compared with their larger counterparts. In addition, a study with cobalt/chromium NP exposure also established particle size–dependent ROS-mediated toxicity (Sohaebuddin et al., 2010; Raghunathan et al., 2013).

ANTIOXIDANT EFFECT OF METAL-BASED NANOPARTICLES

Silver Nanoparticle

Silver NPs are being widely used in the pharmaceutical sector in the production of creams and ointments as they can inhibit infections related to wounds and burns. The silver ion possesses inhibitory effects against several microorganisms (Mohanta and Behera, 2014). Biological synthesis is an alternative and eco-friendly method for the production of silver NPs (Firdhouse and Lalitha, 2015). The silver NPs possess antioxidant and antimicrobial properties and also have potential as a drug carrier in treatment of cancer that has recently gained considerable attention (Nayak et al., 2016). Silver NPs can be produced using a number of physical and chemical methods such as chemical reduction, photochemical reduction (Mallick et al., 2005), electrochemical reduction (Liu and Lin, 2004), and heat vaporization (Smetana et al., 2005). However, a large number of toxic chemicals are required as reducing agents in these processes. Because of increasing use of transition metal NPs in areas of human contact, researchers in the past few years have developed eco-friendly biological systems for NP synthesis that do not require the use of toxic chemicals (Song and Kim, 2008; Tsibakhashvil et al., 2010). Studies have reported that silver NPs biosynthesized using *Chenopodium murale* leaf extract possess antioxidant and antimicrobial property (Abdel-Aziz et al., 2014).

Cerium Oxide (CeO$_2$)

Cerium is one of the most abundant rare-earth elements from the lanthanide series in the periodic table. It exists in both Ce^{3+} and Ce^{4+} oxidation states, which gives it the extraordinary ability as an antioxidant (Das et al., 2013). The ROS-scavenging capability of CeO_2 is based on oxidation-reduction cycling between Ce^{3+} and Ce^{4+} on the NP surface. Studies have shown that treatment of

CeO_2 NPs attenuates tumorigenesis in ovarian cancer mice models and significantly inhibits the metastasis of these cancer cells into the lungs of mice. Angiogenesis in these cancerous cells is also lessened in mice treated with CeO_2 NPs as indicated by CD31-positive staining (Giri et al., 2013). It has also been demonstrated that interaction of CeO_2 NPs with folic acid at the surface along with their delivery with cisplatin further decreases tumor cell formation and angiogenesis (Hijaz et al., 2016). In other studies, CeO_2 NP administration to carbon tetrachloride (CCl_4)–induced liver fibrosis mice was shown to inhibit oxidative stress signaling pathways and reduction in levels of inflammatory cytokines such as TNF-α and IL-1β (Siu et al., 2014; Hirst et al., 2013).

Platinum Nanoparticles

Platinum NPs possess catalytic activity because of very high electron to particle surface ratio, which also contributes to its radical scavenging ability (Katsumi et al., 2014). Studies in SOD1 knockout mice have revealed the therapeutic efficacy of platinum NPs against age-related skin diseases (Shibuya et al., 2014). HaCaT keratinocyte cell lines have shown that platinum NPs have the ability to protect cells from UV-induced apoptosis (Yoshihisa et al., 2010).

Mesoporous Silica Nanoparticles

Mesoporous silica NPs are known to possess ROS, OH$^{\bullet}$, and free radical scavenging ability (Hao et al., 2014; Huang et al., 2010a,b; Morry et al., 2015). Scavenging of ROS by these NPs in vitro can downregulate the expression of mRNA in melanocyte tumor cells. Hence, mesoporous silica NPs are used in the treatment of oxidative stress–induced pathological conditions such as fibrosis and cancer.

Zinc Oxide Nanoparticles

Zinc oxide (ZnO) is a white powder that is barely soluble in water. It has strong antimicrobial activities and is extensively used in food packaging and synthetic textiles. Zinc is a cofactor in several metabolic enzymes. Therefore, it has the potential to enhance phagocytic activity in the body and act upon the defense system. ZnO NPs have the unusual ability to form an excitonic pair (e^-,h^+), thereby acting as ROS scavengers (Sanjay et al., 2014).

CONCLUSION

In the recent years, the production and use of engineered NPs have increased significantly in diverse fields, leading to several folds enhancement in risk of human exposure to NPs. Owing to their small size, NPs acquire many unique physical and chemical properties, which can be hazardous for human health, and hence their potential toxicity and harmful effects cannot be disregarded. However, some metal-based NPs (Ag, Ce, Zn, and Pt) have also shown considerable antioxidant, antimicrobial, and anticancer properties. Therefore, biological and eco-friendly methods of NP synthesis should be encouraged, barring the use of toxic chemicals for optimum utilization of nanotechnology without any adverse effects.

ACKNOWLEDGMENTS

RG, AKS, and RK acknowledge financial support from CSIR-UGC New Delhi in the form of Junior Research Fellowship. AG acknowledges UGC for CRET fellowship. Authors also acknowledge DST-FIST and UGC-SAP facilities of the Department of Biochemistry, University of Allahabad, Allahabad.

REFERENCES

Abdel-Aziz, M.S., Shaheen, M.S., El-Neekety, A.A., et al., 2014. Antioxidant and antibacterial activity of silver nanoparticles biosynthesized using *Chenopodium murale* leaf extract. Journal of Saudi Chemical Society 18 (4), 356–363.

Aust, S.D., Chignell, C.F., Bray, T.M., et al., 1993. Free radicals in toxicology. Toxicology and Applied Pharmacology 120 (2), 168–178.

Avalos, A., Haza, A.I., Mateo, D., et al., 2014. Cytotoxicity and ros production of manufactured silver nanoparticles of different sizes in hepatoma and leukemia cells. Journal of Applied Toxicology 34, 413–423.

Buzea, C., Pacheco II, Robbie, K., 2007. Nanomaterials and nanoparticles: sources and toxicity. Biointerphases 2 (4), 17–71.

Chiang, H.M., Xia, Q., Zou, X., et al., 2012. Nanoscale ZnO induces cytotoxicity and DNA damage in human cell lines and rat primary neuronal cells. Journal of Nanoscience and Nanotechnology 12, 2126–2135.

Coccini, T., Barni, S., Vaccarone, R., 2013. Pulmonary toxicity of instilled cadmium-doped silica nanoparticles during acute and subacute stages in rats. Histology & Histopathology 28 (2), 195–209.

Das, S., Dowding, J.M., Klump, K.E., et al., 2013. Cerium oxide nanoparticles: applications and prospects in nanomedicine. Nanomedicine (London, England) 8 (9), 1483–1508.

Donaldson, K., Stone, V., Tran, C.L., et al., 2004. Nanotoxicology Occupational and Environmental Medicine 61 (9), 727–728.

Donaldson, K., Tran, C.L., 2002. Inflammation caused by particles and fibers. Inhalation Toxicology 14 (1), 5–27.

Dreaden, E.C., Alkilany, A.M., Huang, X., et al., 2012. The golden age: gold nanoparticles for biomedicine. Chemical Society Reviews 41, 2740–2779.

Eom, H.J., Choi, J., 2009. Oxidative stress of ceO2 nanoparticles via p38-Nrf-2 signaling pathway in human bronchial epithelial cell, Beas-2B. Toxicology Letters 187, 77–83.

Fahmy, B., Cormier, S.A., 2009. Copper oxide nanoparticles induce oxidative stress and cytotoxicity in airway epithelial cells. Toxicology In Vitro 23, 1365–1371.

Firdhouse, M.J., Lalitha, P., 2015. Biosynthesis of silver nanoparticles and its applications. Journal of Nanotechnology 2015, 829526. https://doi.org/10.1155/2015/829526.

Fu, P.P., Xia, Q., Hwang, H.-M., Ray, P.C., Yu, H., 2014. Mechanisms of nanotoxicity: generation of reactive oxygen species. Journal of Food and Drug Analysis 22, 64–75.

Fubini, B., Hubbard, A., 2003. Reactive oxygen species (ROS) and reactive nitrogen species (RNS) generation by silica in inflammation and fibrosis. Free Radical Biology and Medicine 34 (12), 1507–1516.

Giri, S., Karakoti, A., Graham, R.P., et al., 2013. Nanoceria: a rare-earth nanoparticle as a novel anti-angiogenic therapeutic agent in ovarian cancer. PLoS One 8 (1), e54578.

Gonzalez, L., Lison, D., Kirsch-Volders, M., 2008. Genotoxicity of engineered nanomaterials: a critical review. Nanotoxicology 2, 252–273.

Hao, N., Yang, H., Li, L., et al., 2014. The shape effect of mesoporous silica nanoparticles on intracellular reactive oxygen species in A375 cells. New Journal of Chemistry 38 (9), 4258–4266.

Hijaz, M., Das, S., Mert, I., et al., 2016. Folic acid tagged nanoceria as a novel therapeutic agent in ovarian cancer. BMC Cancer 16, 220.

Hirst, S.M., Karakoti, A., Singh, S., et al., 2013. Biodistribution and in vivo antioxidant effects of cerium oxide nanoparticles in mice. Environmental Toxicology 28 (2), 107–118.

Huang, C., Aronstam, R.S., Chen, D., et al., 2010a. Oxidative stress, calcium homeostasis, and altered gene expression in human lung epithelial cells exposed to ZnO nanoparticles. Toxicology in Vitro 24 (1), 45–55.

Huang, X., Zhuang, J., Teng, X., et al., 2010b. The promotion of human malignant melanoma growth by mesoporous silica nanoparticles through decreased reactive oxygen species. Biomaterials 31 (24), 6142–6153.

Karlsson, H.L., Cronholm, P., Gustafsson, J., et al., 2008. Copper oxide nanoparticles are highly toxic: a comparison between metal oxide nanoparticles and carbon nanotubes. Chemical Research in Toxicology 21, 1726–1732.

Katsumi, H., Fukui, K., Sato, K., et al., 2014. Pharmacokinetics and preventive effects of platinum nanoparticles as reactive oxygen species scavengers on hepatic ischemia/reperfusion injury in mice. Metallomics 6 (5), 1050–1056.

Khan, I., Saeed, K., Khan, I., 2017 (article in press). Nanoparticles: properties, applications and toxicities. Arabian Journal of Chemistry. https://doi.org/10.1016/j.arabjc.2017.05.011.

Khanna, P., Ong, C., Bay, B.H., et al., 2015. Nanotoxicity: an interplay of oxidative stress, inflammation and cell death. Nanomaterials 5, 1163–1180.

Knaapen, A.M., Borm, P.J.A., Albrecht, C., et al., 2004. Inhaled particles and lung cancer, part A: mechanisms. International Journal of Cancer 109 (6), 799−809.

Kumar, S., Dwivedi, A., Kumar, R., Pandey, A.K., 2015. Preliminary evaluation of biological activities and phytochemical analysis of Syngonium podophyllum leaf. National Academy Science Letters 38 (2), 143−146.

Kumar, S., Pandey, A.K., 2015. Free radicals: health implications and their mitigation by herbals. British Journal of Medicine and Medical Research 7 (8), 438−457.

Lam, C.W., James, J.T., McCluskey, R., et al., 2004. Pulmonary toxicity of single-wall carbon nanotubes in mice 7 and 90 days after intratracheal instillation. Toxicological Sciences 77, 126−134.

Li, J.J., Muralikrishnan, S., Ng, C.T., et al., 2010. Nanoparticle-induced pulmonary toxicity. Experimental Biology and Medicine 235 (9), 1025−1033.

Lin, W., Stayton, I., Huang, Y.-W., et al., 2008. Cytotoxicity and cell membrane depolarization induced by aluminum oxide nanoparticles in human lung epithelial cells A549. Toxicological and Environmental Chemistry 90, 983−996.

Liu, Y.-C., Lin, L.-H., 2004. New pathways Plant mediated synthesis of biomedical silver nanoparticles by using leaf extracts of Citrulluscolosynthis for the synthesis of ultrafine silver nanoparticles from bulk silver substrates in aqueous solutions by sonoelectrochemical methods. Electrochemistry Communications 6, 1163−1168.

Mallick, K., Witcomb, M.J., Scurrell, M.S., 2005. Self-assembly ofsilver nanoparticles in a polymer solvent: formation of a nanochain through nanoscale soldering. Materials Chemistry and Physics 90, 221−224.

Manke, A., Wang, L., Rojanasakul, Y., 2013. Mechanisms of nanoparticle-induced oxidative stress and toxicity. BioMed Research International 2013. Article ID 942916, 15 pages.

Maurya, A., Chauhan, P., Mishra, A., Pandey, A.K., 2012. Surface functionalization of TiO$_2$ with plant extracts and their combined antimicrobial activities against E. faecalis and E. coli. Journal of Research Updates in Polymer Science 1, 43−51.

Medina, C., Santos-Martinez, M.J., Radomski, A., et al., 2007. Nanoparticles: pharmacological and toxicological significance. British Journal of Pharmacology 150, 552−558.

Misawa, M., Takahashi, J., 2011. Generation of reactive oxygen species induced by gold nanoparticles under X-ray and UV irradiations. Nanomedicine 7, 604−614.

Mishra, A., Sharma, A.K., Kumar, S., Saxena, A.K., Pandey, A.K., 2013. Bauhinia variegata leaf extracts exhibit considerable antibacterial, antioxidant, and anticancer activities. BioMed Research International 2013. Article ID 915436, 10 pages.

Mohanta, Y.K., Behera, S.K., 2014. Biosynthesis, characterization and antimicrobial activity of silver nanoparticles by Streptomyces sp. SS2. Bioprocess and Biosystems Engineering 37, 2263−2269.

Morimoto, Y., Horie, M., Kobayashi, N., et al., 2013. Inhalation toxicity assessment of carbon-based nanoparticles. Accounts of Chemical Research 46, 770−781.

Morry, J., Ngamcherdtrakul, W., Gu, S., et al., 2015. Dermal delivery of HSP47 siRNA with NOX4- modulating mesoporous silica-based nanoparticles for treating fibrosis. Biomaterials 66, 41−52.

Nayak, D., Ashe, S., Rauta, P.R., et al., 2016. Bark extract mediated green synthesis of silver nanoparticles: evaluation of antimicrobial activity and antiproliferative response against osteosarcoma. Materials Science and Engineering C58, 44−52. https://doi.org/10.1016/j.msec.2015.08.022.

Nel, A., Xia, T., Madler, L., et al., 2006. Toxic potential of materials at the nanolevel. Science 311, 622−627.

Nel, A., 2005. Air pollution-related illness: effects of particles. Science 308 (5723), 804−806.

Oberdorster, E., 2004. Manufactured nanomaterials (fullerenes, C60) induce oxidative stress in the brain of juvenile largemouth bass. Environmental Health Perspectives 112, 1058−1062.

Oberdorster, G., Maynard, A., Donaldson, K., et al., 2005. Principles for characterizing the potential human health effects from exposure to nanomaterials: elements of a screening strategy. Particle and Fibre Toxicology 2 article 8.

Park, E.-J., Park, K., 2009. Oxidative stress and proinflammatory responses induced by silica nanoparticles in vitro and in vitro. Toxicology Letters 184, 18−25.

Raghunathan, V.K., Devey, M., Hawkins, S., et al., 2013. Influence of particle size and reactive oxygen species on cobalt chrome nanoparticle-mediated genotoxicity. Biomaterials 34 (14), 3559−3570.

Risom, L., Møller, P., Loft, S., 2005. Oxidative stress-induced DNA damage by particulate air pollution. Mutation Research 592 (1−2), 119−137.

Sanjay, S.S., Pandey, A.C., Kumar, S., Pandey, A.K., 2014. Cell membrane protective efficacy of ZnO nanoparticles. SOP Transactions on Nano-Technology 1 (1), 21−29.

Sarkar, A., Ghosh, M., Sil, P.C., 2014. Nanotoxicity: oxidative stress mediated toxicity of metal and metal oxide nanoparticles. Journal of Nanoscience and Nanotechnology 14, 730−743.

Schins, R.P.F., 2002. Mechanisms of genotoxicity of particles and fibers. Inhalation Toxicology 14 (1), 57−78.

Sharma, U.K., Sharma, A.K., Gupta, A., Kumar, R., Pandey, A., Pandey, A.K., 2017. Pharmacological activities of cinnamaldehyde and eugenol: antioxidant, cytotoxic and antileishmanial studies. Cellualr and Molecular Biology 63 (6), 73−78.

Shibuya, S., Ozawa, Y., Watanabe, K., Izuo, N., Toda, T., Yokote, K., Shimizu, T., 2014. Palladium and platinum nanoparticles attenuate aging-like skin atrophy via antioxidant activity in mice. PLoS One 9 (10), e109288.

Sies, H., 1991. Oxidative stress: introduction. In: Sies, H. (Ed.), Oxidative Stress Oxidants and Antioxidants, vols. 15−22. Academic Press, London, UK.

Singh, D., Singh, S., Sahu, J., et al., 2014. Ceramic nanoparticles: Recompense, cellular uptake and toxicity concerns. Artificial Cells Nanomedicine and Biotechnology. https://doi.org/10.3109/21691401.2014.955106.

Sioutas, C., Delfino, R.J., Singh, M., 2005. Exposure assessment for atmospheric Ultrafine Particles (UFPs) and implications in epidemiologic research. Environmental Health Perspectives 113 (8), 947–955.

Siu, K.S., Chen, D., Zheng, X., et al., 2014. Non-covalently functionalized single walled carbon nanotube for topical siRNA delivery into melanoma. Biomaterials 35 (10), 3435–3442.

Smetana, A.B., Klabunde, K.J., Sorensen, C.M., 2005. Synthesis of spherical silver nanoparticles by digestive ripening, stabilization with various agents, and their 3-D and 2-D super lattice formation. Journal of Colloid and Interface Science 284, 521–526.

Smith, K.R., Klei, L.R., Barchowsky, A., 2001. Arsenite stimulates plasma membrane NADPH oxidase in vascular endothelial cells. American Journal of Physiology 280 (3), 442–449.

Sohaebuddin, S.K., Thevenot, P.T., Baker, D., et al., 2010. Nanomaterial cytotoxicity is composition, size, and cell type dependent. Particle and Fibre Toxicology 7 article 22.

Song, J.Y., Kim, B.S., 2008. Biological synthesis of bimetallic Au/Ag nanoparticles using Persimmon (*Diopyros kaki*) leaf extract. Korean Journal of Chemical Engineering 25, 808–811.

Thannickal, V.J., Fanburg, B.L., 2000. Reactive oxygen species in cell signaling. American Journal of Physiology 279 (6), 1005–1028.

Tsibakhashvili, N., Kalabegishvili, T., Gabunia, V., et al., 2010. Synthesis of silver nanoparticles using bacteria. Nano Studies 2, 179–182.

Valko, M., Rhodes, C.J., Moncol, J., et al., 2006. Free radicals, metals and antioxidants in oxidative stress induced cancer. Chemico-Biological Interactions 160 (1), 1–40.

Vallyathan, V., Shi, X., 1997. The role of oxygen free radicals in occupational and environmental lung diseases. Environmental Health Perspectives 105 (1), 165–177.

Wang, J., Sun, P., Bao, Y., et al., 2011. Cytotoxicity of single-walled carbon nanotubes on PC12 cells. Toxicology In Vitro 25, 242–250.

Wilson, M.R., Lightbody, J.H., Donaldson, H., et al., 2002. Interactions between ultrafine particles and transition metals in vivo and in vitro. Toxicology and Applied Pharmacology 184 (3), 172–179.

Xia, T., Kovochich, M., Brant, J., et al., 2006. Comparison of the abilities of ambient and manufactured nanoparticles to induce cellular toxicity according to an oxidative stress paradigm. Nano Letters 6 (8), 1794–1807.

Xia, T., Kovochich, M., Liong, M., et al., 2008. Comparison of the mechanism of toxicity of zinc oxide and cerium oxide nanoparticles based on dissolution and oxidative stress properties. ACS Nano 2, 2121–2134.

Yoshihisa, Y., Honda, A., Zhao, Q.L., et al., 2010. Protective effects of platinum nanoparticles against UV-light-induced epidermal inflammation. Experimental Dermatology 19 (11), 1000–1006.

Yoshikawa, T., Naito, Y., 2002. What is oxidative stress? Japan Medical Association Journal 45 (7), 271–276.

Zhornik, E.V., Baranova, L.A., Strukova, A.M., et al., 2012. ROS induction and structural modification in human lymphocyte membrane under the influence of carbon nanotubes. Biofizika 57, 446–453.

CHAPTER 4

Nanomaterials-Based Next Generation Synthetic Enzymes: Current Challenges and Future Opportunities in Biological Applications

JUHI SHAH[a] • SHRUSTI DAVE[a] • AASHNA VYAS • MAITRI SHAH • HOMICA ARYA • AKHIL GAJIPARA • AKDASBANU VIJAPURA • MALVIKA BAKSHI • PRACHI THAKORE • RUTVI SHAH • VIJAYLAXMI SAXENA • ANMOL SHAMAL • SANJAY SINGH, PHD

INTRODUCTION

Enzymes are macromolecular biological catalysts that accelerate a biochemical reaction by lowering the activation energy, without actually participating in the reaction (Robinson, 2015). Several physiological factors such as pH, temperature, concentration of substrate-enzyme complex, and the presence of inhibitors or activators determine the rate at which an enzymatic reaction proceeds. However, most enzymes function at an optimum temperature, mostly in the range of 30–45°C, and an optimum pH ranging from 4 to 7.5. Most enzymes denature at a temperature higher than 45°C because of the breaking of hydrogen bonds present in their protein structure. Despite the innumerable applications of natural enzymes, they hold certain drawbacks, due to which many efforts have been made toward creating alternative or artificial enzymes. The time- and labor-intensive procedures for enzyme synthesis and purification are some of the important factors, which limits the full potential applications of enzyme technology.

"Nanoscience deals with artificially arranging the atoms, molecules and macromolecules, and studying the properties, which differs significantly from those at bulk counterparts." The materials of small dimension show quantum effects, which is dependent on nanomaterials size and other properties (Filipponi et al., 2010).

Nanoparticles (NPs) exhibiting intrinsic biological enzyme-like characteristics are also termed as nanozymes. In the year 2004, the term "nanozymes" was first coined by Flavio Manea, Florence Bodar Houillon, Lucia Pasquato, and Paolo Scrimin for NPs that mimic the properties of natural enzymes (Manea et al., 2004). Nanozymes offers several advantages over natural enzymes including a higher surface area to volume ratio pertaining to their small size.

Additionally, NPs offer longer shelf life and are relatively stable. Their catalytic activities are resistant to a wide range of pH and temperature conditions. The mass production of NPs can be smoothly facilitated. Nanozymes exhibit size, shape, and pH-dependent catalytic activities that can be easily tuned. A comparison between natural enzymes and nanozymes highlights these unique characteristics of nanozymes along with a large surface area available for modification with biomolecules, which makes them a better candidate for catalysis than natural enzymes and smart response to external stimuli (Wang et al., 2016b). Nanomaterials are mostly reported to mimic the properties of only a few natural enzymes such as peroxidase (Cho et al., 2018; Fu et al., 2017; Yang et al., 2018), oxidase (Cheng et al., 2016; Lu et al., 2015), catalase (Pratsinis et al., 2017; Singh and Singh, 2015; Wang et al., 2016c), superoxide dismutase (SOD) (Jalilov et al., 2016; Naganuma, 2017; Yang et al., 2017), and esterase (Pengo et al., 2005). The intrinsic enzyme-mimicking activities of nanozymes are usually thought to be produced by

[a]Both the authors have contributed equally.

Nanotechnology in Modern Animal Biotechnology. https://doi.org/10.1016/B978-0-12-818823-1.00004-1

atoms that are present on the surface as well as the core. Therefore, the atomic composition of nanozymes is one of the key factors that contributes significantly to their catalytic activity. Various categories of nanomaterials such as bimetallics, trimetallics, and quadrametallics have been reported to exhibit natural enzyme-like characteristics. Nanozymes can also be broadly classified into three categories corresponding to their composition such as metal-based, metal oxide—based, and carbon-based. Metallic NPs such as AuNPs (gold nanoparticles), PtNPs (platinum nanoparticles), CuNPs (copper nanoparticles), AgNPs (silver nanoparticles) as well as bimetallics such as Au-Pt nanorods, Au-Ag nanostructures, Fe-Co alloys, and trimetallics such as Au-Pd-Pt NPs are some of the common examples of metal-based nanozymes. Metal-based nanozymes very often exhibit synergistic effects, which considerably intensifies the overall catalytic performance when coupled with other metallic nanozymes as nanocomposites. Metal oxide—based NPs have been broadly used in the field of biomedical applications such as targeted drug delivery, biosensors, tissue repair, immunoassays, contrast agents in magnetic resonance imaging (MRI), and cell separation. As they are believed to be chemically and biologically inert, additional surface modifications and subsequent conjugation with substances that render functionality is required. Generally, metal oxide—based nanozymes that have undergone significant research are CeNPs (cerium oxide nanoparticles) and IONPs (iron oxide nanoparticles). Carbon-based nanozymes such as fullerenes, graphene oxide, and carbon nanotubes have lured notable interest owing to their unique nanoscale properties. They have been found to bear peroxidase and SOD-mimicking abilities and are largely utilized for detecting analytes in immunoassays, biosensors, and also as signaling agents for signal amplification (Wang et al., 2016b). Nanozymes have gained increasing popularity in recent years and show promising potential as enzyme substitutes. In the following section, we have discussed the different enzyme-like activities shown by different nanomaterials.

NANOMATERIALS EXHIBITING PEROXIDASE-LIKE ACTIVITY

Peroxides, such as hydrogen peroxide (H_2O_2), which are created as by-products in various biochemical reactions within organisms, are potentially dangerous as they act as oxidizing agents. Peroxidases are a large family of protein-based iron porphyrin enzymes that carry out the catalysis of peroxides, most commonly H_2O_2.

Basically, peroxidases utilize H_2O_2 to oxidize a substrate. Nanomaterials exhibiting the peroxidase enzyme-like activities also utilize H_2O_2 to oxidize the peroxidase substrate. IONPs specifically possess an intrinsic peroxidase mimetic activity first reported by Yan and coworkers (Gao et al., 2007). Since this report, nanoscale materials as peroxidase mimetics and their applications in different fields have become a popular subject of research. There has been substantial research on peroxidase mimetic patterns shown by a variety of nanomaterials such as Fe_3O_4 (Fan et al., 2016; Liang et al., 2013), CuO (Deng et al., 2017), Au (Deng et al., 2014; Yokchom et al., 2018), Ag (Sloan-Dennison et al., 2017), AgAu (Han et al., 2015), AuPd (Yang et al., 2016), AgPt (Wu et al., 2016), Zn-CuO (Nagvenkar et al., 2016), and Au-Pt (Peng et al., 2017a).

A general reaction of peroxidase activity can be given by the equation:

$$2AH + H_2O_2 \rightarrow 2A + 2H_2O$$

Or

$$2AH + H_2O_2 \rightarrow 2A + 2\, {}^{\bullet}OH \text{ (Zheng et al., 2016)}$$

IONPs (which include Fe_2O_3 and Fe_3O_4) are known to effectively catalyze the oxidation of peroxide substrates, with very high binding affinity for the substrate 3,3,5,5-tetramethylbenzidine (TMB). IONPs find a wide application as a detection tool or in wastewater treatment (to remove phenols from wastewater) owing to their ability to dilute the toxicity of organic peroxide substrates or cause the color change. Horseradish peroxidase (HRP)—entrapped NPs have been explored for their potential as biocatalysts, bioseparation of DNA and proteins, tissue engineering, MRI, and hyperthermic agents for cancer therapy. The catalytic activity of the Fe_3O_4 NPs depends on the pH of the reaction buffer, temperature, and H_2O_2 concentration, similar to HRP. However, excessive H_2O_2 concentration can hinder the catalytic function of IONPs. Apart from oxidizing TMB, to give a blue color, IONPs also catalyze the oxidation of 3,3'-diaminobenzidine (DAB), which produces a brown color product, and o-phenylenediamine dihydrochloride (OPD) to produce an orange color product. These results demonstrate that the IONPs exhibit peroxidase-like activity similar to the typical peroxidase substrates (Gao et al., 2007).

Cupric oxide NPs (CuONPs) also show intrinsic peroxidase-like catalytic activity by oxidizing TMB in presence of H_2O_2, thus showing the characteristic absorbance at 652 nm. CuONPs are known to oxidize other peroxidase substrates also such as 2,2'-azino-bis

(3-ethylbenzothiazoline-6-sulfonicacid, ABTS) and DAB. Similar to peroxidases, CuONPs also show catalytic activity, which is pH, temperature, and H₂O₂ concentration dependent. With respect to HRP, CuONPs need higher H₂O₂ concentration to achieve the maximum peroxidase activity. This peroxidase-like activity was attributed to the leached copper ions from the NPs in the solution. The K_m value (0.013 mM) of TMB was found to be about 30 times lesser than HRP, whereas K_m value of H₂O₂ was significantly higher than HRP. The stability of CuONPs was investigated and found intact for a wide range of temperatures and pH; therefore, they find many applications in biomedicine and environmental chemistry as biosensors and biocatalysts. They could serve as HRP substitutes for labeling antibodies and for antigen detection through enzyme-linked immunosorbent assay (ELISA) (Chen et al., 2011).

Metallic NPs such as Au, Ag, and Pt also show intrinsic peroxidase activity. Recently, Jv et al. (2010) made a breakthrough discovery that cysteamine-modified AuNPs and bare AuNPs show different peroxidase-like activities. As the surface charge properties of AuNPs directly impact their catalytic property, cysteamine- or citrate-occupied gold surface alters the catalytic activity of AuNPs. Shortly after that discovery, Wang et al. (2012) compared the catalytic activity of amino-modified or citrate-capped AuNPs with unmodified AuNPs where unmodified AuNPs showed significantly higher intrinsic peroxidase activity toward peroxidase substrates, namely, TMB and ABTS. This highlights the role of the charge of NPs as well as the substrate in influencing the magnitude of catalysis (Wang et al., 2012).

Alloy NPs of multiple metals such as AgAu, AgPt, and CuPt are also found to show unique catalytic, electronic, and optical characteristic. He et al. (2010) showed that Ag-related bimetallic alloy NPs (AgAu, AgPd, and AgPt) having a porous structure exhibit peroxidase-like activity. The increased catalytic activity of bimetallic NPs is owed to the synergistic as well as electronic effect. The formation of bimetallic NPs decreases poisoning by CO-like intermediates and retards catalytic degradation. The catalytic performance of AgM alloy NPs can be altered by influencing their morphology, which could be altered by controlling the Ag/M ratio with surfactant concentrations. These NPs can effectively catalyze the oxidation of typical peroxidase reaction substrates in presence of H₂O₂. Apart from size- and shape-dependent regulation, the peroxidase-like activity can also be effectively enhanced by varying the alloy composition as observed in PdAg alloy NPs. The catalytic output of other bimetallic NPs

can be enhanced by extending this approach in future experiments. Bimetallic NPs are better candidates for enzyme mimetics because of their increased stability, low cost of synthesis, and tuneable composition. In near future, such materials can find applications as novel therapeutics for oxidative stress-related diseases such as aging, cancer, and neurodegenerative diseases as well as in environmental monitoring and immunoassays (He et al., 2010).

Nagvenkar et al. confirmed that bimetallic NPs used by doping of CuONPs with zinc (Zn-CuONPs) mimic superior peroxidase-like enzyme activity against various chromogenic substrates such as TMB, OPD, and ABTS in presence of H₂O₂ as shown in Fig. 4.1 (Nagvenkar and Gedanken, 2016). The mechanism obeyed the Michaelis-Menten enzymatic reaction pattern, whereas electron spin resonance (ESR) provided evidence for enhanced enzyme mimetics owing to applications for glucose and antioxidants detection (Nagvenkar and Gedanken, 2016).

Nanohybrids are a class of materials, which typically combine the unique properties of NPs with other functional materials, resulting in entirely new materials with varied and improved applications, especially in sensors and diagnostic devices. Recent research shows that developing nanohybrids can notably heighten the catalytic activity of NPs. Cho et al. (2017) have formulated nanohybrids by combining gold nanoclusters (AuNCs) with IONPs. These constructed nanohybrids greatly enhanced catalytic performance and enabled rapid

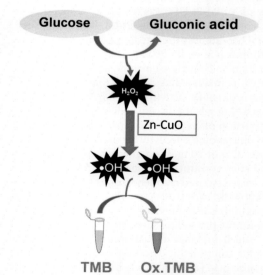

FIG. 4.1 Zn-CuO NPs exhibit the peroxidase-like activity used in the biosensing of glucose.

oxidation of TMB substrate, by virtue of the synergistic effect. Under ideal assay conditions (pH and temperature), K_m values of nanohybrids were found to be significantly lower than those of free IONPs and AuNCs, indicating enhanced substrate affinity. Such nanohybrids are used in biosensors for the colorimetric detection of glucose, owing to their high selectivity and sensitivity toward glucose (Cho et al., 2017).

Wang and coworkers showed that chitosan (CS)-capped NPs of silver halides (AgX, X = Cl, Br, I) (CS-AgX) can exhibit peroxidase-like activities. They oxidized the colorless peroxidase dyes in colored products in the presence of H_2O_2. They could also oxidize the typical substrates in the absence of H_2O_2, when photo activated. Peroxidase mimetics of CS-AgNPs had several advantages over natural peroxidase such as excellent enzyme-like activity over a wide pH range (3.0–7.0), optimal reutilization without much loss of catalytic activity, and also easy regulation of enzymatic activity through light irradiation. These photoactive CS-AgI NPs were used to develop a quick and sensitive colorimetric method for the detection of cancer cells (Wang et al., 2014b).

Ge et al. (2015) explored the potential of trimetallic NPs as candidates for exhibiting intrinsic peroxidase activity. Trimetallic dendritic Au-Pd-Pt NPs catalyze the reduction of H_2O_2 with the help of thionine, an electron mediator, which produces highly amplified electrochemical signals. Therefore, Au-Pt-Pd NPs as nonenzymatic catalysts can be used as signal-amplifying nanoprobes for ultrasensitive and electrochemical sensing of a cancer cell. The electrocatalytic activity of Au-Pd-Pt NPs was by virtue of well-defined AuNP core and a dendritic Pt-Pd bimetallic shell. Au-Pt-Pd NPs amplify the signal as a result of synergetic contribution (Ge et al., 2015). Zhao et al. (2017) reported that Cu@Au-Hg trimetallic amalgam possesses enhanced peroxidase-like activity than bare AuNPs. The peroxidase-like activity of Cu@Au NPs can be improved through the in situ reduction of Hg^{2+}; Au/Hg interaction offers magnified specificity, which leads toward outstanding selectivity for the detection of Hg^{2+} with the LOD of 10 mM in tap water or lake water.

Maji et al. constructed the nanohybrid mesoporous silica—coated reduced graphene oxide NPs surface decorated with folic acids (GSF@ AuNPs), a target for cancer cell detection. GSF@ AuNPs show the peroxidase activity toward the peroxidase substrate TMB in presence of H_2O_2, which provides a dual approach: (1) selective and rapid colorimetric detection of cancer cells and (2) ascorbic acid (AA) and H_2O_2-mediated therapeutics for a cancer cell. In HeLa cells (human cervical cancer cells), cell damage has been activated through the elevated production of $^{\bullet}OH$ radical from exogenous H_2O_2 carried out by GSF@ AuNPs. Endogenous H_2O_2 was generated by AA inside the cell cytoplasm which was also responsible for increased cytotoxicity of cancer cells. In case of normal HEK 293 (human embryonic kidney cells) exposed to hybrid and AA or H_2O_2 shows no clear damage indicating that hybrid has a particular killing impact to cancer cells (Figs. 4.2 and 4.3) (Maji et al., 2015) (Table 4.1)

NANOMATERIALS EXHIBITING OXIDASE-LIKE ACTIVITY

Typically, an oxidase enzyme catalyzes a redox reaction, wherein dioxygen (O_2) acts as an electron acceptor. Oxygen gets reduced to water (H_2O) or H_2O_2 in reactions that involve the donation of a hydrogen atom. Some oxidation reactions, such as the ones that involve xanthine oxidase or monoamine oxidase, typically do not involve free molecular oxygen. Cytochrome oxidase is a key enzyme that assists the body in using oxygen in the generation of energy and is the last component of the electron transfer chain. The oxidase enzymes belong to oxido-reductases classes of an enzyme. Common types of oxidases include glucose oxidase, monoamine oxidase, cytochrome P_{450} oxidase, xanthine oxidase, L-gulonolactone oxidase, NADPH oxidase, lysyl oxidase, and laccase. Nanomaterials such as CeNPs (Cheng et al., 2016), Pt (Yu et al., 2014), and AgNPs (Ni et al., 2015) have found to exhibit oxidase-like catalytic properties. However, as compared with peroxidase mimetics, nanomaterials exhibiting oxidase activity are very few. Some nanomaterials have been studied for their oxidase-mimicking activities. However, the candidates for oxidase mimetics are very rare, as compared with peroxidase. Nanomaterials that have been known to show oxidase-mimicking catalytic properties are CeNPs, Pt, Se, Au-Pt as well as AgNPs. There has been limited research on nanomaterials exhibiting oxidase enzyme-like activity. Asati et al. (2009) have reported that polymer-coated CeNPs show intrinsic oxidase activity at acidic pH values because of their ability to switch reversibly from Ce^{3+} to Ce^{4+}, highlighting their utility in providing protection against oxidative stress and radiation damage. Catalysis by CeNPs is not solely pH-dependent but also depends on the size of the CeNPs, and the thickness of the dextran coating plays pivotal roles. As opposed to peroxidase-catalyzed reactions that require H_2O_2 as the primary electron acceptor or oxidizing agent, CeNPs can quickly catalyze the oxidation of TMB and ABTS as well as dopamine, which is a catecholamine with difficulty in oxidizing at low pH

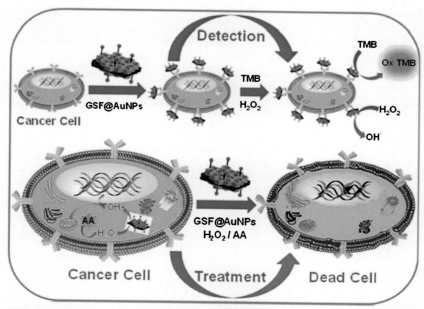

FIG. 4.2 Schematic illustration of peroxidase-like activity of GSF@ AuNPs for cancer cell detection and therapeutic cancer treatment. (Reprinted from Ref. (Maji et al., 2015) with permission, copyright © American Chemical Society (ACS).)

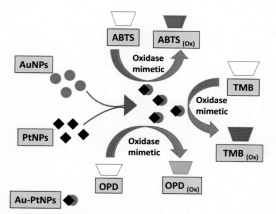

FIG. 4.3 Bimetallic Au-PtNPs possess intrinsic oxidase-like activity towards TMB, ABTS and OPD substrates.

values, in the absence of H_2O_2. CeNPs have been widely used in catalytic converters of automobile exhausts as a fuel cell electrolyte and as an ultraviolet absorber. When conjugated with targeting ligands, CeNPs act as efficient nanocatalyst and detection tools in immunoassays (Asati et al., 2009).

Luo et al. showed for the first time that selenium NPs (Se NPs) possess intrinsic oxidase-like activity by catalyzing the oxidation of TMB by dissolved oxygen (Guo et al., 2016). They exhibited optimum catalytic activity at pH 4 and at 30°C. The Michaelis-Menten constant (K_m) value was 0.0083 mol/L with a maximum reaction velocity (V_{max}) value of 3.042 μmol/L min. Se NPs possessed oxidase-like activity that was directly proportional to their concentration and inversely to their size. The oxidase-mimicking activity of Se NPs was relatively lower than other nanozymes but even then, Se NPs show commendable biocompatibility, rendering them as significant oxidase mimetics. The oxidase mimetic patterns of Se NPs provide new molecular insight into the biological function of Se NPs as observed in cell and animal experiments and expand their application in biosensors, nanomedicine, and biochemistry.

He et al. (2011) showed that the ability of Au-Pt nanostructures to effectively catalyze the oxidation of AA that bears similarity with the functioning of natural oxidase enzyme. Au-Pt nanostructures were used in place of HRP and were evaluated for their oxidase-like activity with TMB and OPD as substrates wherein IL-2 was detected as proof in ELISA. These nanocatalysts are easily modulated by variations in composition and alloy formation, making them unique,

TABLE 4.1
List of NPs Exhibiting Peroxidase-Like Activity.

NPs	Size/Shape	Key Observations	Applications	Reference
IONPs	– 300 nm	A colorimetric detection assay that was composed of IONPs, acetyl cholinesterase (AChE), and choline oxidase catalyzes the production of H_2O_2, which triggers IONPs to oxidize colorimetric substrates. After incubating with the organophosphorus (OP) neurotoxins, the AChE enzymatic activity was inhibited and it produced less H_2O_2, thereby resulting in reduced oxidation of colorimetric substrates.	OP pesticides detection, a rapid method for screening of OP neurotoxins	Liang et al. (2013)
AuNPs	Spherical 30.5 nm	Hg stimulates the peroxidase-mimicking activity of AuNPs very selectively and sensitively. Sodium citrate reduces Hg^{2+} to Hg^0 and Hg^0 deposited on the AuNPs, thus increasing peroxidase-like activity.	Development of biosensors, label-free assay for detection of Hg^{2+}	Long et al. (2011)
Apoferritin-paired gold clusters (Au-Ft)	Oval <2 nm	Apoferritin-paired gold clusters (Au-Ft) catalyzed the oxidation of TMB by H_2O_2 to form a blue color reaction with higher activity near acidic pH and a broad range of temperatures. The kinetic parameters showed a lower K_m value (0.097 mM) and a higher K_{cat} value (5.8×10^{-4} s^{-1}) than HRP.	Glucose detection, biosensing, biomedicine, clinical diagnostics	Jiang et al. (2015)
Dendrimer-encapsulated PtNPs	– 1.25 ± 0.27 nm	PtNPs encapsulated with amine-terminated dendrimers showed intrinsic peroxidase-like activity for sensitive colorimetric analysis. They provide high-density amine groups at the terminals of dendrimers, enabling easy conjugation of graphene oxide (GOx) without the need for coating layers on the PtNPs.	Colorimetric analysis of glucose	Ju and Kim (2015)
CeNPs/NiO nanocomposite	Coral-like shape 2.4 nm	CeNPs/NiO nanocomposites displayed improved peroxidase-like activity compared with that of pure NiO or CeNPs. Their catalytic activity depended on the cerium content with the optimal content being 2.5%. Their heightened catalytic activity was by virtue of their higher ability of electron transfer due to the incorporation of cerium.	Construction of H_2O_2 sensor had a low detection limit with 0.88 μM and a wide linear range from 0.05 to 40 mM	Mu et al. (2017)

AuNPs	Spherical 8.1 ± 1.1 nm	Melamine addition improved the peroxidase-like activity of AuNPs. Melamine's colorimetric detection limit was 0.2 nM. Naked eye detection limit was 0.5 μM.	Melamine detection	Ni et al. (2014)
Pd-Au bimetallic NPs	Rod shaped —	Palladium-gold nanozyme exhibits brilliant peroxidase mimetic activity with OPD in the presence of H_2O_2, with better K_m values compared with HRP. The validation on tap water samples that were spiked with varying malathion concentrations showed decent recovery in the range of 80%—106%.	Highly sensitive detection method for malathion	Singh et al. (2017b)
IONPs	Spherical 350 nm	Glucose oxidase hydrolyzed glucose to H_2O_2 in the presence of oxygen that was followed by the activation of Fe_3O_4 NPs, finally leading to the oxidation of benzoic acid (BA). Fe_3O_4 H_2O_2/BA system's detection limit is 0.008 μM for H_2O_2, 0.025 μM for glucose, and 0.05 μM for p-nitro phenol.	Peroxidase-based method for fluorescent method for detection of glucose and p-nitrophenol	(Shi et al., 2014)
3D graphene-IONPS AuNPs	Crystal structure AuNPs 15 nm Fe_3O_4 NPs 7 nm	A 3D graphene/Fe_3O_4-AuNP displayed flexibly shuttling peroxidase-like activity. As compared with the traditional 2D graphene-based monometallic composite, this 3D structure greatly enhanced the catalytic activity, the catalysis velocity, and the substrate affinity. The catalytic activity could be tuned by the adsorption and desorption of single-stranded DNA molecules.	Label-free sensing of glucose, sequence-specific DNA, mismatched nucleotides and oxy-tetracycline, environmental monitoring, gene detection, and molecular diagnosis	Yuan et al. (2016)
Carbon nanodots (C-dots)	Spherical 2.1 ± 0.7 nM	C-dots showed peroxidase-like activity toward peroxidase substrate TMB, OPD, and pyrogallol in presence of H_2O_2. Detection limit for visual observation of glucose response is 0.001 mM.	The method was used to detect glucose level in the serum sample	(Shi et al., 2011)
Helical carbon nanotubes (HCNTs)	Twin helices 150—200 nM	HCNT exhibit intrinsic peroxidase-like activity with high catalytic efficiency.	HCNTs were used to construct biocatalyst and an amperometric sensor for detection of H_2O_2	(Cui et al., 2011)

Continued

TABLE 4.1
List of NPs Exhibiting Peroxidase-Like Activity.—cont'd

NPs	Size/Shape	Key Observations	Applications	Reference
WO$_x$ QDs (water-soluble quantum dots)	Spherical 1.33 ± 0.5 nm	WO$_x$ QDs have shown the intrinsic peroxidase-like activity for peroxidase substrate ABTS in presence of H$_2$O$_2$ and produce a green color. Two $^\bullet$OH radicals are generated by cleavage of O–O bond of H$_2$O$_2$. By partial electron exchange interaction, the WOx QDs stabilize the $^\bullet$OH radical, which may be a contributing factor in catalytic efficiency of WOx QDs.	Detection of glucose concentration in human serum samples. It can be used for the detection of H$_2$O$_2$	Peng et al. (2017b)

stable and highly effective oxidase mimics (He et al., 2011). PtNPs possess catechol oxidase-mimicking activity, which oxidizes polyphenols into the corresponding o-quinones. This was demonstrated by Liu et al. (2015) using approaches such as UV-Vis spectroscopy, ultrahigh-performance liquid chromatography, and ESR techniques. Furthermore, they established the higher efficiency of PtNPs toward catalyzing the oxidation of polyphenols than the natural enzyme. PtNPs may find extensive utility in cosmetics and dietary supplements as they have the potential of influencing or altering antioxidant activities of polyphenols (Liu et al., 2015).

Wang and coworkers recently established that Hg^{2+} enabled citrate-capped AgNPs (Cit-AgNPs) catalyze the oxidation of TMB by dissolved oxygen. These oxidase mimetics are pH, temperature, and concentration (Hg^{2+}) sensitive and have a higher affinity for Hg^{2+}. Hg–Ag alloys were formed, which activates oxygen to generate superoxide anions, further contributing to the catalytic activity for TMB oxidation. They found that Hg^{2+} can stimulate the oxidase-like activity of Cit-AgNPs which enables a rapid colorimetric assay for Hg^{2+} (Wang et al., 2014a). Cai et al. conjugate aptamer with PtCo bimetallic NPs that possess high oxidase-like activity, which can be used for the development of magnetic-improved colorimetric assay. Deposition of magnetic Co atoms into NPs leads to enhanced oxidase-like activity, which was used for the detection of

cancer cells without the use of destructive H$_2$O$_2$ (Figs. 4.4–4.6) (Cai et al., 2016) (Table 4.2)

NANOMATERIALS EXHIBITING SOD-LIKE ACTIVITY

SOD is an enzyme catalyzing the partitioning or dismutation of the superoxide (O$_2^{\bullet-}$) radicals into either H$_2$O$_2$ or O$_2$. The by-product formed as a result of oxygen metabolism is known as O$_2^{\bullet-}$. Just like peroxides, O$_2^{\bullet-}$ is potentially dangerous to cells and must be regulated or degraded. Hence, the catalytic reaction by SOD enzyme is a key antioxidant defense in almost all living cells that are exposed to oxygen. SOD enzymes remove an electron from the O$_2^{\bullet-}$ molecules, thus converting the O$_2^-$ into less damaging species. There has been substantial research on the SOD-mimicking properties of CeNPs. Other nanomaterials that mimic natural SOD are Cu(OH)$_2$ NPs and certain metals such as Pt and Au. Furthermore, fullerenes also show very efficient SOD-like catalytic activity.

CeNPs mimic SOD activity owing to their ability to switch between two valence states Ce^{3+} and Ce^{4+} coexisting on their surface. An increase in the level of cerium in the 3+ oxidation state, i.e., a higher Ce^{3+}/Ce^{4+} ratio triggers the SOD-like activity of CeNPs, causing the dismutation or portioning of O$_2^{\bullet-}$ to H$_2$O$_2$ or molecular oxygen (Babu et al., 2014; Ceccone and Shard, 2016). CeNPs are advantageous over natural enzymes, offering

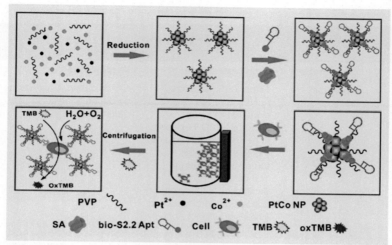

FIG. 4.4 Schematic illustration of synthesis of bimetallic NPs-aptamer nanoconjugates for magnetic separation and detection of target cancer cells. (Reprinted with permission from Ref. (Cai et al., 2016) copyright © Royal Society of Chemistry (RSC).)

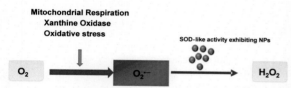

FIG. 4.5 NPs exhibiting SOD-like activity to scavenge $O_2^{\bullet-}$ generated by various processes.

multiple catalytic sites (oxygen vacancies) that are usually not destroyed postreaction with reactive oxygen species.

Korsvik et al. analyzed two CeNPs preparations for size, surface chemistry, and X-ray diffraction pattern, testing them for their ability to react with $O_2^{\bullet-}$ (Ragini Singh et al., 2016). Analysis revealed higher

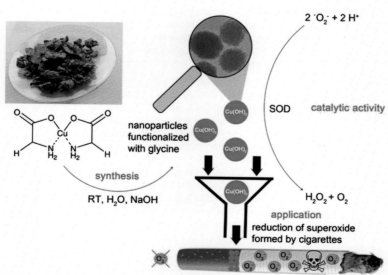

FIG. 4.6 Synthesis, enzymatic activity, and application of Gly-Cu(OH)$_2$ NPs. Cu(Gly)$_2$ was prepared by mixing aqueous solutions of copper(II) acetate and glycine. Addition of NaOH to a solution of Cu(Gly)$_2$ in water at room temperature (RT) led to the formation of Gly-Cu(OH)$_2$ NPs with high intrinsic SOD activity. As-synthesized NPs as part of cigarette 20 filters reduce the concentration of $O_2^{\bullet-}$, produced by commercial cigarettes. (Reprinted with permission from Ref. (Korschelt et al., 2017) © Royal society of Chemistry (RSC).)

TABLE 4.2
List of NPs Exhibiting Oxidase-Like Activity.

NPs	Size/Shape	Key Observations	Applications	References
AuNPs	Spherical 6 nm	Triton X-100-stabilized AuNPs catalyzed the oxidation of NADH. Reported NADH oxidase activity was 0.08 turnover/min in air and 0.32 turnover/min in argon. The catalytic activity diminishes with increasing concentration of gold and increasing size of AuNPs.	CO oxidation, liquid-phase oxidation of glycols and aldehydes.	Kulikova (2005)
BSA-templated MnO_2 NPs	– 4.5 nm	BSA-MnO_2 NPs were utilized as colorimetric immunoassay labels for the detection of goat antihuman IgG instead of HRP. These NPs have good stability, dispersion, and biocompatibility.	BSA-MnO_2 provides the facile and biocompatible labeling technique for the biosensor development, which can be used in biotechnology, bioassay, and biomedicine.	(Liu et al., 2012)
PdNPs	Hexagonal 40 nm	2-D Pd-based nanostructures (e.g., Pd nanosheets, Pd-Au and Pd-Pt nanoplates) bear intrinsic oxidase-like activity. They were capable of activating H_2O_2 or dissolved oxygen for catalyzing the quick oxidation of various organic substrates, and scavenge H_2O_2 to produce oxygen.	Colorimetric detection of glucose, electrocatalytic reduction of H_2O_2, biocatalysis, bioassays, nanobiomedicine.	(Wei et al., 2015)
Citrate-capped PtNPs	Spherical 6 nm	Citrate-capped PtNPs effectively oxidize TMB, dopamine, ABTS, and methylene blue in the presence of O_2, through a four-electron exchange process. The suggested system made the determination of the therapeutic heparin concentration in 1 drop of blood possible.	Clinical diagnosis, colorimetric heparin sensors. Adenosine determination in urine samples.	You et al. (2017)
Iridium NPs (IrNPs)	– 2.5 ± 0.5	Dissolved oxygen (DO) molecules forming $O_2^{\cdot-}$, which in turn will oxidize TMB in presence of IrNPs and produce a blue color product.	Development of colorimetric chemosensor for measurement DO with an LOD of 4.7 μM.	(Cui et al., 2017)

CoNPs/MC (cobalt NPs coated in metal-organic framework—derived carbon)	Quasispherical 200 nm	Metal Co atoms will incorporate into magnetic carbon NPs, which gives high peroxidase-like activity.	Construction of biosensor for detection of glucose in biological samples.	(Dong et al., 2018)
Au-Ag-ICPs hollow core-shell nanostructures	Spherical aggregates 14.6 nm	Au-Ag-ICPs hollow core-shell nanostructures display highly specific and efficient oxidase-like activity without requiring any cosubstrate. Owing to the synergistic effect between Ag-DMcT ICPs and the greatly dispersed tiny gold NPs, the formed 3D nanoassemblies have a brilliant catalytic performance with regard to TMB degradation.	Active precursors and 3D host frameworks for the infiltration of HAuCl₄, wastewater treatment.	(Wang et al., 2015)

intensity peaks in the sample with a smaller particle size, indicating a higher Ce^{3+}/Ce^{4+} ratio. The specific surface areas of the two test samples were found to be 140 and 115 m^2/g, respectively. When compared with ferricytochrome C for $O_2^{\bullet-}$ reduction in a SOD activity assay, the sample with a smaller particle size competed efficiently than CeNPs of larger, providing an inverse co-relation of particle size with catalytic activity. CeNPs with a higher Ce^{3+}/Ce^{4+} ratio exhibit excellent SOD-like activity with a catalytic rate constant that exceeds the one determined for natural SOD. The SOD mimetic activity of CeNPs holds numerous applications such as the treatment of diseases associated with oxidative stress, neurodegenerative disorders, autoimmune diseases, cancer, ischemia, diabetes, lifespan extension by rendering protection against ROS, as catalytic convertors in automobile exhaust systems, as UV absorbents, and as electrolytes in fuel cells (Korsvik et al., 2007). Singh et al. demonstrated that PEGylated CeNPs exhibit the SOD-like activity, which has a capability in ROS scavenging even in the low concentration of natural cellular antioxidant glutathione (GSH). CeNPs inhibit the free-radical generation produced by buthionine sulfoximine and scavenged ROS unvaryingly from the cell cytosol (Singh et al., 2016).

Singh et al. studied the changes occurring in the catalytic properties of CeNPs under different biological conditions in vitro. They showed that the catalytic behavior of CeNPs is altered in the presence of phosphate. The phosphate-treated CeNPs showed an increase in catalase mimetic activity and a subsequent decrease in SOD mimetic activity. CeNPs in 4+ oxidation state accepts the electron and decompose H_2O_2 to produce H_2O, whereas phosphate-treated CeNPs in 3+ oxidation state catalytically reduce H_2O_2, showing catalase mimetic activity. CeNPs were found to be resistant to broad pH change, and CeNPs conjugate forms are easily taken up by cells in culture,

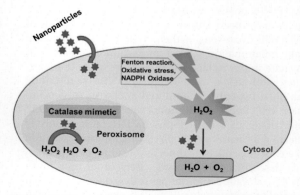

FIG. 4.7 NPs exhibiting catalase mimetic activity in the cytoplasm and peroxisome converting H_2O_2 into H_2O and O_2.

but the presence of phosphate anions can alter their characteristics. The SOD activity of CeNPs was not affected by the acidic or basic environment (Singh et al., 2011).

Korschelt et al. demonstrated that glycine-functionalized copper hydroxide NPs show intrinsic SOD activity (Korschelt et al., 2017). Gly-Cu(OH)$_2$ NPs were incorporated into cigarette filters containing high concentrations of toxic radicals, including ROS and reactive nitrogen species (Fig. 4.7). A significant reduction in the ROS concentrations was observed, showing no toxic side effects. Bulk Cu(OH)$_2$ NPs being insoluble in water are catalytically inactive, whereas the synthesized Gly-Cu(OH)$_2$NPs with good water dispersibility competently catalyzed the decomposition of O$_2^{\bullet-}$. A paramagnetic shift in the NMR signal of water indicates that Cu shuttles between two oxidation states $+2$ and $+1$, during dismutation of the O$_2^{\bullet-}$. A colorimetric UV assay for superoxide dismutation using a superoxide sensitive dye, iodonitrotetrazolium chloride (INT), showed an even high intrinsic SOD activity of Gly-Cu(OH)$_2$NPs than CeNPs with a rate constant = kINT = $3.94 \cdot 104$ M/s. The oxidative damage that results from exposure to ROS and other free radicals is the cause for smoking-induced cancer, cardiovascular, and pulmonary diseases; hence such SOD-like formulations can be incorporated in cigarettes to decrease the free radical–mediated damage in the respiratory tract. Thus, Gly-Cu(OH)$_2$NPs could be used as SOD replacements in cosmetic creams to retard aging, control pigmentation, and provide UV protection. Furthermore, Gly-Cu(OH)$_2$NPs can lower the mutagenicity effects of other SOD substitutes like quercetin (Korschelt et al., 2017).

Metal NPs possess multiple enzyme mimetic properties. PtNPs possess an intrinsic SOD mimetic activity. PtNPs have been known to preserve the viability of HeLa cells by scavenging their UV-induced ROS, reduce lung inflammation, confer protection to mice from cigarette smoke extract, and display antiaging characteristics. Polyacrylic acid–capped PtNPs significantly inhibited hyperthermia-induced apoptosis in two human cell lines—myelomonocytic lymphoma U937 and cutaneous T-cell lymphoma HH cells, negatively affecting all pathways involved in apoptosis execution in a dose-dependent manner and leading to hyperthermia desensitization. However, they could only lower the O$_2^{\bullet-}$ elevation and not peroxides, questioning their durability in scavenging ROS. ROS elevation was selectively suppressed in both the cell lines, which is indicative of the other modes of action expended by PtNPs apart from their SOD-like activity. Citrate and pectin-stabilized bimetallic NPs consisting of alloy-structured

AuPtNPs (CP-Au/PtNPs) quenched H$_2$O$_2$ and O$_2^{\bullet-}$ in a concentration-dependent manner, bringing to light their SOD-like activity, which could be used in the medical treatment of diseases originated due to oxidative stress (Yoshihisa et al., 2011).

Buckminsterfullerene or fullerene, an allotropic crystal form of carbon, is a nanostructure that along with its water-soluble derivatives displays very promising SOD-mimicking activity toward many ROS such as H$_2$O$_2$, nitric oxide, $^\bullet$OH, O$_2^{\bullet-}$, and singlet oxygen. The localized electron-deficient surface portions act as preferred reaction sites where O$_2^{\bullet-}$ is absorbed, contributing to ROS scavenging activity. The fullerene surface remains structurally unaltered, further affirming a catalytic mechanism. Ali and coworkers demonstrated in 2004 that the C3 tris-malonyl-C60-fullerene derivative is a potential SOD mimetic. SOD-deficient mice were used as a test system to determine whether C3 could eliminate O$_2$. It was confirmed that C3 could remove O$_2$ slower than SOD1 and SOD2 by virtue of catalysis and not direct O$_2$ scavenging. It was proposed that C3 electrostatically drives O$_2^{\bullet-}$ toward the electron-deficient areas around the malonic acid groups, eventually allowing dismutation of O$_2$. Quick et al. demonstrated in 2006 that orally administered C3 could reverse alterations associated with aging in the T-lymphocyte CD^{4+}/CD^{8+} ratio and also the age-associated loss of B cells in mice, drawing attention to the role of C3 in reducing age-linked mitochondrial O$_2^{\bullet-}$ production and improvisation of the mitochondrial function (Korschelt et al., 2017). Fullerene, being an excelled SOD mimetic, has found various applications such as protecting human keratinocytes from apoptosis caused by UV exposure, protecting zebrafish embryos from radiation-induced alterations, treating neurological disorders by curbing oxidative stress–induced apoptosis, and providing cardioprotection in reperfusion injuries and coronary occlusion (Ali et al., 2004) (Table 4.3)

NANOMATERIALS EXHIBITING CATALASE-LIKE ACTIVITY

Catalase is a tetrameric, porphyrin containing enzyme present in almost all the living organisms, which are exposed to oxygen (Vainshtein et al., 1981). Catalase triggers the catalytic decomposition of H$_2$O$_2$, which is a potentially harmful by-product of various metabolic processes, to produce water (H$_2$O) and oxygen, mediated by its four porphyrin heme groups. It plays a primary role in providing protection to the cell from oxidative stress caused by ROS. Interestingly, catalase bears the highest turnover number among enzymes;

TABLE 4.3
List of NPs Exhibiting SOD-Mimetic Activity.

NPs	Size/Shape	Key Observations	Applications	Reference
PtNPs	Spherical —	PtNPs show antioxidant activity against X-irradiation-induced apoptosis in the human lymphoma U937 cells, revealing a major reduction in apoptosis. Cells pretreated with PtNPs showed a critical decrease in ROS production, expression of Fas, and loss of mitochondrial membrane potential, whereas inhibition in the activity of caspase-3 was also reported.	Antiapoptotic effects, protection against UV-induced inflammation.	Jawaid et al. (2014)
CeNPs-functionalized polycaprolactone (PCL)-gelatin nanofiber (PGNPNF)	Quasipherical 42 nm	CeNPs exhibiting SOD mimetic activity demonstrate the antioxidant effect of PGNPNF in different buffer systems. SOD activity of CeNPs protects 3T3-L1 cells from the oxidative damage.	Wound healing due to antioxidant property.	Rather et al. (2017)
PtNPs	Spherical 2.4–0.7 nm	PtNPs are known to prolong the lifespan of *Caenorhabditis elegans* by antioxidant treatment. At 0.5 mM concentration, PtNPs greatly extended the lifespans of wild-type N2 nematodes while at 0.25 and 0.5 mM concentrations. 0.5 mM PtNPs majorly decreased lipofuscin accumulation and paraquat-induced ROS, rendering PtNPs as a more potent SOD mimetic compared with EUK-8.	Antiaging, increased survival time against oxidative stress, protection against ROS-mediated diseases.	Kim et al. (2008)
MnO (manganese oxide) NPs	Spherical 7.6 ± 0.7 nm	MnONPs catalyze the decomposition of $O_2^{\bullet-}$ with a greater efficiency compared with native Mn-SOD.	Dual property of MnO includes SOD-like activity with enhanced MRI contrast, which is useful for detection of oxidative stress —promoted tumor progression and metastases or as an MRI imaging probe.	Ragg et al. (2016)

Continued

TABLE 4.3
List of NPs Exhibiting SOD-Mimetic Activity.—cont'd

NPs	Size/Shape	Key Observations	Applications	Reference
Ce-intercalated titanate nanosheets	Spherical Anisotropic 10 nm	Ce-doped titanate nanosheets (Ce-TNS) bear SOD mimetic activity against $O_2^{\bullet-}$. Ce-TNS activity was in excess of CeNPs NPs due to Ce-TNS's high dispersion stability, leading to a huge reaction area. The DNA molecules were protected from UV-induced oxidative damage by Ce-TNS.	UV light screening, ROS annihilation.	Kamada and Soh (2015)
CeNPs	Spherical 5–8 nm	Electron paramagnetic resonance analysis certifies CeNPs reactivity as an SOD mimetic. Decreasing Ce 3+/4+ ratio correlates directly with a fall in SOD mimetic activity. The oxidation state of CeNPs on the surface plays a chief role in its SOD mimetic activity, and its ability to quench $O_2^{\bullet-}$ is directly correlated to the concentrations of Ce^{3+} at the particle surface.	Protection against cellular senescence and inflammation, highly effective antioxidant.	Heckert et al. (2008)
CeNPs	Spherical agglomerates 3–5 nm	CeNPs offer protection to the gastrointestinal epithelium against oxidative damage induced by radiations, hence acting as a free-radical scavenger.	Protection against radiation-induced damage, reduction in inflammation, and fibrosis	Colon et al. (2010)

one molecule of catalase enzyme is reported to quickly convert millions of H_2O_2 molecules to oxygen and H_2O (Goodsell, 2004). Human catalase has an optimum pH of 7 and an optimum temperature of $45°C$ (Chance and Maehly, 1955). Catalase has various applications in the food industry, textile industry, esthetics and cosmetics, as a disinfectant, and so on. Certain NPs are known to show intrinsic catalase-like activity and have diverse uses based on their catalytic behavior. Usually, NPs that mimic catalase activity usually also exhibit SOD-like or peroxidase-like activity (Chen et al., 2012). Commonly found catalase mimetics are CeNPs (Pirmohamed et al., 2010), AgNPs (Krishnaraj et al., 2012), and a few others.

CeNPs exhibit catalase-like activity owing to their ability to shift between the two oxidation states Ce^{3+} and Ce^{4+} coexisting on their surface. An increase in the level of cerium in the 4+ oxidation state, i.e., a higher Ce^{4+}/Ce^{3+} ratio triggers the catalase-like activity of CeNPs, causing the catalytic decomposition of H_2O_2 to H_2O or molecular oxygen. Pirmohamed et al. (2010) proposed that CeNPs show redox state-dependent catalase mimetic activity when detected by the Amplex Red assay at a H_2O_2 concentration of 2 mM. This assay revealed that the CeNP preparation with a higher percentage of cerium in the 4+ state displayed significant catalase-like activity $(9.08 \times 10^{-4} \, nmol/min)$ in contrast to the preparation with a higher level of cerium atoms in the 3+ state with nil catalase-like activity, suggesting that the breakdown of H_2O_2 by CeNPs depended on their 3+/4+ surface charge ratios. Similar results were obtained when H_2O_2 concentrations were altered as well as when changes in the DO levels were measured (Pirmohamed et al., 2010). Singh et al. (2011) demonstrated that the catalase mimetic activity of CeNPs (4+) can resist to the exposure of phosphate ions. CeNPs preserved their catalytic properties even in severe environments, as supported by their stability over different pH solutions. However, upon incubation of CeNPs in phosphate buffer, it was observed that there were almost no changes in their surface oxidation states or catalase mimetic activity when exposed to 50 μM phosphate concentration and thus no change in zeta potential and hydrodynamic size up to 100 μM phosphate concentration, pointing toward the low affinity of CeNPs (4+ oxidation state) toward phosphate concentration. Furthermore, the redox cycling between the 3 + and 4 + oxidation states in these phosphate-incubated CeNPs (4+) was not blocked by the phosphate ions, in contrast to the case of CeNPs (3+). All these findings reveal that exposure to phosphate anions does not bring about any considerable alteration in the surface chemistry of CeNPs (4+) and hence their catalase mimetic activity. As a result of their oxidative stress countering properties, CeNPs have been used in animal and cell models, inclusive of ROS involvement in cardiovascular disease, radiation damage, and retinal degeneration (Singh et al., 2011). Singh and Singh (2015) demonstrated that catalase mimetic activity of CeNPs (4+) did not alter by the interaction of NPs with phosphate anions, pH changes, and cell culture components.

NANOMATERIALS EXHIBITING DUAL ENZYME-LIKE/BIFUNCTIONAL ACTIVITY

Certain nanomaterials exhibit dual enzyme-like activity, i.e., mimic two enzymes at the same time. They often exhibit catalase/SOD-like activity, catalase/peroxidase-like activity, catalase/oxidase-like activity, or peroxidase/oxidase-like activity.

Catalase and peroxidase-like activity: Certain nanomaterials possess catalytic enzymes similar to both natural catalase and peroxidase, as they have the ability to break down H_2O_2. Li Su and coworkers demonstrated that MFe_2O_4 (M = Mg, Ni, Cu) IONPs bore catalytic properties similar to natural catalase and peroxidase. MFe_2O_4 shows peroxidase-like activity by utilizing H_2O_2 to oxidize typical peroxidase substrates, while it also catalyzes the decomposition of H_2O_2 to form H_2O and molecular oxygen. Furthermore, it was observed that the number of $^{\bullet}OH$ and O_2 that were generated had a correlation with the concentration of MFe_2O_4 NPs. $NiFe_2O_4$ NPs were used as colorimetric biosensors for detecting 9.4×10^{-7} to

2.5×10^{-5} mol/L glucose under optimal conditions. This proposed sensor was further successfully utilized to determine glucose in urine samples. Depending on further developments, these NPs can be exploited in different fields such as medical diagnostics, biotechnology, and environmental monitoring (Su et al., 2015). Zhang et al. demonstrated that peroxidase and catalase-like activity of Co_3O_4 nanopolyhedrons, nanocubes, nanoplates, and nanorods and their activity is attributed to the exposure of crystal plane with a higher density of Co^{3+}. Dimercaptosuccinic acid (DMSA)—modified Co_3O_4 NPs of polyhedrons shape were synthesized by coprecipitation method. These NPs were used as a suitable candidate for the immunohistochemical (IHC) detection of epidermal growth factor receptor overexpressed in non—small-cell lung cancer cells/tissues (Zhang et al., 2017).

Catalase and SOD-like activity: A few NPs possess both catalase and SOD-like properties because of their ability to scavenge both $O_2^{\bullet-}$ and H_2O_2 radicals. Yoshihisa et al. proposed that PtNPs possess antitumor activity and are capable of scavenging peroxides and $O_2^{\bullet-}$, which labels them as potential SOD/catalase mimetics. Their dual potential can be manipulated to get protection from oxidative stress—associated disorders. In this study, effect of different concentrations of PtNPs in combination with hyperthermia (44°C, 30 min)-induced apoptosis was studied. Apoptosis can be induced by hyperthermia owing to the formation of intracellular $O_2^{\bullet-}$. Hyperthermia-induced apoptosis was studied by investigating the effect of polyacrylic acid—capped PtNPs on human myelomonocytic lymphoma U937 and cutaneous T cell lymphoma HH cells. However, the results stated that these nanomaterials can inhibit HT-induced apoptosis (Yoshihisa et al., 2011).

Oxidase and SOD-like activity: In recent years, nanozymes have attracted much attention. Many metal-based nanozymes or metallic nanozymes show oxidase as well as SOD-like activity, acting as both prooxidants and antioxidants at the same time. Shen et al. demonstrated that metal and alloy nanomaterials possess oxidase and SOD-like activities. The dissociation of O_2 on metal surfaces accounts for the oxidase-mimicking activities of metals, which can be predicted by density functional theory (DFT)—calculated activation and reaction energies. The activity was found to mainly depend on two factors, metallic compositions and the exposed facets. These oxidase mimetic metallic NPs also display SOD-like activities through the protonation of O_2 and the subsequent adsorption and rearrangement of H_2O on the surfaces of metals. These results grant atom-level insights into the SOD and oxidase-like activities of metals and take the plunge for rational designs of enzyme mimetic metal nanomaterials. Furthermore, the dissociative adsorption mechanism of O_2 helps to comprehend the catalytic abilities of nanomaterials as prooxidants and antioxidants. These bifunctional oxidase/SOD mimetic NPs include metals such as PtNPs, AgNPs, AuNPs, PdNPs, and their alloys (Shen et al., 2015). CeNPs, which are metal oxide—based nanomaterials, also possess oxidase as well as SOD-like activities (Shen et al., 2015).

Peroxidase and oxidase-like activity: A wide range of NPs coexist as oxidase as well as peroxidase mimetics. Very recently, Wang et al. proposed the multienzyme-like functionality of nickel-palladium hollow NPs (NiPdhNPs). NiPdhNPs could rapidly oxidize TMB, ABTS, and OPD, which are typical peroxidase substrates in the presence of H_2O_2, giving different colors, indicating their peroxidase-like activity. Furthermore, they could even catalyze the direct oxidation of TMB, producing a blue color, without the need of H_2O_2, suggesting their oxidase-like activity; the mechanism responsible for multiple enzyme-like activities could be owed to the formation of $^\bullet OH$ during the reactions. Surprisingly, the fluorescence intensity was found to decrease with an increase in NiPdhNPs, indicating that they consumed $^\bullet OH$ instead of generating them, which only suggests that the only generation of $^\bullet OH$ does not give rise to the peroxidase-like activity of NiPdhNPs. It was found that the multienzyme mimetic activity of NiPdhNPs was mainly due to Pd; however, Ni also played a key role in enhancing the enzyme-like activity by providing an active site for the reaction to proceed. NiPdhNPs ($K_m = 0.11$ mM) possessed much higher substrate (TMB) affinity than that of HRP ($K_m = 0.434$ mM). These NPs show promising potential in the field of biotechnology, environmental chemistry, and medical diagnostics, offering advantages such as great stability, simple preparation, and high reproducibility (Wang et al., 2016a). Hu et al. showed that CuONPs behave as an oxidase mimic, which was used for the aerobic oxidation of cysteine to cystine with the formation of H_2O_2 and peroxidase-like activity that catalyzed the H_2O_2-mediated oxidation of terephthalic acid. Dual enzymatic-like activity of CuONPs was a suitable candidate for the fluorescent-based sensor for the detection of cysteine in pharmaceutical and biological samples with the LOD of 6.6 nM (Hu et al., 2017).

NANOMATERIALS WITH MULTIENZYME MIMETIC ACTIVITY

A few nanomaterials have shown to mimic the function of all three important cellular antioxidant enzymes, namely, SOD, peroxidase as well as catalase. Zhang and coworkers synthesized Prussian Blue NPs (PBNPs), which exhibits multiple enzyme activities such as peroxidase, SOD, and catalase. PB is commonly used as a dye in iron staining of biopsy specimens. PB is also approved by the Food and Drug Administration as an antidote for thallium in 2010. PBNPs are reported to inhibit $^\bullet OH$ generation and also quench $O^{2\bullet-}$ and H_2O_2. ROS scavenging potential of PBNPs makes them as an efficient antioxidants and have several applications in radical scavenging (Zhang et al., 2016). Dong et al. demonstrated that Co_3O_4 possesses peroxidase, catalase, and SOD-like activity, which is dependent on the surrounding pH. At alkaline and neutral environment, Co_3O_4 will trigger the breakdown of H_2O_2 and $O_2^{\bullet-}$, which indicates that the Co_3O_4 can work as a catalase and SOD mimetics. At acidic environment, Co_3O_4 will catalyze the substrate TMB in presence of H_2O_2 and act as peroxidase mimetics. Conjugation of avastin antibody on the surface of Co_3O_4 leads to the development of IHC assay attributed to the peroxidase-like activity of Co_3O_4. This assay was used in the detection of vascular endothelial growth factor (VEGF), which found to be overexpressed in cancer tissues (Dong et al., 2014). Wang et al. constructed nickel-palladium hollow NPs (Ni-Pd) possess triple enzyme-like activity, including oxidase, peroxidase, and catalase mimetic activity. Peroxidase mimetic activity of Ni-Pd NPs used glucose detection with an LOD of 4.2 μM (Wang et al., 2016a). Bhagat et al. showed that gold (core)-CeNPs (shell) (Au/CeNPs CSNPs) exhibit catalase, peroxidase, and SOD mimetic activity because the proximity of the redox potentials of Au+/Au and Ce(3+)/Ce (4+) may result in redox couple, stimulating the multienzyme-like activity. The peroxidase-like activity of core-shell NPs was utilized in the glucose detection, which gave a linear range of detection limit between 100 μM and 1 mM (Bhagat et al., 2017). Singh and coworkers reported that Mn_3O_4 nanozyme with flower-like morphology (Mnf) can mimic major antioxidant enzymes, i.e., catalase, glutathione peroxidase (GPx), and SOD. The oxidative state of Mn plays a crucial role in the enzyme-like mimetic activity. The presence of Mn in both 3+ and 2+ oxidation state was confirmed from XPS data. Mn-based nanomaterial having higher Mn^{3+}/Mn^{2+} ratio (oxidized Mnf) shows enhanced catalase and GPx-like activities as compared with Mnf. Mn-based nanomaterials protect the cells from MPP+-induced cytotoxicity in a Parkinson disease—like cellular model. Thereby, Mn_3O_4 nanozyme can be a potential therapeutic measure to prevent diseases caused due to $^\bullet OH$ (Singh et al., 2017a).

NUCLEOTIDES AND NUCLEOSIDE AS A COCATALYST

Many metal ions show high affinity toward nucleotide bases, which makes DNA oligonucleotides a good stabilizers in the synthesis of small metal nanoclusters (Liu, 2012). Nucleic acids have unique properties such as tunable conformation and biocompatibility, which makes them attractive templates for synthesis of constructing nanomaterial. Wang et al. have developed methods to prepare H_2O soluble and fluorescent Au/ AgNCs and AuNCs using nucleotide templates such as cytidine and adenosine (Zhang et al., 2014). Fang Pu and coworkers have developed a review report on the strategies for the synthesis of nanomaterials using nucleosides, nucleobases, and nucleotides as templates or modulators and their application. They showed that the generation of such coordination polymer NPs (CPNs) can be synthesized because of the self-assembly of nucleotides and metal ion (Pu et al., 2018). Xu et al. recently demonstrated that nucleoside triphosphates (NTPs) can also enhance the oxidase mimetic activity of CeNPs. They reported that improved efficiency is related to the type of NTPs. As CeNPs exhibit phosphatase-like activity, the improving effect of NTPs can be linked with the coupling of oxidation reactions of NTPs hydrolysis. They observed that the energy released during the hydrolysis of NTPs increases the oxidation reactions of CeNPs. Based on different enhancement capabilities of NTPs, colorimetric assays for single-nucleotide polymorphism typing were also reported (Xu et al., 2014). Additionally, Shah et al. have demonstrated that AuNPs show size and concentration-dependent, strong peroxidase-like activity in the presence of ATP as opposed to the expected outcome that NPs exposed to biomolecules may be deprived of their catalytic properties. However, an addition of free phosphate, carbonate, and sulfate anions did not alter the catalytic activity, whereas the addition of AA to the peroxidase reaction decreased the enzyme mimetic activity of AuNPs in the presence of ATP (Shah et al., 2015). In another report, Shah et al. further demonstrated that the peroxidase-like activity of AuNPs is dependent on the surface charge and capping molecule present on the surface of NPs. Uncharged (polyethylene glycol—coated AuNPs, PEG AuNPs) was reported to exhibit maximum enzyme-like activity than negatively charged citrate and positively charged N-cetyl'N,N,N' trimethylammonium bromide AuNPs (CTAB AuNPs). It was observed that the peroxidase-like activity of PEG AuNPs and citrate AuNPs is dependent on the hydroxyl radicals, whereas CTAB AuNPs did not display any appreciable activity under the same

experimental condition. Authors reported that peroxidase-like activity of PEG AuNPs and citrate AuNPs was enhanced in the presence of ATP. However, ATP did not enhance the generation of $^{\bullet}OH$ radicals. It was proposed that ATP offers stabilization to the oxidized TMB because of the oxidation by PEG AuNPs, citrate AuNPs as well as HRP (Shah and Singh, 2018). Vallabani et al. reported that in presence of ATP, Fe_3O_4 NPs possess peroxidase-like activity at physiological pH. They demonstrated that ATP constitutes stable complex with Fe_3O_4 in presence of H_2O_2 turns in the production of H_2O_2. Furthermore, via promoting single electron transfer, ATP will stabilize the oxidized TMB radical cation. Peroxidase-like activity at a physiological pH provides a way for single-step glucose detection (0.05—4 mM) with the limit of detection at 50 μM (Vallabani et al., 2017).

CONCLUSIONS AND FUTURE PROSPECTS

Phenomenal progress has been made in the field of nanozymes because of their unique properties as compared with natural enzymes, which makes them potential candidates for next-generation artificial enzymes (Wei and Wang, 2013). Nanozymes propose excellent advantages over natural enzymes such as high stability, easy production in bulk, low cost, synthesis, longer shelf life, robustness, stability over various temperature, and pH ranges and mainly a larger surface area to volume ratio, which provides a greater enzymatic activity. Their aforementioned beneficiary edges over natural enzymes make them promising enzyme substitutes and have found diverse applications in biosensors, biocatalysts, bioseparation, bioassays, immunoassays, glucose detection, H_2O_2 detection, therapeutic applications, tumor diagnosis, environmental engineering, pollutant detection and removal, drug and gene targeting, tissue engineering, and treatments for oxidative stress-related diseases (Shin et al., 2015; Wei and Wang, 2013). However, despite all these advantages and wide range of applications, nanozymes also face certain challenges, which can be looked into near future. Firstly, most nanozymes show somewhat lower activity as compared with natural enzymes. It has been observed that even if the NPs are itself highly active, its catalytic performance is compromised because of the presence of surface modifications or additional surface coatings. Surface modifications definitely enhance the substrate affinity, but there is an evident fall in the selectivity. It would therefore be ideal to develop nanozymes bearing increased activity and also use fitting techniques for surface modifications.

Future research can be made more meaningful by using a strategy that includes an analytical screening of enzyme-mimicking activities that depend on the atomic composition of NPs, instead of current primitive research, which is based on novel nanozyme development through random screening of the enzyme-mimicking properties of the existing unknown nanomaterials. Furthermore, the strategy may also be developed and can try to increase the catalytic activity of nanomaterials by exploiting the synergistic effect of these particles in facilitating the transfer of electrons during redox reactions (Shin et al., 2015).

Nanozymes have been found to show relatively low selectivity toward target molecules because of the absence of active sites for the substrate to bind and undergo an effective catalytic reaction as in the case of natural enzymes. This can often hinder the detection and biosensing mechanisms in therapeutic applications. Substantial nanozyme designs and surface modifications using nucleic acids, antibodies, and polymers can be put to use in an attempt to develop more selective and specific nanozymes. Advancements toward creating fresh surface engineering techniques that can make nanozymes more selective toward their target molecules and substrates will lead to a critical development in this field. The toxicity pertaining to nanozymes poses as a threat to humans as well as to the ecosystem and has become a critical issue, which needs to be addressed. This can be resolved by adopting a biologically inspired synthesis of nanozymes and thereby producing nontoxic nanozymes by efficiently limiting the usage of noxious chemicals, thereby increasing their utility in wider therapeutic applications (Shin et al., 2015). For nanozymes to be placed as a novel technology, a better comprehension of the fundamental underlying mechanisms is required, and certain unknown questions need to be answered such as why different types of NPs such as AuNPs, GOx, AgNPs, IONPs, and carbon nanotubes show similar enzymatic activity. The mechanism behind dual or triple functionality of nanozymes is also unknown. For instance, IONPs show peroxidase-like activity at an acidic pH but catalase-like activity at a neutral pH. There is also immense potential for exploring a wider range of catalytic activities in nanozymes such as transferases, ligases, oxidoreductases, lyases, hydrolases, and isomerases. The enzymatic activity of nanozymes can be coupled with their other indigenous nanoscale properties such as electronic, optic, and magnetic to broaden their applications (Gao and Yan, 2016).

With the aforementioned developments, nanozymes are expected to be largely engaged in even more diverse applications in the near future. In conclusion, nanozymes have already been established as an exceptionally robust interdisciplinary field overarching and linking biology and nanotechnology.

ACKNOWLEDGMENTS

J Shah thanks the Department of Science and Technology (DST), New Delhi, for providing INSPIRE Junior Research Fellowship (JRF). The financial assistance for the Centre for Nanotechnology Research and Applications (CENTRA) by The Gujarat Institute for Chemical Technology (GICT) is acknowledged. The funding from the Department of Science and Technology, Science and Engineering Research Board (SERB) (Grant No.: ILS/SERB/2015-16/01), to Dr Sanjay Singh under the scheme of Start-Up Research Grant (Young Scientists) in Life Sciences is also gratefully acknowledged.

Compliance with ethical standards: Conflict of interest: The authors declare that they have no conflict of interest.

REFERENCES

Ali, S.S., Hardt, J.I., Quick, K.L., Sook Kim-Han, J., Erlanger, B.F., Huang, T.-t., Dugan, L.L., 2004. A biologically effective fullerene (C_{60}) derivative with superoxide dismutase mimetic properties. Free Radical Biology and Medicine 37, 1191–1202.

Asati, A., Santra, S., Kaittanis, C., Nath, S., Perez, J.M., 2009. Oxidase activity of polymer-coated cerium oxide nanoparticles. Angewandte Chemie (International Ed. in English) 48, 2308–2312.

Babu, K.S., Anandkumar, M., Tsai, T.Y., Kao, T.H., Inbaraj, B.S., Chen, B.H., 2014. Cytotoxicity and antibacterial activity of gold-supported cerium oxide nanoparticles. International Journal of Nanomedicine 9, 5515–5531.

Bhagat, S., Srikanth Vallabani, N.V., Shutthanandan, V., Bowden, M., Karakoti, A.S., Singh, S., 2017. Gold core/ceria shell-based redox active nanozyme mimicking the biological multienzyme complex phenomenon. Journal of Colloid and Interface Science 513, 831–842.

Cai, S., Qi, C., Li, Y., Han, Q., Yang, R., Wang, C., 2016. PtCo bimetallic nanoparticles with high oxidase-like catalytic activity and their applications for magnetic-enhanced colorimetric biosensing. Journal of Materials Chemistry B 4, 1869–1877.

Ceccone, G., Shard, A.G., 2016. Preface: in focus issue on nanoparticle interfaces. Biointerphases 11, 04B101.

Chance, B., Maehly, A., 1955. [136] Assay of catalases and peroxidases. Methods in Enzymology 2, 764–775.

Chen, W., Chen, J., Liu, A.L., Wang, L.M., Li, G.W., Lin, X.H., 2011. Peroxidase-like activity of cupric oxide nanoparticle. ChemCatChem 3, 1151–1154.

Chen, Z., Yin, J.-J., Zhou, Y.-T., Zhang, Y., Song, L., Song, M., Gu, N., 2012. Dual enzyme-like activities of iron oxide nanoparticles and their implication for diminishing cytotoxicity. ACS Nano 6, 4001–4012.

Cheng, H., Lin, S., Muhammad, F., Lin, Y.-W., Wei, H., 2016. Rationally modulate the oxidase-like activity of nanoceria for self-regulated bioassays. ACS Sensors 1, 1336–1343.

Cho, S., Lee, S.M., Shin, H.Y., Kim, M.S., Seo, Y.H., Cho, Y.K., Kim, M.I., 2018. Highly sensitive colorimetric detection of allergies based on an immunoassay using peroxidase-mimicking nanozymes. Analyst 143, 1182–1187.

Cho, S., Shin, H.Y., Kim, M.I., 2017. Nanohybrids consisting of magnetic nanoparticles and gold nanoclusters as effective peroxidase mimics and their application for colorimetric detection of glucose. Biointerphases 12, 01A401.

Colon, J., Hsieh, N., Ferguson, A., Kupelian, P., Seal, S., Jenkins, D.W., Baker, C.H., 2010. Cerium oxide nanoparticles protect gastrointestinal epithelium from radiation-induced damage by reduction of reactive oxygen species and upregulation of superoxide dismutase 2. Nanomedicine: Nanotechnology, Biology and Medicine 6, 698–705.

Cui, M., Zhao, Y., Wang, C., Song, Q., 2017. The oxidase-like activity of iridium nanoparticles, and their application to colorimetric determination of dissolved oxygen. Microchimica Acta 1–7.

Cui, R., Han, Z., Zhu, J.J., 2011. Helical carbon nanotubes: intrinsic peroxidase catalytic activity and its application for biocatalysis and biosensing. Chemistry 17, 9377–9384.

Deng, H.H., Li, G.W., Hong, L., Liu, A.L., Chen, W., Lin, X.H., Xia, X.H., 2014. Colorimetric sensor based on dual-functional gold nanoparticles: analyte-recognition and peroxidase-like activity. Food Chemistry 147, 257–261.

Deng, H.H., Zheng, X.Q., Wu, Y.Y., Shi, X.Q., Lin, X.L., Xia, X.H., Hong, G.L., 2017. Alkaline peroxidase activity of cupric oxide nanoparticles and its modulation by ammonia. Analyst 142, 3986–3992.

Dong, J., Song, L., Yin, J.-J., He, W., Wu, Y., Gu, N., Zhang, Y., 2014. Co₃O₄ nanoparticles with multi-enzyme activities and their application in immunohistochemical assay. ACS Applied Materials and Interfaces 6, 1959–1970.

Dong, W., Zhuang, Y., Li, S., Zhang, X., Chai, H., Huang, Y., 2018. High peroxidase-like activity of metallic cobalt nanoparticles encapsulated in metal–organic frameworks derived carbon for biosensing. Sensors and Actuators B: Chemical 255, 2050–2057.

Fan, K., Wang, H., Xi, J., Liu, Q., Meng, X., Duan, D., Yan, X., 2016. Optimization of Fe₃O₄ nanozyme activity via single amino acid modification mimicking an enzyme active site. Chemical Communications 53, 424–427.

Filipponi, L., Sutherland, D., Center, I.N., 2010. Introduction to Nanoscience and Nanotechnologies.

Fu, S., Wang, S., Zhang, X., Qi, A., Liu, Z., Yu, X., Li, L., 2017. Structural effect of Fe3O4 nanoparticles on peroxidase-like activity for cancer therapy. Colloids and Surfaces B: Biointerfaces 154, 239–245.

Gao, L., Yan, X., 2016. Nanozymes: an emerging field bridging nanotechnology and biology. Science China Life Sciences 59, 400.

Gao, L., Zhuang, J., Nie, L., Zhang, J., Zhang, Y., Gu, N., Perrett, S., 2007. Intrinsic peroxidase-like activity of ferromagnetic nanoparticles. Nature Nanotechnology 2, 577–583.

Ge, S., Zhang, Y., Zhang, L., Liang, L., Liu, H., Yan, M., Yu, J., 2015. Ultrasensitive electrochemical cancer cells sensor based on trimetallic dendritic Au@PtPd nanoparticles for signal amplification on lab-on-paper device. Sensors and Actuators B: Chemical 220, 665–672.

Goodsell, D., 2004. Catalase. Molecule of the Month.

Guo, L., Huang, K., Liu, H., 2016. Biocompatibility selenium nanoparticles with an intrinsic oxidase-like activity. Journal of Nanoparticle Research 18, 74.

Han, L., Li, C., Zhang, T., Lang, Q., Liu, A., 2015. Au@Ag heterogeneous nanorods as nanozyme interfaces with peroxidase-like activity and their application for one-pot analysis of glucose at nearly neutral pH. ACS Applied Materials and Interfaces 7, 14463–14470.

He, W., Liu, Y., Yuan, J., Yin, J.J., Wu, X., Hu, X., Guo, Y., 2011. Au@Pt nanostructures as oxidase and peroxidase mimetics for use in immunoassays. Biomaterials 32, 1139–1147.

He, W., Wu, X., Liu, J., Hu, X., Zhang, K., Hou, S., Xie, S., 2010. Design of AgM bimetallic alloy nanostructures (M= Au, Pd, Pt) with tunable morphology and peroxidase-like activity. Chemistry of Materials 22, 2988–2994.

Heckert, E.G., Karakoti, A.S., Seal, S., Self, W.T., 2008. The role of cerium redox state in the SOD mimetic activity of nanoceria. Biomaterials 29, 2705–2709.

Hu, A.L., Deng, H.H., Zheng, X.Q., Wu, Y.Y., Lin, X.L., Liu, A.L., Hong, G.L., 2017. Self-cascade reaction catalyzed by CuO nanoparticle-based dual-functional enzyme mimics. Biosensors and Bioelectronics 97, 21–25.

Jalilov, A.S., Zhang, C., Samuel, E.L., Sikkema, W.K., Wu, G., Berka, V., Tour, J.M., 2016. Mechanistic study of the conversion of superoxide to oxygen and hydrogen peroxide in carbon nanoparticles. ACS Applied Materials and Interfaces 8, 15086–15092.

Jawaid, P., Rehman, M., Yoshihisa, Y., Li, P., Zhao, Q., Hassan, M.A., Kondo, T., 2014. Effects of SOD/catalase mimetic platinum nanoparticles on radiation-induced apoptosis in human lymphoma U937 cells. Apoptosis 19, 1006–1016.

Jiang, X., Sun, C., Guo, Y., Nie, G., Xu, L., 2015. Peroxidase-like activity of apoferritin paired gold clusters for glucose detection. Biosensors and Bioelectronics 64, 165–170.

Ju, Y., Kim, J., 2015. Dendrimer-encapsulated Pt nanoparticles with peroxidase-mimetic activity as biocatalytic labels for sensitive colorimetric analyses. Chemical Communications 51, 13752–13755.

Jv, Y., Li, B., Cao, R., 2010. Positively-charged gold nanoparticles as peroxidase mimic and their application in hydrogen peroxide and glucose detection. Chemical Communications 46, 8017–8019.

Kamada, K., Soh, N., 2015. Enzyme-mimetic activity of Ce-intercalated titanate nanosheets. Journal of Physical Chemistry B 119, 5309–5314.

Kim, J., Takahashi, M., Shimizu, T., Shirasawa, T., Kajita, M., Kanayama, A., Miyamoto, Y., 2008. Effects of a potent antioxidant, platinum nanoparticle, on the lifespan of

Caenorhabditis elegans. Mechanism of Ageing and Development 129, 322–331.

Korschelt, K., Ragg, R., Metzger, C.S., Kluenker, M., Oster, M., Barton, B., Tremel, W., 2017. Glycine-functionalized copper(ii) hydroxide nanoparticles with high intrinsic superoxide dismutase activity. Nanoscale 9, 3952–3960.

Korsvik, C., Patil, S., Seal, S., Self, W.T., 2007. Superoxide dismutase mimetic properties exhibited by vacancy engineered ceria nanoparticles. Chemical Communications 1056–1058.

Krishnaraj, C., Jagan, E., Ramachandran, R., Abirami, S., Mohan, N., Kalaichelvan, P., 2012. Effect of biologically synthesized silver nanoparticles on *Bacopa monnieri* (Linn.) Wettst. plant growth metabolism. Process Biochemistry 47, 651–658.

Kulikova, V., 2005. NADH oxidase activity of gold nanoparticles in aqueous solution. Kinetics and Catalysis 46, 373–375.

Liang, M., Fan, K., Pan, Y., Jiang, H., Wang, F., Yang, D., Yan, X., 2013. Fe_3O_4 magnetic nanoparticle peroxidase mimetic-based colorimetric assay for the rapid detection of organophosphorus pesticide and nerve agent. Analytical Chemistry 85, 308–312.

Liu, J., 2012. Adsorption of DNA onto gold nanoparticles and graphene oxide: surface science and applications. Physical Chemistry Chemical Physics 14, 10485–10496.

Liu, X., Wang, Q., Zhao, H., Zhang, L., Su, Y., Lv, Y., 2012. BSA-templated MnO_2 nanoparticles as both peroxidase and oxidase mimics. Analyst 137, 4552–4558.

Liu, Y., Wu, H., Chong, Y., Wamer, W.G., Xia, Q., Cai, L., Yin, J.J., 2015. Platinum nanoparticles: efficient and stable catechol oxidase mimetics. ACS Applied Materials and Interfaces 7, 19709–19717.

Long, Y.J., Li, Y.F., Liu, Y., Zheng, J.J., Tang, J., Huang, C.Z., 2011. Visual observation of the mercury-stimulated peroxidase mimetic activity of gold nanoparticles. Chemical Communications 47, 11939–11941.

Lu, W., Shu, J., Wang, Z., Huang, N., Song, W., 2015. The intrinsic oxidase-like activity of Ag_2O nanoparticles and its application for colorimetric detection of sulfite. Materials Letters 154, 33–36.

Maji, S.K., Mandal, A.K., Nguyen, K.T., Borah, P., Zhao, Y., 2015. Cancer cell detection and therapeutics using peroxidase-active nanohybrid of gold nanoparticle-loaded mesoporous silica-coated graphene. ACS Applied Materials and Interfaces 7, 9807–9816.

Manea, F., Houillon, F.B., Pasquato, L., Scrimin, P., 2004. Nanozymes: gold-nanoparticle-based transphosphorylation catalysts. Angewandte Chemie International Edition in English 43, 6165–6169.

Mu, J., Zhao, X., Li, J., Yang, E.C., Zhao, X.J., 2017. Coral-like CeO_2/NiO nanocomposites with efficient enzyme-mimetic activity for biosensing application. Mater Sci Eng C Mater Biol Appl 74, 434–442.

Naganuma, T., 2017. Shape design of cerium oxide nanoparticles for enhancement of enzyme mimetic activity in therapeutic applications. Nano Research 10, 199–217.

Nagvenkar, A.P., Gedanken, A., 2016. $Cu_{0.89}Zn_{0.11}O$, a new peroxidase-mimicking nanozyme with high sensitivity for glucose and antioxidant detection. ACS Applied Materials and Interfaces 8, 22301–22308.

Ni, P., Dai, H., Wang, Y., Sun, Y., Shi, Y., Hu, J., Li, Z., 2014. Visual detection of melamine based on the peroxidase-like activity enhancement of bare gold nanoparticles. Biosensors and Bioelectronics 60, 286–291.

Ni, P., Sun, Y., Dai, H., Hu, J., Jiang, S., Wang, Y., Li, Z., 2015. Highly sensitive and selective colorimetric detection of glutathione based on Ag [I] ion-3,3′,5,5′-tetramethylbenzidine (TMB). Biosensors and Bioelectronics 63, 47–52.

Peng, C.F., Pan, N., Zhi-Juan, Q., Wei, X.L., Shao, G., 2017a. Colorimetric detection of thiocyanate based on inhibiting the catalytic activity of cystine-capped core-shell Au@Pt nanocatalysts. Talanta 175, 114–120.

Peng, H., Lin, D., Liu, P., Wu, Y., Li, S., Lei, Y., Xia, X., 2017b. Highly sensitive and rapid colorimetric sensing platform based on water-soluble WOx quantum dots with intrinsic peroxidase-like activity. Analytica Chimica Acta 128–134.

Pengo, P., Polizzi, S., Pasquato, L., Scrimin, P., 2005. Carboxylate-imidazole cooperativity in dipeptide-functionalized gold nanoparticles with esterase-like activity. Journal of the American Chemical Society 127, 1616–1617.

Pirmohamed, T., Dowding, J.M., Singh, S., Wasserman, B., Heckert, E., Karakoti, A.S., Self, W.T., 2010. Nanoceria exhibit redox state-dependent catalase mimetic activity. Chemical Communications 46, 2736–2738.

Pratsinis, A., Kelesidis, G.A., Zuercher, S., Krumeich, F., Bolisetty, S., Mezzenga, R., Sotiriou, G.A., 2017. Enzyme-mimetic antioxidant luminescent nanoparticles for highly sensitive hydrogen peroxide biosensing. ACS Nano 11, 12210–12218.

Pu, F., Ren, J., Qu, X., 2018. Nucleobases, nucleosides, and nucleotides: versatile biomolecules for generating functional nanomaterials. Chemical Society Reviews 47, 1285–1306.

Ragg, R., Schilmann, A., Korschelt, K., Wieseotte, C., Kluenker, M., Viel, M., Frey, H., 2016. Intrinsic superoxide dismutase activity of MnO nanoparticles enhances the magnetic resonance imaging contrast. Journal of Materials Chemistry B 4, 7423–7428.

Rather, H.A., Thakore, R., Singh, R., Jhala, D., Singh, S., Vasita, R., 2017. Antioxidative study of Cerium Oxide nanoparticle functionalised PCL-Gelatin electrospun fibers for wound healing application. Bioactive Materials.

Robinson, P.K., 2015. Enzymes: principles and biotechnological applications. Essays in Biochemistry 59, 1–41.

Shah, J., Purohit, R., Singh, R., Karakoti, A.S., Singh, S., 2015. ATP-enhanced peroxidase-like activity of gold nanoparticles. Journal of Colloid and Interface Science 456, 100–107.

Shah, J., Singh, S., 2018. Unveiling the role of ATP in amplification of intrinsic peroxidase-like activity of gold nanoparticles. 3 Biotech 8, 67.

Shen, X., Liu, W., Gao, X., Lu, Z., Wu, X., Gao, X., 2015. Mechanisms of oxidase and superoxide dismutation-like activities of gold, silver, platinum, and palladium, and their alloys: a general way to the activation of molecular oxygen. Journal of the American Chemical Society 137, 15882–15891.

Shi, W., Wang, Q., Long, Y., Cheng, Z., Chen, S., Zheng, H., Huang, Y., 2011. Carbon nanodots as peroxidase mimetics and their applications to glucose detection. Chemical Communications 47, 6695–6697.

Shi, Y., Su, P., Wang, Y., Yang, Y., 2014. Fe_3O_4 peroxidase mimetics as a general strategy for the fluorescent detection of H_2O_2-involved systems. Talanta 130, 259–264.

Shin, H.Y., Park, T.J., Kim, M.I., 2015. Recent research trends and future prospects in nanozymes. Journal of Nanomaterials.

Singh, N., Savanur, M.A., Srivastava, S., D'Silva, P., Mugesh, G., 2017a. A redox modulatory Mn_3O_4 nanozyme with multienzyme activity provides efficient cytoprotection to human cells in a Parkinson's disease model. Angewandte Chemie International Edition in English 56, 14267–14271.

Singh, R., Karakoti, A.S., Self, W., Seal, S., Singh, S., 2016. Redox-sensitive cerium oxide nanoparticles protect human keratinocytes from oxidative stress induced by glutathione depletion. Langmuir 32, 12202–12211.

Singh, R., Singh, S., 2015. Role of phosphate on stability and catalase mimetic activity of cerium oxide nanoparticles. Colloids and Surfaces B: Biointerfaces 132, 78–84.

Singh, S., Dosani, T., Karakoti, A.S., Kumar, A., Seal, S., Self, W.T., 2011. A phosphate-dependent shift in redox state of cerium oxide nanoparticles and its effects on catalytic properties. Biomaterials 32, 6745–6753.

Singh, S., Tripathi, P., Kumar, N., Nara, S., 2017b. Colorimetric sensing of malathion using palladium-gold bimetallic nanozyme. Biosensors and Bioelectronics 92, 280–286.

Sloan-Dennison, S., Laing, S., Shand, N.C., Graham, D., Faulds, K., 2017. A novel nanozyme assay utilising the catalytic activity of silver nanoparticles and SERRS. Analyst 142, 2484–2490.

Su, L., Qin, W., Zhang, H., Rahman, Z.U., Ren, C., Ma, S., Chen, X., 2015. The peroxidase/catalase-like activities of MFe_2O_4 (M= Mg, Ni, Cu) MNPs and their application in colorimetric biosensing of glucose. Biosensors and Bioelectronics 63, 384–391.

Vainshtein, B., Melik-Adamyan, W., Barynin, V., Vagin, A., Grebenko, A., 1981. Three-dimensional structure of the enzyme catalase. Nature 293, 411.

Vallabani, N.V., Karakoti, A.S., Singh, S., 2017. ATP-mediated intrinsic peroxidase-like activity of Fe_3O_4-based nanozyme: one step detection of blood glucose at physiological pH. Colloids and Surfaces B: Biointerfaces 153, 52–60.

Wang, G.-L., Xu, X.-F., Cao, L.-H., He, C.-H., Li, Z.-J., Zhang, C., 2014a. Mercury (II)-stimulated oxidase mimetic activity of silver nanoparticles as a sensitive and selective mercury (II) sensor. RSC Advances 4, 5867–5872.

Wang, G.L., Xu, X.F., Qiu, L., Dong, Y.M., Li, Z.J., Zhang, C., 2014b. Dual responsive enzyme mimicking activity of AgX (X=Cl, Br, I) nanoparticles and its application for cancer cell detection. ACS Applied Materials and Interfaces 6, 6434–6442.

Wang, L., Zeng, Y., Shen, A., Zhou, X., Hu, J., 2015. Three dimensional nano-assemblies of noble metal nanoparticle–infinite coordination polymers as specific oxidase mimetics for degradation of methylene blue without adding any cosubstrate. Chemical Communications 51, 2052–2055.

Wang, Q., Zhang, L., Shang, C., Zhang, Z., Dong, S., 2016a. Triple-enzyme mimetic activity of nickel–palladium hollow nanoparticles and their application in colorimetric biosensing of glucose. Chemical Communications 52, 5410–5413.

Wang, S., Chen, W., Liu, A.L., Hong, L., Deng, H.H., Lin, X.H., 2012. Comparison of the peroxidase-like activity of unmodified, amino-modified, and citrate-capped gold nanoparticles. ChemPhysChem 13, 1199–1204.

Wang, X., Guo, W., Hu, Y., Wu, J., Wei, H., 2016b. Introduction to Nanozymes Nanozymes: Next Wave of Artificial Enzymes. Springer, pp. 1–6.

Wang, X., Yang, Q., Cao, Y., Hao, H., Zhou, J., Hao, J., 2016c. Metallosurfactant ionogels in imidazolium and protic ionic liquids as precursors to synthesize nanoceria as catalase mimetics for the catalytic decomposition of H_2O_2. Chemistry 22, 17857–17865.

Wei, H., Wang, E., 2013. Nanomaterials with enzyme-like characteristics (nanozymes): next-generation artificial enzymes. Chemical Society Reviews 42, 6060–6093.

Wei, J., Chen, X., Shi, S., Mo, S., Zheng, N., 2015. An investigation of the mimetic enzyme activity of two-dimensional Pd-based nanostructures. Nanoscale 7, 19018–19026.

Wu, L.-L., Wang, L.-Y., Xie, Z.-J., Pan, N., Peng, C.-F., 2016. Colorimetric assay of l-cysteine based on peroxidase-mimicking DNA-Ag/Pt nanoclusters. Sensors and Actuators B: Chemical 235, 110–116.

Xu, C., Liu, Z., Wu, L., Ren, J., Qu, X., 2014. Nucleoside triphosphates as promoters to enhance nanoceria enzyme-like activity and for single-nucleotide polymorphism typing. Advanced Functional Materials 24, 1624–1630.

Yang, L., Liu, X., Lu, Q., Huang, N., Liu, M., Zhang, Y., Yao, S., 2016. Catalytic and peroxidase-like activity of carbon based-AuPd bimetallic nanocomposite produced using carbon dots as the reductant. Analytica Chimica Acta 930, 23–30.

Yang, Z., Luo, S., Zeng, Y., Shi, C., Li, R., 2017. Albumin-mediated biomineralization of shape-controllable and biocompatible ceria nanomaterials. ACS Applied Materials and Interfaces 9, 6839–6848.

Yang, Z., Zhu, Y., Chi, M., Wang, C., Wei, Y., Lu, X., 2018. Fabrication of cobalt ferrite/cobalt sulfide hybrid nanotubes with enhanced peroxidase-like activity for colorimetric detection of dopamine. Journal of Colloid and Interface Science 511, 383–391.

Yokchom, R., Laiwejpithaya, S., Maneeprakorn, W., Tapaneeyakorn, S., Rabablert, J., Dharakul, T., 2018. Paper-based immunosensor with signal amplification by enzyme-labeled anti-p16(INK4a) multifunctionalized gold nanoparticles for cervical cancer screening. Nanomedicine.

Yoshihisa, Y., Zhao, Q.-L., Hassan, M.A., Wei, Z.-L., Furuichi, M., Miyamoto, Y., Shimizu, T., 2011. SOD/catalase mimetic platinum nanoparticles inhibit heat-induced apoptosis in human lymphoma U937 and HH cells. Free Radical Research 45, 326–335.

You, J.-G., Liu, Y.-W., Lu, C.-Y., Tseng, W.-L., Yu, C.-J., 2017. Colorimetric assay of heparin in plasma based on the inhibition of oxidase-like activity of citrate-capped platinum nanoparticles. Biosensors and Bioelectronics 92, 442–448.

Yu, C.J., Chen, T.H., Jiang, J.Y., Tseng, W.L., 2014. Lysozyme-directed synthesis of platinum nanoclusters as a mimic oxidase. Nanoscale 6, 9618–9624.

Yuan, F., Zhao, H., Zang, H., Ye, F., Quan, X., 2016. Three-Dimensional graphene supported bimetallic nanocomposites with DNA regulated-flexibly switchable peroxidase-like activity. ACS Applied Materials and Interfaces 8, 9855–9864.

Zhang, W., Dong, J., Wu, Y., Cao, P., Song, L., Ma, M., Zhang, Y., 2017. Shape-dependent enzyme-like activity of Co_3O_4 nanoparticles and their conjugation with his-tagged EGFR single-domain antibody. Colloids and Surfaces B: Biointerfaces 154, 55–62.

Zhang, W., Hu, S., Yin, J.-J., He, W., Lu, W., Ma, M., Zhang, Y., 2016. Prussian blue nanoparticles as multienzyme mimetics and reactive oxygen species scavengers. Journal of the American Chemical Society 138, 5860–5865.

Zhang, Y., Jiang, H., Ge, W., Li, Q., Wang, X., 2014. Cytidine-directed rapid synthesis of water-soluble and highly yellow fluorescent bimetallic AuAg nanoclusters. Langmuir 30, 10910–10917.

Zhao, Y., Qiang, H., Chen, Z., 2017. Colorimetric determination of Hg (II) based on a visually detectable signal amplification induced by a Cu@ Au-Hg trimetallic amalgam with peroxidase-like activity. Microchimica Acta 184, 107–115.

Zheng, C., Ke, W., Yin, T., An, X., 2016. Intrinsic peroxidase-like activity and the catalytic mechanism of gold@ carbon dots nanocomposites. RSC Advances 6, 35280–35286.

CHAPTER 5

Nanoparticle-Based Drug Delivery for Chronic Obstructive Pulmonary Disorder and Asthma: Progress and Challenges

RIDHIMA WADHWA • TARU AGGARWAL • NOOPUR THAPLIYAL •
DINESH KUMAR CHELLAPPAN • GAURAV GUPTA • MONICA GULATI, PHD •
TRUDI COLLET, PHD • BRIAN OLIVER, PHD • KYLIE WILLIAMS, PHD •
PHILIP MICHAEL HANSBRO, PHD • KAMAL DUA, PHD •
PAWAN KUMAR MAURYA, PHD

INTRODUCTION

Nanoparticle-based drug system is a powerful technique, which can be customized for targeting and causing a therapeutic effect. Numerous nanoparticle-based drug system has experimented, which includes different types of nanoparticles such as nanocrystals, dendrimers, polymeric micelles, protein-based, carbon nanotube, and liposome (Singh and Lillard Jr, 2009). When compared with conventional drug systems, these nanoparticle-based drug delivery systems offer many advantages, which include improving the pharmacokinetic behavior of drug molecule inside a biological entity such as rapid absorption of the nanodrug owing to its high surface area (Magenheim et al., 1993). The most common obstruction faced by conventional drug formulation is rapid immunoclearance by immune cells, which accounts for an inefficient drug system, but in case of this nanoparticle drug design, it seems to negotiate with this barrier, as there is the usage of biocompatible molecules, which prolong the circulation time of a drug molecule inside the target system. Some cases like doxorubicin and paclitaxel involved an increase in safety and morbidity of patients. If correct formulation has taken place, drug molecules have capabilities such as enhanced solubility and adhesion, which accounts for an early response (Muller et al., 2018). When the entire formulation is more effective, it obviously reduces the dosage of drug administered to achieve the same level of benefit as earlier, thereby making it safer and cost-effective. Nanocarrier systems can provide the advantage of sustained release in the lung tissue, resulting in reduced dosing frequency and improved patient compliance. Local delivery of inhalable nanocarriers appears to be a promising alternative to oral or intravenous administration, thus decreasing the incidence of side effects associated with a high drug serum concentration. When compared with the actual therapeutic system in terms of outcome, this nanoparticle-based drug system offers modest improvement over conventional systems, some disadvantages associated are accumulation of nanoparticle in organs, which are specifically involved in immune response such as spleen and liver as when delivered intravenously macrophages opsonize this nanodrug, and thus, these nanotherapeutics are targeted to nonspecific healthy organs. Despite being biocompatible and having increased half-life, they are restricted by series of biological barriers which eventually leads to limited availability within the specific target organ, thus lowering the achievement of nanodrug. Immense tasking is involved like complex mathematical modeling is required to modulate the fluid dynamics of a nanoformulation by altering its geometry, which enhances its margination. This chapter will discuss the importance and use of nanoparticles for respiratory diseases such as asthma and COPD.

NANOPARTICLES FOR CHRONIC OBSTRUCTIVE PULMONARY DISEASE AND ASTHMA

Chronic obstructive pulmonary disorder (COPD) and asthma are the two leading respiratory diseases, with the increase in pollution, and decreasing air quality,

Nanotechnology in Modern Animal Biotechnology. https://doi.org/10.1016/B978-0-12-818823-1.00005-3

the disease prominence in the population is growing in the population. According to a recent study by GBD 2015 Chronic Respiratory Disease Collaborators, 2015 COPD ranked 8th and asthma ranked 23rd for global burden among the population (Soriano et al., 2017). Because of this increasing burden, scientists are working for the development of novel therapeutics and diagnostics for improved disease management. Hence research is being reduced to nanoscale for an efficient system of targeting and curing respiratory disease. A large population is affected by these respiratory diseases including individuals from all age groups from young fetus to elders. Therefore, for the treatment and diagnosis of respiratory disease, lungs are an attractive target due to the following reasons:

- Lungs carry a large surface area, i.e., approximately 50 m^2 for an average adult (Hasleton, 1972).
- Drug absorption in the lungs is enhanced because of the presence of extensive vascularization providing an ideal environment (Gaul et al., 2018).
- Lungs prove to be an important drug delivery target considering that they are able to bypass the systematic delivery route (oral administration) conditions such as the digestive system and help in avoiding first-pass metabolism (Kuzmov and Minko, 2015).

The lungs also exhibit low enzymatic activity, compared with the digestive system. Inhalation is considered to be a more comfortable and patient-friendly route of drug delivery than subcutaneous and intravenous injections. The continued development of inhalable therapeutics has been encouraged by the wide variety of medications that have been successfully delivered through the pulmonary route. The advantages of local delivery to the lung are numerous as it allows the targeting of diseased cells, leading to improved efficacy and fewer systemic side effects (Kuzmov and Minko, 2015). Particle size is a critical parameter for inhalation therapy as it determines drug delivery and deposition. Particle size is also crucial for uptake, as they need to be in the low nanometer range to be taken up by cells. Respirable nanoparticles are an interesting prospect for drug delivery to the lung. Nanoparticles may act in increasing life quality through better therapeutics and diagnosis. There are several routes of drug administration such as oral, local, inhalation, and parenteral using injections. Because of suitable anatomy and physiology of lungs, inhalation route of drug administration is preferred in respiratory disease; types of pulmonary drug delivery devices are mentioned in Table 5.1. In pulmonary drug targeting, easy, noninvasive, and safe administration of the drug can be achieved by inhalation of aerosols or inhalation of dry powder. This results in direct delivery of drug at the target site, i.e., airways where local drug action and systematic drug absorption can be achieved owing to the availability of lavage surface area (Bahadori and Mohammadi, 2012).

LUNG PHYSIOLOGY AND PARTICLE ABSORPTION

Gas exchange between the blood and the external environment acts as the main function for lungs. They also maintain the homeostatic, i.e., systematic pH. The lung includes trachea, which diverges into bronchi, and leads to the alveolar sac. The leading airways are such that they are fixed with ciliated columnar epithelium, which

TABLE 5.1
Pulmonary Drug Delivery Devices Used in COPD and Asthma.

S.No.	Device	Model	Therapeutic Container	Reference
1	Pressurized metered dose inhaler (pMDI)	Modeled on perfume bottles and consist of a pressurized canister	Canister contains the therapeutic suspended in a propellant	
2	Dry powder inhaler	Inhaling plastic device with an inhaling mouth piece and capsule holder	Dry powder transportable capsule	Ibrahim et al. (2015)
3	Soft mist inhaler	pMDIs and nebulizers	Medication is formulated as a solution	Dalby et al. (2004)
4	Nebulizers	Group of devices that generate an aerosol from a solution	Medication is formulated as a solution	(Ibrahim et al., 2015; Gaul et al., 2018)

changes to a cuboidal shape drawing closer to the distal airways. The entrapment of aerosolized particles occurs in the lumen of bronchial airways, which is known to have a floating layer of mucus, which helps in capturing also. This mucous layer is on the constant move owing to the rhythmic beating of cilia. Therefore, particles present in the periphery of lungs are reported to be of 24 h approximately in a healthy individual (Patton and Byron, 2007). The alveoli are composed of type I pneumocytes and type II pneumocytes, where type I pneumocytes include the surface of alveoli and basement membrane with the pulmonary capillaries and type II pneumocytes are responsible for clearing large particle by secreting surfactants and macrophages. A number of alveoli present in a lung are approximately 300 million with a combined surface area greater than 100 m^2 and 0.1 mm as the combined thickness of alveolar epithelium (Brain, 2007; Patton and Byron, 2007). For efficient transfer of particles, these alveoli serve as target providing a large surface area, consisting of thin barrier between the pulmonary lumen and the capillaries.

In the lungs, particles are deposited by diffusion, sedimentation, or inertial impact. Particles more than 10 μm in diameter in size are for the most part subject to inertial impaction in the oropharyngeal region, or sedimentation in the bronchial area; at this location, this target for the drug serves as possessing less systemic therapeutic effect (Brain, 2007; Edwards and Dunbar, 2002; Sung et al., 2007). Particles smaller than 1 μm are generally exhaled and are unsuitable for pulmonary delivery, although it has been reported that particles around 100 nm in diameter can settle in the lower alveolar region. Particles between 1 and 5 μm are the optimum size for deposition in the small airways and alveoli; these particles are expected to bypass deposition in mouth and throat, resulting in efficient drug delivery by depositing in the periphery region of lungs (Edwards and Dunbar, 2002; Musante et al., 2002; Patton and Byron, 2007; Sung et al., 2007). After deposition of these particles, conformational change may bring change to release the drug molecule. This released drug molecule firstly will encounter biological and physicochemical barriers, such as the mucous barrier, macrophages in the alveolar region, and catabolic enzymes in the tracheobronchial region. In the peripheral region of the lung, these particles must release the drug molecule or diffuse drug at the target via epithelial barrier directly into the bloodstream, whereas large particles due to slow drug release may get in contact with alveolar macrophage resulting in phagocytosis. Therefore, to overcome such situation penetration, enhancers

can be combined with the particle formulation for the systematic availability of drug and to increase the long-time safety of these particles.

The lungs are suitable for both local and systemic drug delivery. There are several lung diseases that are key acceptors for pulmonary therapy, such as asthma, emphysema, chronic obstructive pulmonary disease (COPD), cystic fibrosis, primary pulmonary hypertension, and cancer (Patton et al., 2010). The local drug delivery is advantageous, as the drug avoids first-pass metabolism and deposits directly at the disease site. This type of topical application of the drug to the lung epithelium also eliminates potential side effects caused by the high systemic concentrations typical of conventional delivery methods and can reduce costs because smaller doses can be used (Patton et al., 2010). The benefits of systemic drug delivery through the lungs typically include a rapid onset of action and an increased bioavailability over oral formulations, especially for peptide drugs (Bailey and Berkland, 2009; Shoyele and Slowey, 2006).

THE BEHAVIOR OF NANOPARTICLE IN VIVO

Targeted delivery is a mode of transfer in which the therapeutic formulation is dispensed to a precise target with the help of carriers; in case of nanotargeting, nanoparticles-based formulations are taken in account (Cheng et al., 2007). This strategy shows great potential for diseases such as cancer, and a huge number of studies have been published stating the importance of this targeted delivery approach. Although there are numerous successful examples of preclinical studies, their translation to corresponding clinical studies is limited because of multiple challenges. For advancement toward clinical approvals, a deep understanding of the underlying principle is essential to overcome associated issues and devise a novel strategy. Nanoparticle upon entry inside a living moiety shows a complex behavior, as its dynamics are governed by multiple factors such as the route of administration, disease involved, and the intensity of dosage, which in totality decides its cellular and metabolic fate (Albanese et al., 2012). If taking a generalized overview in which only interaction of a naïve nanoparticle and cell is considered, it is easy to understand and improvise on nanoparticle-based drug system. A cell recognizes nanoparticle in an energy-dependent process and identifies specific feature associated with it, whereas nanoparticle interacts with biomolecules, which are freely and abundantly found in circulation such as proteins and form a biological corona around its surface; this interaction depends on the surface charge, size, and composition of a nanoparticle (Albanese et al., 2012).

To obtain targeted delivery in lung diseases such as COPD and asthma, nanoparticle system should overcome several challenges such as overcoming resistance by respiratory mucus and alveolar fluids, reticuloendothelial system evasion of nanoparticles, cell-specific targeting of nanoparticles for intranasal delivery of nanoparticles, cell entry, and endosomal escape of nanoparticles. It is achieved by designing nanoparticle by taking the nanoparticle size and surface in special considerations. Another strategy, which can be accepted to evade the mucosal barrier, is by fabricating nanoparticles along with doped magnetic properties, resulting in gliding across the mucosal barrier with the help of external magnetic force; it also utilizes mucolytic agents, for example, hydrolyzing enzymes for the disruption of the mucosal network (Colina et al., 1996; Connaris and Greenwell, 1997; Gersting et al., 2004). In lungs owing to the inflammatory response, several macrophages and neutrophils are found to be present in the airways; therefore, these macrophages and neutrophils can be targeted by conjugating nanoparticle with antibodies against macrophage or neutrophil markers (Fajac et al., 1999; Wiewrodt et al., 2002). Bioconjugating nanoparticle loaded with the drug is the most widely accepted strategy for targeted drug delivery in lung disorders; refer to Fig. 5.1 for the drug delivery and absorption representation.

There are two major requirements accompanying an efficient targeted drug delivery system, i.e., the drug should reach specifically to the target with maximum activity and minimum loss in terms of drug amount and the second is limiting nonspecific targeting or off-targeting holding minimum side effects. To achieve these two basic requirements, active and passive modes of targeting have been devised.

Passive Targeting

It depends on the microenvironmental difference between the normal and diseased cells exploiting the unique physiopathological condition of the cell-like increased vascularization, enhanced expression of surface receptors, increased porosity, varied morphology, and many more (Bazak et al., 2014; Maeda and Matsumura, 1989). This gradient helps in the rapid targeting of nanodrug to the site of infection without the aid of any external energy so it is an energy independent method.

In respiratory disorders such as COPD and asthma, passive targeting gains attraction when combined with pulmonary drug delivery. In some specific situations, for example, inflammation and tumor development, the blood vessel wall containing endothelial lining develops additional permeability than in healthy tissues (Shaji and Lal, 2013). Therefore, nanoparticles going from 10 to 500 nm in size are able to store deep in these tissues (Shaji and Lal, 2013). This spontaneous

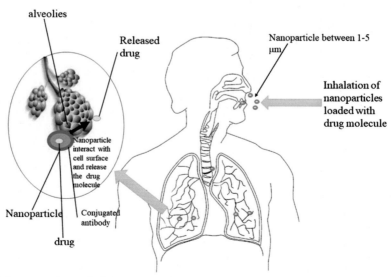

FIG. 5.1 **Pulmonary drug administration.** Drug release from antibiotic-conjugated nanoparticle ranging from 1 to 5 μm in diameter in the airways. Antibiotic-conjugated particle deposits in the lumen of alveoli where macrophages and neutrophils are present.

accumulation is known as the enhanced permeability and retention (EPR) phenomenon. EPR effects of inflamed tissues and tumors create a significant positive impact on passive targeting (Bai et al., 2016; Suer and Bayram, 2017). For example, liposomal systems (nano-based) mainly have discrete benefits because of an important feature, i.e., the ability of liposomes to be internalized later in the lungs, creating passive targeting a reliable method. A mixture of surfactants, which comprise increased levels of dipalmitoylphosphatidyl-choline (DPPC), is secreted by the lungs, which is one of the key phospholipids utilized in liposomal mem-brane; hence, the mechanism of surfactant acceptance facilitates the internalization of liposomes (Mosgoeller et al., 2012; Paturi et al., 2012). Saari et al. (1999) demonstrated that Bec-L created by DPPC accumulates certainly in the lung, representing slower clearance and providing a sustained release.

Active Targeting

It is a well-coordinated and controlled process; this approach was developed as a complementary method to the existing approach in improving targeting and drug retention at the site of infection. It assisted in easy delivery of high-molecular-weight biomolecules such as DNA and RNA by overcoming rejection by phys-iological barrier; one approach to make it more target specific is by using receptor-oriented method in which ligands are designed, which have high affinity toward surface receptor present on the target cell such as small peptides, aptamers, and other small molecules (Naseri et al., 2015). This type of conjugation enables receptor-mediated endocytosis followed by internaliza-tion in the form of the endosome and delivering the package at a specific site of action and lastly release of drug component by enzymatic activity or change in pH retaining its biological activity to maximum (Naseri et al., 2015). The choice for precise delivery vehicle and its modification to achieve high specificity are manda-tory steps in this strategy. Itraconazole in a lipid nano-carrier form as dry powder has shown enhanced itraconazole solubility, which is used to treat an inva-sive form of pulmonary aspergillosis. Nanostructured lipid carrier (NLC) has been widely used in active target-ing, which is a combination of liquid and solid in def-inite ratio forming a less ordered structure providing more space for the drug to be incorporated; it is found in activated forms by PEGylating, attachment of ligand, and adding multifunctional groups on its structure. PEGylation involves addition of a hydrophilic nonionic polymer called polyethylene glycol, and additional attachment of ligand like NAG makes it an active targeting component on whole, which persist properties such as improved permeability and half-life (Jawahar and Reddy, 2012; Paranjpe and Müller-Goymann, 2014; Souto and Müller, 2010; Weber et al., 2014). Li-gands can be attached for specific binding such as trans-ferrin and hyaluronic acid, for treatment of lung cancer and TB. In case of COPD out of many failed attempts, one of the NLC proved its efficacy both in vivo and in vitro formulated by combination of ultrasonication and freeze-drying, named rosuvastatin (RV)-loaded NLCs with Sf/Si ratio 1.02 (Weber et al., 2014).

APPLICATIONS OF NANOPARTICLES IN CHRONIC OBSTRUCTIVE PULMONARY DISEASE AND ASTHMA
Disease Diagnosis

The diagnosis of disease at the molecular level in its initial stages is one of the major goals. Nanoparticles act as probes in in vitro and in vivo conditions to pro-duce a signal at the cellular level in the disease state. These probes are designed such that they cross the cellular barriers and help in analysis from sample prep-aration to diagnosis onto a miniaturized device. They are highly robust for the use by patients and practi-tioners. The progress in nanotechnology has laid the way for disease diagnosis and drug development, thus providing scope to pharmaceutical development. Nanotechnology-based diagnosis will have a great impact in future in the diagnosis of disease.

Nanoparticles have prospective ability to enhance X-ray-based pulmonary diagnosis (Omlor et al., 2015). Imaging probes based on folic acid–modified dendrimer-entrapped gold nanoparticles have been used in targeted computed tomography scanning. Such nanoparticles have been used for in vitro and in vivo studies; entrapped in lysosomes of the folic acid receptor, they have been reported to express lung adenocarcinomas (SPC-A1), allowing tumor cell detec-tion by microcomputed tomography scanning proceed-ing to nanoparticle uptake. This proves to have better biocompatibility, unaffected cell morphology, cell cy-cle, viability, and apoptosis (Wang et al., 2013). Nano-particles have the potential to improve magnetic resonance imaging (MRI) for the lungs. It has been demonstrated in mice where gadolinium-DOTA-based nanoparticle was administered, enhanced signal in several organs was observed, including lungs by ultrashort-echo-time-proton-MRI. The produced signals changed over the time during passage of nanoparticles through the body from lungs to blood, and then to kid-neys and bladder (Bianchi et al., 2013).

Targeted Drug Delivery

Drug delivery for treatment of pulmonary diseases has gained considerable attention because of high diffusion ability across the alveolar epithelium, target specificity especially across the airways, lower toxic effect, and better acquiescence in patients (Dua et al., 2017). Several different types of nanoparticles including proteins, lipids, polymers, and metals have been used for this purpose. However, clearance by macrophages (Storm and van Etten, 1997) and mucociliary defense (Soane et al., 1999) are the major barriers to effective drug delivery. Therefore, to overcome these barriers, biodegradable polymers with better half-life (Madan et al., 2014; Sharma et al., 2014) and high surface area to volume ratio such as chitosan, alginates, gelatin, dextran, poly(lactic-co-glycolic acid) (PGLA), and poly(ε-caprolactone) (PCL) can be used as drug carrier (Lyn et al., 2011), obtaining long-term therapeutic effect as represented in Fig. 5.2.

Liposomes

Liposome constitutes an outer hydrophobic layer and an inner hydrophilic layer with size varying from 50 to 100 nm depending on the cholesterol and phospholipid composition (Yhee et al., 2016). Because of the presence of phospholipids, liposomes are ideal for encapsulation of hydrophilic drugs such as chemotherapeutic drugs, i.e., paclitaxel and doxorubicin, and this nanoparticle-dependent drug delivery system has been approved by the FDA (Paranjpe and Müller-Goymann, 2014). The first liposome-based product was extracted from bovine surfactant against acute respiratory distress syndrome in infants via pulmonary instillation (Adler-Moore and Proffitt, 2002). Till date, inhaled liposome therapy faces limitations. It is essential to maintain the physical properties of liposome after inhalation to get the desired product. Thus, at present, dry inhaled powder, Arikace and Pulmaquin are at clinical development phases (Paranjpe and Müller-Goymann, 2014). Arikace is liposomal Amikacin containing dipalmitoyl-phosphatidylcholine and cholesterol against *Pseudomonas aeroginosa* causing a lung infection, leading to cystic fibrosis. Pulmaquin is ciprofloxacin liposome for treatment of lung infections. Several other liposome-based formulations containing antioxidants are under study for acute oxidant-related lung injury and hypoxia (Hood et al., 2014).

Dendrimers

Dendrimers are defined as hyperbranched polymeric macromolecules, having well-defined radical branching (Kesharwani et al., 2014). Dendrimers of small size,

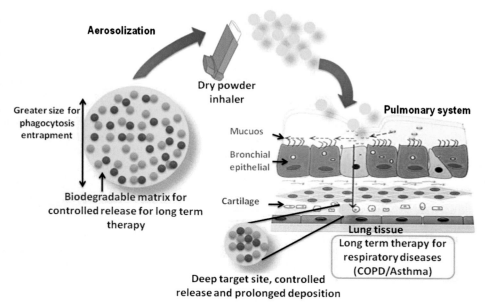

FIG. 5.2 **Mechanism of dry powder-based drug delivery.** Nanoparticle possessing biodegradable matrix and in size such that it can entrap phagocytes. This dry powder can be administered through pulmonary route crossing the mucosal and bronchial epithelial barrier, resulting in deep targeting of nanoparticle for prolonged deposition and long-term therapy of COPD and asthma.

spherical shape, and lipophilicity are better than linear polymers, as they have better penetration to the cell membrane. The loading on dendrimers depends on type and number of active sites, loading capacity, external functional groups, and lipophilicity. Dendrimers-based formulations are useful in treating endothelium dysfunction and preventing inflammation and metastasis in cancer (Khan et al., 2015). PEGylation of dendrimers is highly useful in the therapy of pulmonary diseases after inhalation (Ryan et al., 2013).

Inorganic nanoparticles

Typically, inorganic nanoparticles comprise noble metals such as gold and silver, iron oxide, and silica. Depending on distinctive properties of magnetism and plasmon resonance, especially gold and iron are used for diagnostics via CT, MRI, and positron emission tomography (Xie et al., 2011). These cationic metallic nanoparticles are also being used for gene delivery owing to the fact that they can easily bind to anionic DNA/RNA (Ding et al., 2014). Based on this principle, gene delivery has been successful to COPD alveolar epithelium in mice model, but cytotoxicity remains a major concern. As well as, cationic metal nanoparticles can also interact with negatively charged serum proteins leading to agglomeration (Geiser et al., 2013).

Silver nanoparticles are also budding because of high conductivity, lower toxicity, stability, and therapeutic effect. They are nontoxic at a low dose (Seong et al., 2006). The therapeutic effect has been demonstrated in H1299 cell line for lung cancer; once silver nanoparticles have been injected, a suppression in proliferation was observed (Seong et al., 2006).

Chitosan

Chitosan is derived from alkaline deacetylation of chitin and is a naturally occurring linear biopolyaminosaccharide (Hu et al., 1999). It is suitable as a carrier molecule for anticancer (Ignjatović et al., 2016), antiinflammatory (Friedman et al., 2013), antibiotics (Huang et al., 2016), and pulmonary (Teijeiro-Osorio et al., 2009) drugs because of controlled drug release properties. The only drawback of chitosan nanoparticles is agglomeration during synthesis (Dua et al., 2017). Chitosan derivatives such as trimethyl chitosan, carboxymethyl chitosan, N-succinyl chitosan, PEGylated chitosan, and thiolated chitosan have been used as a preferable excipient for safe and controlled drug delivery. Trimethyl chitosan is highly mucoadhesive, which is responsible for better absorption, leading to the opening of tight junctions between the adjoining epithelial (Sieval et al., 1998).

Carboxymethyl chitosan has an affinity in both basic and neutral solutions (Anitha et al., 2011). Use of rifampicin-based nano-oleoyl-carboxymethyl chitosan has been reported by Li et al. (2011) for pulmonary tuberculosis. N-succinyl chitosan is polar in nature and thus proves to be biocompatible in nature (Hou et al., 2010). PEGylation of chitosan increases solubility and biocompatibility helping to escape host immune defense with prolonged circulation time in the blood (Andrade et al., 2011). Thiolated chitosan is formed by introduction of cysteine disulfide bonds in mucous glycoproteins as a result of which it has increased mucoadhesive properties up to 140 times (Sieval et al., 1998).

Gelatin

It is obtained as a product from denatured protein after acid or alkaline hydrolysis of collagen (Lee et al., 2015). Gelatin consists of polypeptides with hydrophobic, cationic, and anionic amino acids in 1:1:1 ratio. Polypeptides are as ~13% positively charged, ~12% negatively charged, and ~11% hydrophobic in nature (Elzoghby, 2013). Hydrophobic- and hydrophilic-based drugs including resveratrol (Karthikeyan et al., 2013), curcumin (Cao et al., 2011), cisplatin (Tseng et al., 2009), and methotrexate (Cascone et al., 2002) are delivered gelatin-based nanoparticles (GNPs). Continuous drug release and low toxic effect to cells are important factors for GNPs-based drug delivery (Elzoghby, 2013). Along with, gelatin is highly reproducible and cost-effective. For instance, GNP-based cisplatin-loaded biotinylated EGF is more efficient as compared with cisplatin in solution against A549 lung adenocarcinoma in vitro (Tseng et al., 2009). In addition, studies have revealed that it had a high concentration of cisplatin nanoformulation as that of cisplatin solution in mice (Lee et al., 2015).

Other polymeric nanoparticles

Natural and synthetic polymers are commonly used for nanoparticle synthesis. These are macromolecules with reoccurring monomer units, and the chemical groups are important for functionalization and conjugation to drugs. They are preferable because of high drug encapsulation, surface modification, conserved drug deterioration, better shelf life, and prolonged delivery of the drug. For instance, PEGylation is bioinert, brings about surface modification of nanoparticle, and diminishes opsonization from immune cells and much-inert nature (Jokerst et al., 2011). Similarly, polyoxalate with PGLA composition attenuates inflammatory response and downregulates proinflammatory cytokines. Thus, hydroxybenzyl alcohol (HBA) integrated with polyoxalate—PGLA-based nanoparticle has been studied for the treatment of asthma and pulmonary inflammation (Yoo et al., 2013). Similarly, cationic polymers such as polyethyleneimine (PEI) can bind to nucleotides because of electrostatic forces so they are efficient in gene delivery for pulmonary diseases (Seong et al., 2006).

CURRENT NANOPARTICLE-BASED MEDICINE IN CHRONIC OBSTRUCTIVE PULMONARY DISEASE

COPD is the fourth major cause of death worldwide, and by the year 2020, it is expected to be the third cause of motility (To, 2018). It is a chronic disease designated by an abnormal inflammatory response because of inhalation of noxious particles directing airflow limitation. The pathophysiology of the disease is marked by chronic airways inflammation and degradation of lung parenchyma, as a result of which, alveolar macrophages release inflammatory moderators such as tumor necrosis factor (TNF-α), interleukins (IL-6, IL-8), monocyte chemotactic peptide-1 (MCP-1), leukotriene LTB4, and reactive oxygen species/reactive nitrogen species (ROS/RNS). Recent therapeutic choices are anticholinergics, β2-agonists, and inhaled corticosteroids, which are responsible to govern the symptoms rather than to cure the disease (To, 2018). The challenge faced by nanotherapeutics is airways immune response, mucous hypersecretion, and inflammation. A study by Vij (2011) elaborates upon the application of versatile polymer vesicles of poly(ethylene glycol) and PLGA for integrated delivery of corticosteroids and antiinflammatory bronchodilators theophylline along with molecular probes for theranostics administration in lung obstruction. Muralidharan et al. (2016) proclaimed the treatment of pulmonary inflammation by inhalable dry powder with dimethyl fumarate, an antioxidant Nrf2 activator (Muralidharan et al., 2016). Through in vitro studies for lung deposition model, the authors reported that the aerosol particles could reach lower pulmonary tract and treat inflammation during COPD. Phosphoinositide 3-kinase inhibitor (PI3K) associated with differentiation of alveolar epithelial stem cells had been used for alveoli repair in COPD-induced animal model (Horiguchi et al., 2015). Combinations of the inhibitor drug with patient-friendly nanoparticles are predicted to have effective treatment of COPD. Gold nanoparticles can be used for successful target delivery in COPD epithelial cells of alveoli and macrophages (Geiser et al., 2013). Supermagnetic iron oxide nanoparticles provide an opportunity for imaging of COPD at the molecular level. This has been proved by conjugation of antibody to iron oxide nanoparticles, which were used for MRI of macrophages. Anti-CD206 and anti-CD86 antibody conjugated with iron oxide nanoparticles in COPD lipopolysaccharide-induced mouse model showed different affinity to the proinflammatory and resolute macrophages (Al Faraj etal., 2014). Modified surface of target allows specific and selective delivery of nanoparticles to the target site, thus providing high contrast imaging and targeted therapy. Refer Table 5.2 for examples of nanoparticle used in COPD.

Current Nanoparticle-Based Medicine in Asthma

Asthma is a heterogeneous disease distinguished by chronic inflammation of airways. It has been reported as the 14th significant disorder worldwide concerning to extent and duration (Celik et al., 2012). The disorder is characterized by airways inflammation, which causes bronchial hyperresponsiveness, which in turn results in recurrent reversible airway obstruction. In patients with asthma, Th2 type of inflammation is common to chronic allergic inflammatory responses at multiple tissue sites. IgE, mast cells, eosinophils, macrophages, and cytokines, the local epithelial, mesenchymal, vascular, and neurologic events are involved in directing the Th2 phenotype to the lung and through aberrant injury repair mechanisms to remodeling of the airway wall (Holgate, 2008). The commonly used therapeutics for asthma short- or long-acting β-adrenergic agonist inhaled corticosteroids and bronchodilators that reduce the symptoms of the disease. Long-term use of corticosteroids leads to side effects, decreased symptomatic relief, loss of lung function, and inability to revert remodeling, further causing immune suppression.

Matsuo et al. (2009) showed that sheath steroid with nanoparticles has better efficiency compared with free steroid in airways inflammation. Salbutamol encapsulated in nanoparticles has better efficacy to lung membrane during bronchospasm as a result of drug concentration at the target site. Liposomal-based salbutamol has high retention time as well as concentration and thus better therapeutic effect (Ahmad et al., 2009). Wang et al. (2012) reported increased curcumin bioavailability and efficacy in solid lipid nanoparticles as an anti-inflammatory antioxidant for bronchial asthma. Another class of medication called montelukast, a leukotriene receptor antagonist used for prevention of wheezing, bronchospasm, cough, and tightness of chest when encapsulated in nanolipid, increases its

TABLE 5.2
Relevant Studies Focusing on Nanoparticle-Based Therapy for COPD and Asthma.

S.No	Nanocarrier	Drug	Class	Condition	Route of Administration	Method of Preparation	Mode of Action	Reference
1.	Polyethylene glycol and phosphatidylethanolamine (PEG-DSPE)	Budesonide	Glucocorticoid	Asthma and COPD	Dry inhaled powder	Coprecipitation and reconstitution	Inhibited airway inflammation	Jacobs and Müller (2002)
3.	PLGA LPMs	Oridonin	Antiinflammatory	Asthma and COPD	Aerosol	Freeze-drying	Autophagy concomitantly with loss of AML1-ETO	Yadollahi et al. (2015)
4.	Solid-lipid nanoparticles	Salbutamol sulfate	β_2 agonist	Asthma	Dry inhaled powder	Precipitation	Activation of Gs adenylyl cyclase cyclic AMP pathway	Mehta (2016)
5.	Polyaspartamide	Fluticasone propionate/albuterol sulfate	β_2 agonist/antiinflammatory	Asthma and COPD	Dry inhaled powder	Spray drying and freeze drying	Stimulates glucocorticoid receptor	Craparo et al. (2017)
6.	—	CpG-oligonucleotide (ODN)	Antigen	Asthma	Nasal	—	Trigger Th1 immune response and increase IgG2 and IFN-γ cytokine levels	Givens et al. (2018)

bioavailability and helps to bypass hepatic metabolism, thereby reducing toxicity (Patil-Gadhe and Pokharkar, 2014). Furthermore, telodendrimers have high loading capacity and good stability and provide a gradual release of hydrophobic drugs such as dexamethasone that target lungs inflammation, thereby reducing eosinophils and cytokines further boosting hyperresponsiveness of airways (Kenyon et al., 2013). Nanoparticles have been used for gene therapy for inhibition of Th2 transcription factors, cytokines, and overexpression of Th2 antagonists. Poly(ethylene imines) (PEIs) are the most eminently used polymeric nanoparticles for gene delivery. Chitosan, dendrimers, and poly(lactic-co-glycolic acid) (PLGA) are also being used as nanocarriers of gene therapy (da Silva et al., 2017). Chitosan interferon (IFN)-γ-pDNA (CIN) has been proved to lower hyperresponsiveness of airways against methacholine and ovalbumin-induced asthma mouse model. CIN has also been reported to reduce inflammatory cytokine, CD8+ T lymphocytes ruling downregulation of dendritic cells and block Th2 cytokine production. Nanoparticles composed of poly-L-lysine and polyethylene glycol-linked by a cysteine residue (CK30PEG) are nontoxic and immune-compatible to human lungs (Da Silva et al., 2014; Kong et al., 2008). This nanoparticle system is efficient to carry and deliver thymulin, resulting in an antiinflammatory response, antifibrotic effects, collagen deposition, and smooth muscle hypertrophy in an animal model. Furthermore, cytosine-phosphate-guanine (CpG) adjuvant—loaded biodegradable nanoparticles are successful in the treatment of dust allergies and being used for suppression of Th2 asthmatic response (Salem, 2014). Therefore, prospective molecular targets mediated by nanoparticles are transcription factors, immune cells of airways, kinases, and their receptors along with knock in and knock out of related genes. Refer Table 5.2 for examples of nanoparticle used in asthma.

TOXICITY IN CHRONIC OBSTRUCTIVE PULMONARY DISEASE AND ASTHMA

Inhalable nanoparticles have attained a major concern regarding safety and toxicity. Deposition of insoluble nanoparticles has resulted in the local reaction such as elevated oxidative stress, macrophages, and inflammation. Administration of insoluble nanoparticles such as carbon and titanium oxide induced inflammatory response caused by leukocytes and cytokines especially in bronchoalveolar lavage. Size of the nanoparticle plays a critical role in lung deposition. Greater sized nanoparticles are deposited in the upper airways,

whereas the smaller ones deposit in distilled parts of the pulmonary tract (Kreyling et al., 2009, 2013; Upadhyay et al., 2010, 2014). Ultrafine (<100 nm) particles are taken up by cells and allow crossing of the epithelial and endothelial layer via transcytosis into the circulation leading to distribution throughout the body which are responsible for particle-induced fibrosis, pneumoconiosis. The extent of the toxic effect depends on the exposure time, concentration, and ROS/RNS production (Roy and Vij, 2010). However, chronic lung diseases are being investigated for localized and systematic drug delivery, including small molecules, peptides, siRNA for specific genes. Also, many times, nanovehicle is deposited or entrapped in mononuclear phagocytotic system, which may seep into circulation, reach other organs, and result into other major diseases (Nygaard et al., 2004) and thus causes local and systemic cytotoxicity, allergic reaction, or inflammation. Eventually, biodegradable nanoparticles are being used as a vehicle for drug delivery such as polymers, liposomes, and lipids, which have significantly lower inflammatory response in comparison with nonbiodegradable nanovehicles of the same size (Dailey et al., 2006). Therefore, retention time and permeability should be taken into consideration for pulmonary drug designing for a better understanding of side effects and drug absorption at different sites (Olsson et al., 2011).

CONCLUSION

Asthma and COPD are the chronic respiratory diseases facing an increase in global burden. This indicates the constant requirement of novel therapeutics and diagnostics for these diseases. The nanoparticle is one of the major drug delivery agents, which can be used to fulfill the need. Nanotechnology proposes several exhilarating opportunities for the treatment of respiratory diseases. Systematic and local deliveries are the two modes for drug delivery. In COPD and asthma, we can adopt local drug delivery directly to the airways. While fabricating appropriate nanoparticle for COPD and asthma, it is essential that the nanoparticle size and shape be taken into account. Specific and targeted delivery can be obtained using nanoparticle, which would improve the efficacy of existing therapies and reduce side effects. By overcoming the current challenges of crossing barriers and accurate targeting, an efficient system can be developed. The review also suggests that pulmonary route of drug delivery for nanotechnology-based therapeutics can prove to be a more significant approach.

REFERENCES

Adler-Moore, J., Proffitt, R.T., 2002. AmBisome: liposomal formulation, structure, mechanism of action and preclinical experience. Journal of Antimicrobial Chemotherapy 49 (1), 21–30.

Ahmad, F.J., Mittal, G., Jain, G.K., Malhotra, G., Khar, R.K., Bhatnagar, A., 2009. Nano-salbutamol dry powder inhalation: a new approach for treating broncho-constrictive conditions. European Journal of Pharmaceutics and Biopharmaceutics 71 (2), 282–291.

Al Faraj, A., Shaik, A.S., Afzal, S., Al Sayed, B., Halwani, R., 2014. MR imaging and targeting of a specific alveolar macrophage subpopulation in LPS-induced COPD animal model using antibody-conjugated magnetic nanoparticles. International Journal of Nanomedicine 9, 1491.

Albanese, A., Tang, P.S., Chan, W.C., 2012. The effect of nanoparticle size, shape, and surface chemistry on biological systems. Annual Review of Biomedical Engineering 14, 1–16.

Andrade, F., Goycoolea, F., Chiappetta, D.A., Das Neves, J., Sosnik, A., Sarmento, B., 2011. Chitosan-grafted copolymers and chitosan-ligand conjugates as matrices for pulmonary drug delivery. International Journal of Carbohydrate Chemistry 2011.

Anitha, A., Maya, S., Deepa, N., Chennazhi, K., Nair, S., Tamura, H., Jayakumar, R., 2011. Efficient water soluble O-carboxymethyl chitosan nanocarrier for the delivery of curcumin to cancer cells. Carbohydrate Polymers 83 (2), 452–461.

Bahadori, M., Mohammadi, F., 2012. Nanomedicine for respiratory diseases. Tanaffos 11 (4), 18.

Bai, J., Wang, J.T.-W., Rubio, N., Protti, A., Heidari, H., Elgogary, R., Shah, A.M., 2016. Triple-modal imaging of magnetically-targeted nanocapsules in solid tumours in vivo. Theranostics 6 (3), 342.

Bailey, M.M., Berkland, C.J., 2009. Nanoparticle formulations in pulmonary drug delivery. Medicinal Research Reviews 29 (1), 196–212.

Bazak, R., Houri, M., El Achy, S., Hussein, W., Refaat, T., 2014. Passive targeting of nanoparticles to cancer: a comprehensive review of the literature. Molecular and clinical oncology 2 (6), 904–908.

Bianchi, A., Lux, F., Tillement, O., Crémillieux, Y., 2013. Contrast enhanced lung MRI in mice using ultra-short echo time radial imaging and intratracheally administered Gd-DOTA-based nanoparticles. Magnetic Resonance in Medicine 70 (5), 1419–1426.

Brain, J.D., 2007. Inhalation, deposition, and fate of insulin and other therapeutic proteins. Diabetes Technology and Therapeutics 9 (S1), S14–S15.

Cao, F., Ding, B., Sun, M., Guo, C., Zhang, L., Zhai, G., 2011. Lung-targeted delivery system of curcumin loaded gelatin microspheres. Drug Delivery 18 (8), 545–554.

Cascone, M.G., Lazzeri, L., Carmignani, C., Zhu, Z., 2002. Gelatin nanoparticles produced by a simple W/O emulsion as delivery system for methotrexate. Journal of Materials Science: Materials in Medicine 13 (5), 523–526.

Celik, M., Tuncer, A., Soyer, O.U., Saçkesen, C., Tanju Besler, H., Kalayci, O., 2012. Oxidative stress in the airways of children with asthma and allergic rhinitis. Pediatric Allergy and Immunology 23 (6), 556–561.

Cheng, J., Teply, B.A., Sherifi, I., Sung, J., Luther, G., Gu, F.X., Farokhzad, O.C., 2007. Formulation of functionalized PLGA–PEG nanoparticles for in vivo targeted drug delivery. Biomaterials 28 (5), 869–876.

Colina, A.-R., Aumont, F., Deslauriers, N., Belhumeur, P., de Repentigny, L., 1996. Evidence for degradation of gastrointestinal mucin by Candida albicans secretory aspartyl proteinase. Infection and Immunity 64 (11), 4514–4519.

Connaris, S., Greenwell, P., 1997. Glycosidases in mucin-dwelling protozoans. Glycoconjugate Journal 14 (7), 879–882.

Craparo, E.F., Ferraro, M., Pace, E., Bondì, M.L., Giammona, G., Cavallaro, G., 2017. Polyaspartamide-based nanoparticles loaded with fluticasone propionate and the in vitro evaluation towards cigarette smoke effects. Nanomaterials 7 (8), 222.

da Silva, A.L., Cruz, F.F., Rocco, P.R.M., Morales, M.M., 2017. New perspectives in nanotherapeutics for chronic respiratory diseases. Biophysical Reviews 9 (5), 793–803.

Da Silva, A.L., Martini, S.V., Abreu, S.C., Samary, C. d. S., Diaz, B.L., Fernezlian, S., Goya, R.G., 2014. DNA nanoparticle-mediated thymulin gene therapy prevents airway remodeling in experimental allergic asthma. Journal of Controlled Release 180, 125–133.

Dailey, L., Jekel, N., Fink, L., Gessler, T., Schmehl, T., Wittmar, M., Seeger, W., 2006. Investigation of the proinflammatory potential of biodegradable nanoparticle drug delivery systems in the lung. Toxicology and Applied Pharmacology 215 (1), 100–108.

Dalby, R., Spallek, M., Voshaar, T., 2004. A review of the development of Respimat® soft Mist™ inhaler. International Journal of Pharmaceutics 283 (1–2), 1–9.

Ding, Y., Jiang, Z., Saha, K., Kim, C.S., Kim, S.T., Landis, R.F., Rotello, V.M., 2014. Gold nanoparticles for nucleic acid delivery. Molecular Therapy 22 (6), 1075–1083.

Dua, K., Bebawy, M., Awasthi, R., Tekade, R.K., Tekade, M., Gupta, G., Hansbro, P.M., 2017. Application of chitosan and its derivatives in nanocarrier based pulmonary drug delivery systems. Pharmaceutical Nanotechnology 5 (4), 243–249.

Edwards, D., Dunbar, C., 2002. Therapeutic aerosol bioengineering. Annual Review of Biomedical Engineering 4, 93–107.

Elzoghby, A.O., 2013. Gelatin-based nanoparticles as drug and gene delivery systems: reviewing three decades of research. Journal of Controlled Release 172 (3), 1075–1091.

Fajac, I., Briand, P., Monsigny, M., Midoux, P., 1999. Sugar-mediated uptake of glycosylated polylysines and gene transfer into normal and cystic fibrosis airway epithelial cells. Human Gene Therapy 10 (3), 395–406.

Friedman, A.J., Phan, J., Schairer, D.O., Champer, J., Qin, M., Pirouz, A., Modlin, R.L., 2013. Antimicrobial and anti-inflammatory activity of chitosan–alginate nanoparticles: a targeted therapy for cutaneous pathogens. Journal of Investigative Dermatology 133 (5), 1231–1239.

Gaul, R., Ramsey, J.M., Heise, A., Cryan, S.-A., Greene, C.M., 2018. Nanotechnology approaches to pulmonary drug delivery: targeted delivery of small molecule and gene-based therapeutics to the lung. Design of Nanostructures for Versatile Therapeutic Applications 221–253.

Geiser, M., Quaile, O., Wenk, A., Wigge, C., Eigeldinger-Berthou, S., Hirn, S., Mall, M.A., 2013. Cellular uptake and localization of inhaled gold nanoparticles in lungs of mice with chronic obstructive pulmonary disease. Particle and Fibre Toxicology 10 (1), 19.

Gersting, S.W., Schillinger, U., Lausier, J., Nicklaus, P., Rudolph, C., Plank, C., Rosenecker, J., 2004. Gene delivery to respiratory epithelial cells by magnetofection. The Journal of Gene Medicine 6 (8), 913–922.

Givens, B.E., Geary, S.M., Salem, A.K., 2018. Nanoparticle-based CpG-oligonucleotide therapy for treating allergic asthma. Immunotherapy (0).

Hasleton, P., 1972. The internal surface area of the adult human lung. Journal of Anatomy 112 (Pt 3), 391.

Holgate, S.T., 2008. Pathogenesis of asthma. Clinical and Experimental Allergy 38 (6), 872–897.

Hood, E.D., Chorny, M., Greineder, C.F., Alferiev, I.S., Levy, R.J., Muzykantov, V.R., 2014. Endothelial targeting of nanocarriers loaded with antioxidant enzymes for protection against vascular oxidative stress and inflammation. Biomaterials 35 (11), 3708–3715.

Horiguchi, M., Oiso, Y., Sakai, H., Motomura, T., Yamashita, C., 2015. Pulmonary administration of phosphoinositide 3-kinase inhibitor is a curative treatment for chronic obstructive pulmonary disease by alveolar regeneration. Journal of Controlled Release 213, 112–119.

Hou, Z., Han, J., Zhan, C., Zhou, C., Hu, Q., Zhang, Q., 2010. Synthesis and evaluation of N-succinyl-chitosan nanoparticles toward local hydroxycamptothecin delivery. Carbohydrate Polymers 81 (4), 765–768.

Hu, K.J., Yeung, K.W., Ho, K.P., Hu, J.L., 1999. Rapid extraction of high-quality chitosan from mycelia of *Absidia glauca*. Journal of Food Biochemistry 23 (2), 187–196.

Huang, W.-T., Larsson, M., Lee, Y.-C., Liu, D.-M., Chiou, G.-Y., 2016. Dual drug-loaded biofunctionalized amphiphilic chitosan nanoparticles: enhanced synergy between cisplatin and demethoxycurcumin against multidrug-resistant stem-like lung cancer cells. European Journal of Pharmaceutics and Biopharmaceutics 109, 165–173.

Ibrahim, M., Verma, R., Garcia-Contreras, L., 2015. Inhalation drug delivery devices: technology update. Medical Devices (Auckland, NZ) 8, 131.

Ignjatović, N.L., Penov-Gaši, K.M., Wu, V.M., Ajduković, J.J., Kojić, V.V., Vasiljević-Radović, D., Uskoković, D.P., 2016. Selective anticancer activity of hydroxyapatite/chitosan-poly (d, l)-lactide-co-glycolide particles loaded with an androstane-based cancer inhibitor. Colloids and Surfaces B: Biointerfaces 148, 629–639.

Jacobs, C., Müller, R.H., 2002. Production and characterization of a budesonide nanosuspension for pulmonary administration. Pharmaceutical Research 19 (2), 189–194.

Jawahar, N., Reddy, G., 2012. Nanoparticles: a novel pulmonary drug delivery system for tuberculosis. Journal of Pharmaceutical Sciences and Research 4 (8), 1901.

Jokerst, J.V., Lobovkina, T., Zare, R.N., Gambhir, S.S., 2011. Nanoparticle PEGylation for imaging and therapy. Nanomedicine 6 (4), 715–728.

Karthikeyan, S., Prasad, N.R., Ganamani, A., Balamurugan, E., 2013. Anticancer activity of resveratrol-loaded gelatin nanoparticles on NCI-H460 non-small cell lung cancer cells. Biomedicine & Preventive Nutrition 3 (1), 64–73.

Kenyon, N.J., Bratt, J.M., Lee, J., Luo, J., Franzi, L.M., Zeki, A.A., Lam, K.S., 2013. Self-assembling nanoparticles containing dexamethasone as a novel therapy in allergic airways inflammation. PLoS One 8 (10), e77730.

Kesharwani, P., Jain, K., Jain, N.K., 2014. Dendrimer as nanocarrier for drug delivery. Progress in Polymer Science 39 (2), 268–307.

Khan, O.F., Zaia, E.W., Jhunjhunwala, S., Xue, W., Cai, W., Yun, D.S., Pelet, J.M., 2015. Dendrimer-inspired nanomaterials for the in vivo delivery of siRNA to lung vasculature. Nano Letters 15 (5), 3008–3016.

Kong, X., Hellermann, G.R., Zhang, W., Jena, P., Kumar, M., Behera, A., Mohapatra, S.S., 2008. Chitosan interferon-γ nanogene therapy for lung disease: modulation of T-cell and dendritic cell immune responses. Allergy Asthma & Clinical Immunology 4 (3), 95.

Kreyling, W.G., Hirn, S., Möller, W., Schleh, C., Wenk, A., Celik, G. l., Johnston, B.D., 2013. Air–blood barrier translocation of tracheally instilled gold nanoparticles inversely depends on particle size. ACS Nano 8 (1), 222–233.

Kreyling, W.G., Semmler-Behnke, M., Seitz, J., Scymczak, W., Wenk, A., Mayer, P., Oberdörster, G., 2009. Size dependence of the translocation of inhaled iridium and carbon nanoparticle aggregates from the lung of rats to the blood and secondary target organs. Inhalation Toxicology 21 (Suppl. 1), 55–60.

Kuzmov, A., Minko, T., 2015. Nanotechnology approaches for inhalation treatment of lung diseases. Journal of Controlled Release 219, 500–518.

Lee, W.-H., Loo, C.-Y., Traini, D., Young, P.M., 2015. Inhalation of nanoparticle-based drug for lung cancer treatment: advantages and challenges. Asian Journal of Pharmaceutical Sciences 10 (6), 481–489.

Li, Y., Zhang, S., Meng, X., Chen, X., Ren, G., 2011. The preparation and characterization of a novel amphiphilic oleoyl-carboxymethyl chitosan self-assembled nanoparticles. Carbohydrate Polymers 83 (1), 130–136.

Lyn, L.Y., Sze, H.W., Rajendran, A., Adinarayana, G., Dua, K., Garg, S., 2011. Crystal modifications and dissolution rate of piroxicam. Acta Pharmaceutica 61 (4), 391–402.

Madan, J.R., Khude, P.A., Dua, K., 2014. Development and evaluation of solid lipid nanoparticles of mometasone furoate for topical delivery. International Journal of Pharmaceutical Investigation 4 (2), 60.

Maeda, H., Matsumura, Y., 1989. Tumoritropic and lymphotropic principles of macromolecular drugs. Critical Reviews in Therapeutic Drug Carrier Systems 6 (3), 193–210.

Magenheim, B., Levy, M., Benita, S., 1993. A new in vitro technique for the evaluation of drug release profile from colloidal carriers-ultrafiltration technique at low pressure. International Journal of Pharmaceutics 94 (1−3), 115−123.

Matsuo, Y., Ishihara, T., Ishizaki, J., Miyamoto, K.-i., Higaki, M., Yamashita, N., 2009. Effect of betamethasone phosphate loaded polymeric nanoparticles on a murine asthma model. Cellular Immunology 260 (1), 33−38.

Mehta, P., 2016. Dry powder inhalers: a focus on advancements in novel drug delivery systems. Journal of Drug Delivery 2016.

Mosgoeller, W., Prassl, R., Zimmer, A., 2012. Nanoparticle-mediated treatment of pulmonary arterial hypertension. Methods in Enzymology 508, 325−354.

Muller, M., Vehlow, D., Torger, B., Urban, B., Woltmann, B., Hempel, U., 2018. Adhesive drug delivery systems based on polyelectrolyte complex nanoparticles (PEC NP) for bone healing. Current Pharmaceutical Design 24 (13), 1341−1348.

Muralidharan, P., Hayes, D., Black, S.M., Mansour, H.M., 2016. Microparticulate/nanoparticulate powders of a novel Nrf2 activator and an aerosol performance enhancer for pulmonary delivery targeting the lung Nrf2/Keap-1 pathway. Molecular systems design & engineering 1 (1), 48−65.

Musante, C.J., Schroeter, J.D., Rosati, J.A., Crowder, T.M., Hickey, A.J., Martonen, T.B., 2002. Factors affecting the deposition of inhaled porous drug particles. Journal of Pharmaceutical Sciences 91 (7), 1590−1600.

Naseri, N., Valizadeh, H., Zakeri-Milani, P., 2015. Solid lipid nanoparticles and nanostructured lipid carriers: structure, preparation and application. Advanced Pharmaceutical Bulletin 5 (3), 305.

Nygaard, U.C., Samuelsen, M., Aase, A., Løvik, M., 2004. The capacity of particles to increase allergic sensitization is predicted by particle number and surface area, not by particle mass. Toxicological Sciences 82 (2), 515−524.

Olsson, B., Bondesson, E., Borgström, L., Edsbäcker, S., Eirefelt, S., Ekelund, K., Hegelund-Myrbäck, T., 2011. Pulmonary drug metabolism, clearance, and absorption. Controlled Pulmonary Drug Delivery 21−50.

Omlor, A.J., Nguyen, J., Bals, R., Dinh, Q.T., 2015. Nanotechnology in respiratory medicine. Respiratory Research 16 (1), 64.

Paranjpe, M., Müller-Goymann, C., 2014. Nanoparticle-mediated pulmonary drug delivery: a review. International Journal of Molecular Sciences 15 (4), 5852−5873.

Patil-Gadhe, A., Pokharkar, V., 2014. Montelukast-loaded nanostructured lipid carriers: part I oral bioavailability improvement. European Journal of Pharmaceutics and Biopharmaceutics 88 (1), 160−168.

Patton, J.S., Brain, J.D., Davies, L.A., Fiegel, J., Gumbleton, M., Kim, K.-J., Ehrhardt, C., 2010. The particle has landed—characterizing the fate of inhaled pharmaceuticals. Journal of Aerosol Medicine and Pulmonary Drug Delivery 23 (S2), S71−S87.

Patton, J.S., Byron, P.R., 2007. Inhaling medicines: delivering drugs to the body through the lungs. Nature Reviews Drug Discovery 6 (1), 67.

Paturi, D., Patel, M., Mitra, R., Mitra, A.K., 2012. Nasal and pulmonary delivery of macromolecules to treat respiratory and nonrespiratory diseases. Pulmonary Nanomedicine 45.

Roy, I., Vij, N., 2010. Nanodelivery in airway diseases: challenges and therapeutic applications. Nanomedicine 6 (2), 237−244.

Ryan, G.M., Kaminskas, L.M., Kelly, B.D., Owen, D.J., McIntosh, M.P., Porter, C.J., 2013. Pulmonary administration of PEGylated polylysine dendrimers: absorption from the lung versus retention within the lung is highly size-dependent. Molecular Pharmaceutics 10 (8), 2986−2995.

Saari, M., Vidgren, M.T., Koskinen, M.O., Turjanmaa, V.M., Nieminen, M.M., 1999. Pulmonary distribution and clearance of two beclomethasone liposome formulations in healthy volunteers. International Journal of Pharmaceutics 181 (1), 1−9.

Salem, A.K., 2014. A promising CpG adjuvant-loaded nanoparticle-based vaccine for treatment of dust mite allergies. Immunotherapy 6 (11), 1161−1163.

Seong, J.H., Lee, K.M., Kim, S.T., Jin, S.E., Kim, C.K., 2006. Polyethylenimine-based antisense oligodeoxynucleotides of IL-4 suppress the production of IL-4 in a murine model of airway inflammation. The Journal of Gene Medicine 8 (3), 314−323.

Shaji, J., Lal, M., 2013. Nanocarriers for targeting in inflammation. Asian Journal of Pharmaceutical and Clinical Research 6 (3), 3−12.

Sharma, A., Prasad, A., Dua, K., Singh, G., 2014. Effect of combination of acrylic polymers on the release of nevirapine formulated as extended release matrix pellets using extrusion and spheronization technique. Current Drug Delivery 11 (5), 643−651.

Shoyele, S.A., Slowey, A., 2006. Prospects of formulating proteins/peptides as aerosols for pulmonary drug delivery. International Journal of Pharmaceutics 314 (1), 1−8.

Sieval, A., Thanou, M., Kotze, A., Verhoef, J., Brussee, J., Junginger, H., 1998. Preparation and NMR characterization of highly substituted N-trimethyl chitosan chloride. Carbohydrate Polymers 36 (2−3), 157−165.

Singh, R., Lillard Jr., J.W., 2009. Nanoparticle-based targeted drug delivery. Experimental and Molecular Pathology 86 (3), 215−223.

Soane, R., Carney, A., Jones, N., Frier, M., Perkins, A., Davis, S., Illum, L., 1999. The effect of the nasal cycle on mucociliary clearance. Clinical Otolaryngology and Allied Sciences 24 (4), 377−383.

Soriano, J.B., Abajobir, A.A., Abate, K.H., Abera, S.F., Agrawal, A., Ahmed, M.B., Alam, K., 2017. Global, regional, and national deaths, prevalence, disability-adjusted life years, and years lived with disability for chronic obstructive pulmonary disease and asthma, 1990−2015: a systematic analysis for the Global Burden of Disease Study 2015. The lancet Respiratory medicine 5 (9), 691−706.

Souto, E.B., Müller, R.H., 2010. Lipid nanoparticles: effect on bioavailability and pharmacokinetic changes. Drug Delivery 115–141.

Storm, G., van Etten, E., 1997. Biopharmaceutical aspects of lipid formulations of amphotericin B. European Journal of Clinical Microbiology & Infectious Diseases 16 (1), 64–73.

Suer, H., Bayram, H., 2017. Liposomes as potential nanocarriers for theranostic applications in chronic inflammatory lung diseases. Biomedical and Biotechnology Research Journal (BBRJ) 1 (1), 1.

Sung, J.C., Pulliam, B.L., Edwards, D.A., 2007. Nanoparticles for drug delivery to the lungs. Trends in Biotechnology 25 (12), 563–570.

Teijeiro-Osorio, D., Remuñán-López, C., Alonso, M.J., 2009. Chitosan/cyclodextrin nanoparticles can efficiently transfect the airway epithelium in vitro. European Journal of Pharmaceutics and Biopharmaceutics 71 (2), 257–263.

To, P. G. (2018). Global Initiative for Chronic Obstructive Lung.

Tseng, C.-L., Su, W.-Y., Yen, K.-C., Yang, K.-C., Lin, F.-H., 2009. The use of biotinylated-EGF-modified gelatin nanoparticle carrier to enhance cisplatin accumulation in cancerous lungs via inhalation. Biomaterials 30 (20), 3476–3485.

Upadhyay, S., Ganguly, K., Stoeger, T., Semmler-Bhenke, M., Takenaka, S., Kreyling, W.G., Eickelberg, O., 2010. Cardiovascular and inflammatory effects of intratracheally instilled ambient dust from Augsburg, Germany, in spontaneously hypertensive rats (SHRs). Particle and Fibre Toxicology 7 (1), 27.

Upadhyay, S., Stoeger, T., George, L., Schladweiler, M.C., Kodavanti, U., Ganguly, K., Schulz, H., 2014. Ultrafine carbon particle mediated cardiovascular impairment of aged spontaneously hypertensive rats. Particle and Fibre Toxicology 11 (1), 36.

Vij, N., 2011. Nano-based theranostics for chronic obstructive lung diseases: challenges and therapeutic potential. Expert Opinion on Drug Delivery 8 (9), 1105–1109.

Wang, H., Zheng, L., Peng, C., Shen, M., Shi, X., Zhang, G., 2013. Folic acid-modified dendrimer-entrapped gold nanoparticles as nanoprobes for targeted CT imaging of human lung adencarcinoma. Biomaterials 34 (2), 470–480.

Wang, W., Zhu, R., Xie, Q., Li, A., Xiao, Y., Li, K., Wang, S., 2012. Enhanced bioavailability and efficiency of curcumin for the treatment of asthma by its formulation in solid lipid nanoparticles. International Journal of Nanomedicine 7, 3667.

Weber, S., Zimmer, A., Pardeike, J., 2014. Solid lipid nanoparticles (SLN) and nanostructured lipid carriers (NLC) for pulmonary application: a review of the state of the art. European Journal of Pharmaceutics and Biopharmaceutics 86 (1), 7–22.

Wiewrodt, R., Thomas, A.P., Cipelletti, L., Christofidou-Solomidou, M., Weitz, D.A., Feinstein, S.I., Muzykantov, V.R., 2002. Size-dependent intracellular immunotargeting of therapeutic cargoes into endothelial cells. Blood 99 (3), 912–922.

Xie, J., Liu, G., Eden, H.S., Ai, H., Chen, X., 2011. Surface-engineered magnetic nanoparticle platforms for cancer imaging and therapy. Accounts of Chemical Research 44 (10), 883–892.

Yadollahi, R., Vasilev, K., Simovic, S., 2015. Nanosuspension technologies for delivery of poorly soluble drugs. Journal of Nanomaterials 2015, 1.

Yhee, J.Y., Im, J., Nho, R.S., 2016. Advanced therapeutic strategies for chronic lung disease using nanoparticle-based drug delivery. Journal of Clinical Medicine 5 (9), 82.

Yoo, D., Guk, K., Kim, H., Khang, G., Wu, D., Lee, D., 2013. Antioxidant polymeric nanoparticles as novel therapeutics for airway inflammatory diseases. International Journal of Pharmaceutics 450 (1–2), 87–94.

FURTHER READING

Dua, K., Chellappan, D.K., Singhvi, G., de Jesus Andreoli Pinto, T., Gupta, G., Hansbro, P.M., 2018 Dec. Targeting microRNAs using nanotechnology in pulmonary diseases. Panminerva Med 60 (4), 230–231.

Dua, K., Gupta, G., Awasthi, R., Chellappan, D.K., 2018 Sep. Why is there an emerging need to look for a suitable drug delivery platform in targeting and regulating microbiota? Panminerva Medica 60 (3), 136–137.

Dua, K., Gupta, G., Chellapan, D.K., Bebawy, M., Collet, T., 2018 Dec. Nanoparticle-based therapies as a modality in treating wounds and preventing biofilm. Panminerva medica 60 (4), 237–238.

Dua, K., Gupta, G., Chellappan, D.K., Shukla, S., Hansbro, P.M., 2019 Jun. Targeting bacterial biofilms in pulmonary diseases in pediatric population. Minerva Pediatrica 71 (3), 309–310.

Dua, K., Gupta, G., Koteswara Rao, N., Bebawy, M., 2018 Oct. Nano-antibiotics: a novel approach in treating P. aeruginosa biofilm infections. Minerva Medica 109 (5), 400.

Dua, K., Madan, J.R., Chellappan, D.K., Gupta, G., 2018 Sep. Nanotechnology in drug delivery gaining new perspectives in respiratory diseases. Panminerva Medica 60 (3), 135–136.

Dua, K., Malyla, V., Singhvi, G., Wadhwa, R., Krishna, R.V., Shukla, S.D., Shastri, M.D., Chellappan, D.K., Maurya, P.K., Satija, S., Mehta, M., Gulati, M., Hansbro, N., Collet, T., Awasthi, R., Gupta, G., Hsu, A., Hansbro, P.M., 2019 Feb 1. Increasing complexity and interactions of oxidative stress in chronic respiratory diseases: An emerging need for novel drug delivery systems. Chemico-Biological Interactions 299, 168–178.

Dua, K., Hansbro, N.G., Foster, P.S., Hansbro, P.M., 2018. Targeting MicroRNAs: promising future therapeutics in the treatment of allergic airway disease. Critical Reviews in Eukaryotic Gene Expression 28 (2), 125–127.

Dua, K., Rapalli, V.K., Shukla, S.D., Singhvi, G., Shastri, M.D., Chellappan, D.K., Satija, S., Mehta, M., Gulati, M., Pinto, T.J.A., Gupta, G., Hansbro, P.M., 2018 Nov. Multidrug resistant Mycobacterium tuberculosis & oxidative stress complexity: Emerging need for novel drug delivery approaches. Biomedicine & Pharmacotherapy 107, 1218–1229.

Mehta, M., Deeksha, Sharma, N., Vyas, M., Khurana, N., Maurya, P.K., Singh, H., Andreoli de Jesus, T.P., Dureja, H., Chellappan, D.K., Gupta, G., Wadhwa, R., Collet, T., Hansbro, P.M., Dua, K., Satija, S., 2019 May 1. Interactions with the macrophages: An emerging targeted approach using novel drug delivery systems in respiratory diseases. Chemico-Biological Interactions 304, 10–19. https://doi.org/10.1016/j.cbi.2019.02.021.

Ng, Z.Y., Wong, J.Y., Panneerselvam, J., Madheswaran, T., Kumar, P., Pillay, V., Hsu, A., Hansbro, N., Bebawy, M., Wark, P., Hansbro, P., Dua, K., Chellappan, D.K., 2018 Dec 1. Assessing the potential of liposomes loaded with curcumin as a therapeutic intervention in asthma. Colloids and Surfaces. B, Biointerfaces 172, 51–59.

Riese, P., Sakthivel, P., Trittel, S., Guzmán, C.A., 2014. Intranasal formulations: promising strategy to deliver vaccines. Expert Opinion on Drug Delivery 11 (10), 1619–1634.

Shastri, M.D., Shukla, S.D., Chong, W.C., Dua, K., Peterson, G.M., Patel, R.P., Hansbro, P.M., Eri, R., O'Toole, R.F., 2018 Oct 11. Role of Oxidative Stress in the Pathology and Management of Human Tuberculosis. Oxidative Medicine and Cellular Longevity 2018, 7695364.

Wadhwa, R., Aggarwal, T., Malyla, V., Kumar, N., Gupta, G., Chellappan, D.K., Dureja, H., Mehta, M., Satija, S., Gulati, M., Maurya, P.K., Collet, T., Hansbro, P.M., Dua, K., 2019 Mar 25. Identification of biomarkers and genetic approaches toward chronic obstructive pulmonary disease. Journal of Cellular Physiology. https://doi.org/10.1002/jcp.28482.

CHAPTER 6

Biosensors in Animal Biotechnology

ADITYA ARYA, PHD • ANAMIKA GANGWAR, MSC • AMIT KUMAR, PHD

INTRODUCTION

The term biosensor is used to describe a biological sensing device made up of a transducer and a biological element that may be an enzyme, an antibody, or a nucleic acid, wherein during the sensing process, the biological element interacts with the analyte being tested and the responses are converted into an electrical signal by the transducer. A standard definition of biosensor provided by regulatory committee states that "biosensor is an analytic device that consists of a biologic component in intimate contact with a physical transducer component that converts the signal generated by the biologic component into a measurable electrical or optical signal" (Bhalla et al., 2016). Also, A. Turner, the editor-in-chief of the Biosensors and Bioelectronics journal, defines a biosensor as "a compact analytical device incorporating a biological or biologically derived sensing element either integrated within or intimately associated with a physicochemical transducer" (Palchetti and Mascini, 2008). To call a senor as biosensor, it must have a biological component that acts as the sensor and an electronic component that detects and transmits the signal. In other words, the biological material is immobilized and a contact is made between the immobilized biological material and the transducer. The analyte binds to the biological material to form a bound analyte, which in turn produces the electronic response that can be measured. Sometimes the analyte is converted to a product that could be associated with the release of heat, gas (oxygen), electrons, or hydrogen ions. The transducer then converts the product-linked changes into electrical signals, which can be amplified and measured. We will discuss more about the operating principle and types of the biosensors in subsequent sections.

The beginning of biosensors dates back to the year 1906 when M. Cremer noted that electric potential arising between parts of the fluid located on opposite sides of a glass membrane (Cremer, 1906). Later in 1909 based on this hypothesis, Søren Sørensen developed the concept of pH and later W.S. Hughes developed a pH measurement device in the year 1922 (Hughes, 1922). Meanwhile, Griffin and Nelson demonstrated the immobilization of the enzyme invertase on aluminum hydroxide and charcoal that opened the doors for the development of true biosensors. In 1956, first true biosensor was developed by Leland C. Clark, which was meant to be used for detection of oxygen. The sensor is still known as Clark's electrode in his honor. Furthermore, in the year 1962, Clark developed another biosensor for glucose based on the amperometric detection, which was followed by a stroke for the detection of glucose by Leland Clark in 1962. Eventually in 1975, the first commercial biosensor was developed by Yellow Springs Instrument (YSI) (Palchetti and Mascini, 2008). Since then, hundreds of different types of biosensors have been developed. In 1983, first immunosensor was developed by Roederer and Bastiaans, based on piezoelectric detection (Roederer and Bastiaans, 1983). More recently advanced classes of biosensors including nanobiosensors and lab-on-chip biosensors have also been developed. Fig. 6.1 illustrates a brief timeline of biosensor development.

Animal biotechnology is a branch of biotechnology in which molecular biology techniques are used to create improved animals with their suitability for human use, in particular pharmacological value, commercial value, and productivity. Although in classical terms the animal improvements have been done from thousands of years and are a part of animal breeding programs, use of biotechnology nowadays primarily aims at use of recombinant DNA technology, in particular, to produce genetically modified animal. Genetically modified animals have been used to produce therapeutic proteins, resistant to disease and improved traits for commercial value.

Nanotechnology in Modern Animal Biotechnology. https://doi.org/10.1016/B978-0-12-818823-1.00006-5

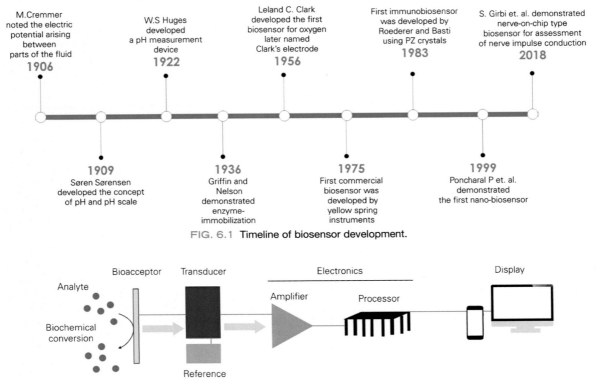

FIG. 6.1 Timeline of biosensor development.

FIG. 6.2 Basic design of a biosensor.

BASIC DESIGN AND OPERATING PRINCIPLE

A biosensor typically consists of three different parts: a signal receptor, a signal transducer, and an electronic signal processor. These components are connected through electronic or optical circuitry and must be able to communicate with each other in unidirectional (most often) or bidirectional manner. A basic outline design of a typical biosensor is provided in Fig. 6.2. Depending on the mechanisms of sensing, signal transduction, and signal processing, there can be a variety of formats of sensors, even for sensing the same analyte. We will discuss the type of sensors later in subsequent sections.

Bioreceptors

The first key component of a sensor is receptor that perceives the exogenous signal; in case of biosensor, the receptor is a biomolecule. A variety of biomolecules are known that can act as bioreceptors such as enzymes, antibodies, nucleic acid, carbohydrates, hormone, intact cells, or even tissues. Its role is to interact specifically with the target analyte, and as a result of biochemical reaction, it is consequently transformed through

transducer to a measurable signal. A variety of bioreceptors are frequently used, such as enzymes, antibodies, nucleic acids, and microbial cells. More recently, nanomaterials have begun to be used as bioacceptors, leading to the development of an emerging class of biosensors called nanobiosensors (discussed in Section 7). Fig. 6.3 represents outline classification of various bioacceptors used in biosensors development.

Following are some commonly used bioreceptors and their pros and cons for the usage in biosensors.

Enzymes

Enzymes are biocatalysts that are capable of enhancing the rate of biological reactions. There are six designated classes of enzymes according to the nature of reactions catalyzed by them. Among these, most commonly used enzymes are oxidoreductase, because of the involvement of electrons, which could be used for easy transduction into electronic signal. However, other classes of enzyme that could couple with colorimetric, thermometric, or fluorimetric signals are also used. During these reactions, interaction between substrate and

Five classes of bio-acceptors used in biosensor

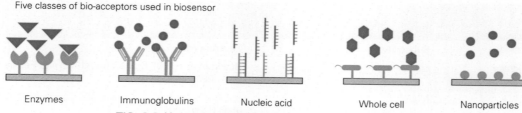

| Enzymes | Immunoglobulins | Nucleic acid | Whole cell | Nanoparticles |

FIG. 6.3 Various bioacceptors used in biosensor development.

enzyme leads to generation of electrons. Rocchitta et al. have recently reviewed the enzyme biosensors and their biomedical applications (Rocchitta et al., 2016). Table 6.1 summarizes the common enzymes that have been widely used in various biosensor activities.

Antibodies

Antibodies are another excellent choice of bioreceptors, primarily because of their highly specific reactions toward the analyte molecule. This unique property of, that is, specificity toward the analyte of interest (antigen) fitting into the antibody-binding site is crucial to their usefulness in immunosensors (Vo-Dinh and Cullum, 2000). Unlike enzymes, binding of antibody to analyte does not provoke an electrochemical reaction but is detected using fluorimetric or colorimetric means by using secondary antibodies. A number of immunoglobulin-based biosensors have been successfully commercialized, for example, pregnancy test cards. Table 6.2 enlists some of the common biosensors based on antibodies as bioreceptors. (Table 6.2)

Nucleic acids

Nucleic acid as a transducer has high sensitivity and selectivity by virtue of its very strong base pair affinity between complementary base pairing in nucleotide strands of bioreceptor and analyte (Borgmann et al., 2011). Both naturally existing and synthetic oligodeoxyribonucleotides (ODNs) are used as probes in the DNA hybridization sensors. End labels, such as thiols, disulfides, amines, or biotin, are incorporated to immobilize ODNs to transducer surfaces. A long flexible spacer is usually added by means of hydrocarbon linkers to provide sufficient accessibility for surface attachment (Labuda et al., 2010). Although most DNA-based sensors develop signals by means of a labeled complementary DNA (part of bioreceptor), which are often fluorescent or chromogenic in nature. However, researchers have recently gathered more interest in electrochemical sensors based on nucleic acids where complementary base pairing generates a current, which is detected using suitable transducer. Unlike

enzymes or antibody transducers, nucleic acid recognition layers can be readily synthesized and regenerated for multiple uses, which is one of the greatest advantages of nucleic acid transducers.

A subtype of nucleic acid bioacceptors is aptamers. Aptamers are artificial single-stranded DNA or RNA oligonucleotides (typically <100 mer), which are often selected from randomized oligonucleotide libraries by SELEX (systematic evolution of ligands by exponential enrichment). These aptamers are promising tools in biosensor development as they specifically bind with various biomolecules, such as proteins, viruses, and bacteria, as well as small molecules, such as organic dyes, metal ions and amino acids (Tan et al., 2011). A comprehensive perspective of pros and cons of aptamers-based biosensors is provided by Lakhin et al. (2013).

Whole cells

Whole cells have also been used as a potential transducer in the development of biosensors. Bacterial cells, animal cells, and plant cells are known to be used (D'Souza, 2001). Microbial cells are known to react with a large number of substrates and show electrochemical response, which can be recorded and transmitted using a transducer, forming the principle of whole cell–based biosensors (Corcoran and Rechnitz, 1985). Immobilized *Azotobacter vinelandii* coupled with ammonia electrode shows sensitivity range between 10 and 8×10 mol/dm^3. It measures the concentration of nitrate within 5–10 min. Examples of a few important microbial biosensors are given in Table 6.3.

Besides bacterial cells, bacteriophages and other viruses have also been explored for their use in development of biosensors. Bacteriophages provide an opportunity to be used as bioreceptors, because of their ability to interact with specific strains of bacteria. Although the use of bacteriophage as biosensing component is more relevant in food technology, where detection of foodborne bacterial pathogens can be performed based on viruses infecting bacterial cells (Zhang et al., 2015), animal biotechnology also makes use of

TABLE 6.1
Common Enzymes Used in Biosensor Development.

Enzyme Class	Substrate	References
Oxidoreductase	Lactate	Huang et al. (2008), Katrlík et al. (1999) and Pereira et al. (2007)
	Malate	Arif et al. (2002), Monošík et al. (2012), Prodromidis et al. (1997), Wang et al. (2008)
	Ascorbate	Vermeir et al. (2007), Wang et al. (2008), Pollegioni et al. (2007) and Sacchi et al. (1998)
	Alcohol	Katrlík et al. (1999), Pena et al. (2002), Smutok et al. (2006), Tkač et al. (2003)
	Cholesterol	(Lia and Gub, 2006), Umar et al. (2009), Vidal et al. (2004)
	Glycerol	Alvarez-Gonzalez et al. (2000), dMonošik et al. (2012), Niculescu et al. (2003)
	Fructose	Tkač et al. (2001), Tkač et al. (2002)
Transferase	Acetic acid	Mieliauskiene et al. (2006), Mizutani et al. (2003)
	Captan	Choi et al. (2003)
	Atrazine	Andreou and Clonis (2002)
Hydrolase	Sucrose	Soldatkin et al. (2008), Surareungchai et al. (1999)
Lyase	Citric acid	Maines et al. (2000), Prodromidis et al. (1997)
Ligase		Pang et al. (2006)
Isomerase		Sheu et al. (2008)
SPECIFIC EXAMPLES		
Glucose oxidase	β-D-Glucose	Reviewed in Rocchitta et al. (2016)
Glutamate oxidase	L-Glutamate	''
Alcohol oxidase	Ethanol	''
Lactate oxidase	L-Lactate	''
Ascorbate oxidase	L-Ascorbic acid	''
Cholesterol oxidase	Cholesterol	''
Choline oxidase	Choline, acetylcholine	''
Laccase	Polyphenols	''
Tyrosinase	Monophenols, dihydroxyphenols, bisphenol A	''
Alcohol dehydrogenase	Ethanol	''
Glutamate dehydrogenase	Glutamate	''
Glucose dehydrogenase	Glucose	''
Lactate dehydrogenase	Lactate	''

such sensors for detection of microbial contamination. Bacteriophage-based biosensors are known to work on two different approaches: one involves unmodified lysing, which can lyse bacteria specifically and release specific components to facilitate detection, and the second approach involves modified phages that carry reporter genes to facilitate the detection of their growth or amplification inside bacterial cells.

TABLE 6.2
List of Some Common Examples of Antibodies as BioAcceptors.

Application	Analyte	Antibody Type	Transducer	Reference
Nitroaromatic explosives detection	2,4,6-Trinitrotoluene (TNT)	scAb	Chemoresistive	Reviewed in (Sharma et al., 2016)
Deep vein thrombosis (DVT) disorders	D-dimer	scAb	Electrochemical	"
Human immunodeficiency virus (HIV)	HIV-1 virion infectivity factor	scFv	Piezoelectric	"
Toxic metabolite detection	Aflatoxin B1	scFv	SPR	"
Disease diagnosis	Fc receptors	scFv	Piezoelectric	"
Listeriosis diagnosis	Listeria monocytogenes	scFv	Electrochemical	"
Pesticide concentration detection	Atrazine	scAb	Electrochemical	"
Encephalomyelitis diagnosis	Venezuelan equine encephalitis virus	scFv	Resonant mirror	"
Entamoeba histolytica diagnosis	*Entamoeba histolytica* antigens	scFv	Amperometric	"
Detection of doping with hormone by	Somatotropin	Half-sized Ab fragment	Surface plasmon resonance (SPR)	"
Detection of inflammation	C-reactive protein (CRP)	Engineered Ab fragment	SPR	"
Antibody aimed at foot-and-mouth disease		Anti-3ABC antibodies	Electrochemical	Reviewed in (Moina and Ybarra, 2012)
Various antibodies aimed at Chagas disease			Electrochemical and fluorescence	"

Nanoparticles

Besides above four categories of bioacceptors, a new class of bioacceptors, nanomaterials, has been added in the recent decade. Advent of nanotechnology has led to the development of various nanomaterials, including nanoparticles and nanofibers as bioacceptors. It must be noted here that nanoparticles provide more diverse application in biosensor technology than any other class of biomolecule. They can act as both bioacceptors and transducers. Nanomaterials such as cerium oxide, which are known to have biomimetic activity, are suitable as bioacceptors, whereas other nanomaterials such as graphene and carbon nanotube—based materials, gold nanoparticles, quantum dots of

various inorganic materials have been successfully used as transducers because of their excellent transduction abilities. Holzinger et al. have described various aspects of use of nanomaterials in biosensing abilities in their review (Holzinger et al., 2014).

Immobilization of Bioreceptors

Bioreceptors, which we discussed above, are often immobilized to the surface of a biosensor; they have several advantages such as loss of wastage during a biochemical reaction, stability of signal due to physical disturbances, and cost optimization. Immobilization approaches can be broadly grouped into two broad categories: physical and chemical. Physical approaches are

TABLE 6.3
Microbial Biosensors Containing Oxygen Electrode.

Microorganisms	Analyte	Type of Biosensors
Escherichia coli	Glutamate	Potentiometric (CO_2)
Proteus morgani	Cysteine	Potentiometric (H_2S)
Nitrosomonas sp.	Ammonia	Amperometric (O_2)
Lactobacillus fermenti	Thiamine	Amperometric (mediated)
Lactobacillus arabinosus	Nicotinic acid	Potentiometric (H^+)
Cyanobacteria	Herbicides	Amperometric (mediated)
Trichosporum cutaneum	BOD	Miniature oxygen electrode
Breibacterium sp.	Acrylamide; acrylic acid	Oxygen electrode
Saccharomyces cerevisiae	Cyanide	Oxygen electrode
Synechococcus sp. *PCC 7942*	Pollutants such as diuron, mercuric chloride	Photochemical
Candida ini	Alcohol	Oxygen electrode
Aspergillus niger	Glucose	Oxygen electrode
Bacillus subtillis	Peptides	Oxygen electrode

based on simple affinity interactions (weak noncovalent), or entrapments using films, or matrices, whereas chemical methods are based on strong covalent link formation between bioacceptor and biosensor surface. Fig. 6.4 illustrates common immobilization approaches. Various immobilization methods have been reviewed by various authors, and detailed description exists elsewhere (Holzinger et al., 2014).

Physical methods for immobilization

Physical methods for immobilization are based on weak physical interactions between the biosensor substratum or immobilization media and bioacceptor.

Physiosorption. This method does not involve any kind of entrapment or covalent linkage, rather attachment of bioacceptor molecules on the outer surface of the inert matrix. This can be preferred in case of enzyme biosensors, as this is relatively mild approach and does not affect the enzyme activity. Moreover, the matrix can also be regenerated after single or multiple use. As a disadvantage, this method of immobilization is affected most by the change in environmental conditions, as the linkage is primarily through very weak interactions such as van der Waals interactions or ionic interactions. Several biosensors, including one of the glucose biosensors based on sol–gel-derived platforms, involve the physiosorption (Narang et al., 1994).

Retention by gel matrix. In this method, the inert support of agarose or dextran is used for trapping the molecules of bioacceptor, so that they do not flow away during the analysis. The advantage of this method is that this is very gentle and no chemical modification of bioacceptor occurs; hence, specificity is not at all affected. However, it has a high diffusion barrier and is not suitable for a prolonged analysis. Among various methods of developing gel matrices, sol–gel matrices are more common and advantageous. These are biocompatible porous structures, which provide space for integrating nanostructured particles, which in turn play an important role in accelerating electron transfer. A well-known electrochemical biosensor developed by Parra-Alfambra et al. (2011) for lactate detection was based on integrating gold nanoparticles (AuNPs) and enzyme lactate oxidase (LOx) into a sol–gel polymeric matrix obtained from (3-mercaptopropyl)-trimethoxysilane (MPTS) (Rathee et al., 2016).

Retention by gel membranes. Gel membranes are ultrathin surfaces, which are permeabilized with the bioacceptor to form an immobile substratum for the reaction. Biocompatible membrane with different porosities serves as a common immobilization matrix. The gel membrane also works as an antifouling electrode and thus protects the enzyme structure and may lead to increase linear response. The thickness

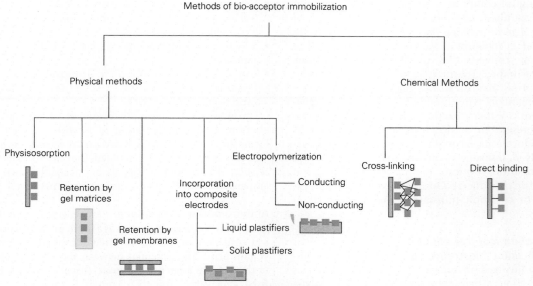

FIG. 6.4 Various immobilization approaches used in biosensor development.

of the membrane also affects the rate of electron transfer, which in turn can be adjusted during the deposition of the membrane layer by the amount of charge applied. Recently, researchers have fabricated a lactate oxidase—based amperometric sensor immobilized on a mucin/albumin hydrogel matrix with nafion membrane, which served as a protecting layer on the electrode surface (Romero et al., 2010; Rathee et al., 2016).

Incorporation into composite electrode. Composite electrodes are relatively more stable and advanced approach, where plastifier agents are used for immobilization of bioacceptor. Two different types of plastifiers are used: liquid and solid. Liquid plastifiers primarily involve concentrated carbon paste, whereas solid plastifiers use carbon, ink, and composite material or carbon composites. A more elaborate discussion on plasticizers is provided by Luqman (2012).

Electropolymerization. Electropolymerization is a promising route to design new functional surfaces. This process involves electrochemically controlled formation of thin polymer films at electrode surfaces. Electropolymerizing is an advanced approach that also helps in miniaturization of biosensors. Sasso et al. (1990) noted that electropolymerized 1, 2-diaminobenzene as a means to prevent interferences and fouling and to stabilize immobilized enzyme in electrochemical biosensors.

Chemical methods for immobilization
Chemical methods for immobilization involve formation of a covalent bond between the biosensor substratum and the bioacceptor. However, the surface used for covalent bond formation can be of various types. In this method, the acceptor is chemically conjugated to the inert matrix by chemical modification and hence covalent linkage. Although, this method provides stronger interaction and hence prolonged detection; also the effect of pH and temperature is least on analysis, but process may involve harsh chemicals and matrix cannot be regenerated once used. Covalent linkage can be direct or in cross-linking format as explained below.

Direct covalent binding. Direct covalent binding involves a strong covalent linkage directly with electrode surface or with an immobilized inert support such as beads or gels. Polymer membranes have also been used for the direct immobilization of the bioacceptor to the biosensor.

Covalent cross-linking. Cross-linking is a mild method of covalently linking acceptor to matrix, without the use of harsh chemicals. Usually, the linkage is performed through some proteins. Cross-linking is often conjugated with entrapment methods.

Transducers
The second important component of the biosensor is a transducer that transmits the signal generated from

the biological component to the electronics system. As per standard text books, *"Transduction of the biosensor signal is a process that is concurrent, and within the special environment of the biosensing element."* The transducer converts the biochemical interactions into measurable electronic signals. Electrochemical, electrooptical, acoustical, and mechanical transducers are among many types used in biosensors. The transducer works either directly or indirectly. The transducers present in a biosensor are also known as biotransducers. Based on the basic principle and physics involved, transducers can be broadly grouped into electrochemical, optical, thermal, field-effect, and gravimetric transducers. Outline classification and examples of common transducers are represented in Fig. 6.5.

Electrochemical transducers

Electrochemical biosensors are used with the bioacceptor involved in an electronic reaction such as redox reactions. There are a large number of analyte-bioacceptor reactions that involve the redox reactions. Sensing of redox equivalents of the cell such as glucose, NADH, calcium, and ATP can be conveniently and economically performed with electrochemical biosensors. There are several approaches that are used to detect electrochemical changes during biorecognition.

Some known electrochemical reactions in biological systems that form basis of electrochemical basis:

- Glucose biosensor (glucose oxidase based): D-Glucose + O_2 → D-glucono-1,5-lactone + H_2O_2 → D-gluconate + H^+
- Urea biosensor (urease based): Urea + H_2O + H^+ → ammonia + CO_2
- Amino acid biosensor (amino acid oxidase based): L-Amino acid + O_2 + H_2O → keto acid + ammonia + H_2O_2

In aforesaid reactions, involvement of protons or electrons leads to generation of electric current that forms the basis of analysis. Based on the sensing devices and intensity of signal generated, various formats of sensing electrochemical signals are available, such as amperometric, voltametric, and potentiometric.

Amperometric transducer. Amperometric transducers detect change in current as a result of electrochemical oxidation or reduction, often measured by cyclic voltammetry. The potential between the working electrode and the reference electrode (usually Ag/AgCl) is fixed at a value, and then current is measured with respect to time. When a potential of −0.6 V, relative to the Ag/AgCl electrode, is applied to the platinum cathode, a current proportional to the oxygen concentration is produced. Normally, both electrodes are bathed in a solution of saturated potassium chloride and separated from the bulk solution by an oxygen-permeable plastic membrane (e.g., Teflon, polytetrafluoroethylene). The applied potential is the driving force for the electron

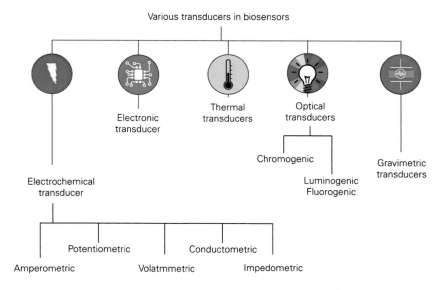

FIG. 6.5 Outline classification and example of types of transducers used in biosensors.

transfer reaction. The current produced is a direct measure of the rate of electron transfer. Following general reaction occurs at each electrode:

- Ag anode $4Ag + 4Cl^- \rightarrow 4AgCl + 4e^-$
- Pt cathode $O_2 + 4H^+ + 4e^- \rightarrow 2H_2O$

The change in current over a cycle is an indicative of reaction occurring between the bioreceptor molecule and analyte. The drawback of this sensor is that it is limited by the mass transport rate of the analyte to the electrode.

measure electrochemical impedance spectroscopy arising from the biochemical reaction in analyte. To measure impedance, a small amplitude sinusoidal electrical stimulus is applied, and the frequency is varied over a range to obtain the impedance spectrum. The resistive and capacitive components of impedance are determined from in-phase and out-of-phase current responses. An impedance analyzer is used to control and apply the stimulus as well as measure the impedance changes.

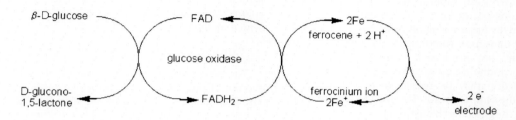

Potentiometric transducers. These are very similar to amperometric transducers, in design; however, unlike amperometric transducers, potentiometric sensors measure a potential or charge accumulation of occurring due to analyte-bioacceptor reaction. The transducer typically comprises an ion selective electrode (ISE) and a reference electrode. The ISE contains a membrane that selectively interacts with the charged ion of interest, causing the accumulation of a charge potential compared with the reference electrode. The reference electrode is used to draw a constant half-cell potential irrespective of the analyte concentration. A high impedance voltmeter is used to measure the electromotive force or potential between the two electrodes.

Conductometric transducers. Conductometric transducers work on the principle of measuring the change in conductive properties of the analyte, in particular the media in which analyte is present. The electrochemical reaction occurring at the bioaccceptor interphase leads to a change in the solution electrical conductivity or current flow, which is detected using two metal electrodes, which are separated at a certain distance and supplied with an alternating current (AC).

Impedometric transducers. The effective resistance of an electric circuit or component to alternating current obtained as a cumulative result of resistance and reactance is called impedance; these transducers

Electronic transducers (field-effect transistor based)

Electronic transducers (transistors) are primarily based on the field effect, hence known as field-effect transistor or FET in short. FET is a three-terminal device that utilizes an electric field to control the current flowing through the device. FET transducers have attracted recent attention of biosensor community because of their ability to directly translate the interactions between analyte and the FET surface. These semiconductor-based transducers have a high input impedance, which is used in sensing the chemical change due to analyte and bioacceptor reaction.

In a field emission transducer, current flows along a channel, which is connected to the source on one side and drain on the other side. FET-based sensing is also known as electronic biosensing; it is advantageous over other biosensing methods because of high sensitivity and high spatial resolution. In addition to ease of on-chip integration of device arrays and the cost-effective device fabrication, the surface ultrasensitivity of FET-based biosensors makes it an attractive alternative to existing biosensor technologies.

Thermal transducers

Thermal transducers, also known as calorimetric sensors, are based on measurement of the heat produced by the molecular recognition reaction, and the amount of heat produced is correlated to the reactant concentration. Calorimetric principles can be used for direct measurement of heat changes associated with thermo-chemical processes (Grime, 1985). The metabolic activity of the biocomponent causes an increase in temperature, which is transformed into a detectable electrical signal. Enzyme-catalyzed reactions exhibit the same enthalpy changes as spontaneous chemical reactions. Considerable heat evolution is noted (5−100 kJ/mol). In most enzymatic reactions, calorimetric transducers are universally applicable in enzyme sensors.

Thermistors are the most commonly used devices for sensing the heat. A thermistor (a thermometer with high sensitivity) can monitor the enthalpy change of an enzyme-based reaction. If the enthalpy change in the biocatalytic process is significant, then the temperature of the transducer (thermistor) is changed and this change can be monitored. Under experimentally controlled conditions, up to 80% of the heat generated in the reaction may be registered as a temperature change in the sample stream. This may be simply calculated from the enthalpy change and the amount reacted. If a 1 mM reactant is completely converted to product in a reaction generating 100 kJ/mol, then each mL of solution generates 0.1 J of heat. At 80% efficiency, this will cause a change in temperature of the solution amounting to approximately $0.02°C$ (Chaplin and Bucke, 1990).

Apart from thermistors, biometallic strips, liquid gas expansion, pyroelectric systems, and metal resistance have also been used as thermal transducers. Thermocouples are also an excellent choice for detecting temperature change with high sensitivity. Platinum/metal thermocouples are of interest as thermal transducer because of significant EMP generation with different metals. A detailed description of various thermal transducers and enthalpy of various biochemical reactions is reviewed by Danielsson (1991).

Recently, pyroelectric transducers have also gained attention because of their higher sensitivity. Pyroelectric sensors are based on generation of an electrical signal because of a change in temperature. A receiver chip of a pyroelectric infrared detector is made up of single-crystalline lithium tantalite, which provides an excellent long-term stability of the signal voltage because of its low temperature coefficient.

Optical biotransducers

Optical biotransducers, used in optical biosensors for signal transduction, use photons to collect information about analyte. These are highly sensitive, highly specific, small in size, and cost-effective. There are two main approaches used for detection in optical biosensors. First involves determining changes in light absorption between the reactants and products of a reaction. In the second approach, the light output is measured by a luminescent process. The first approach is widely established and used in cost-effective colorimetric test strips. Test strips are disposable single-use cellulose pads impregnated with enzyme and reagents. A very common example of such biosensors used in biotechnology is glucose biosensor, where the strips contain glucose oxidase, horseradish peroxidase (EC 1.11.1.7), and a chromogen (e.g., o-toluidine or 3,3′,5,5′-tetramethyl-benzidine). The hydrogen peroxide, produced by the aerobic oxidation of glucose, oxidizing the weakly colored chromogen to a highly colored dye, is represented by the following equation:

$$\text{Chromogen (2H)} + H_2O_2 \rightarrow \text{dye} + 2H_2O$$

The evaluation of the dyed strips is best achieved by the use of portable reflectance meters, although direct visual comparison with a colored chart is often used. A wide variety of test strips involving other enzymes are commercially available at the present time.

Another most promising biosensor involving the second approach, luminescence, uses firefly luciferase (photinus-luciferin 4-monooxygenase (ATP hydrolysing), EC 1.13.12.7) to detect the presence of bacteria in food or clinical samples. Bacteria are specifically lysed and the ATP released (roughly proportional to the number of bacteria present) reacted with D-luciferin and oxygen in a reaction, which produces yellow light in high quantum yield.

The detection mechanism of optical biotransducers depends on the enzyme system that converts analyte into products, which are either oxidized or reduced at the working electrode.

Luciferase

$$ATP + \text{D-luciferin} + O_2 \rightarrow \text{oxyluciferin} + AMP + \text{pyrophosphate} + CO_2 + \text{light (562 nm)}$$

The light produced may be detected photometrically by use of high-voltage, and expensive, photomultiplier tubes or low-voltage cheap photodiode systems. The sensitivity of the photomultiplier-containing systems is, at present, somewhat greater ($<10^4$ cells/mL, $<10^{-12}$ M ATP) than the simpler photon detectors, which use photodiodes. Firefly luciferase is a very expensive enzyme, only obtainable from the tails of wild fireflies. Use of immobilized luciferase greatly reduces the cost of these analyses.

Besides photomultiplier tube, evanescent field detection principle is also commonly used in an optical biosensor system as the transduction principle. This principle is one of the most sensitive detection methods. It enables the detection of fluorophore exclusively in the close proximity of the optical fiber.

Gravimetric or piezoelectric transducers

Gravimetric biosensors use the basic principle of a response to a change in mass. In case of bimolecular interactions, in context to sensing in animal cell culture and animal biotechnology, binding of proteins or antibodies to its target leads to a detectable change in the mass, generating a measureable signal, which is used to sense the analyte properties. Most gravimetric biosensors use thin piezoelectric quartz crystals (quartz crystal microbalance, QCM), either as resonating crystals or as bulk/surface acoustic wave (SAW) devices. A system is described in which acoustic waves are launched in very thin (25 microns) polymer films to produce an oscillatory resonant device. In the majority of these, the mass response is inversely proportional to the crystal thickness. The use of thin quartz crystals for the detection of small additions of bound mass to its surface has been reported over a period of more than 40 years. Because of its piezoelectricity, the crystal can be made to oscillate, using simple electric circuitry, in a shear mode and at a natural frequency, which is inversely proportional to the crystal thickness. The detailed description on acoustic transducers and associated biosensors has been provided by Fogel et al. (2016) in their review.

OUTLINE CLASSIFICATION SCHEMES AND QUALITY INDICATORS OF BIOSENSORS

There are various criteria on which biosensors may be classified. First and most common classification scheme of biosensors is based on nature of transducers used. As discussed above, the transducers can fall into major categories such as electrochemical, calorimetric (thermal), colorimetric (optical), and mass-based or piezoelectric; hence this becomes the first most popular basis of sensors' classification. Second classification scheme is based on the type of bioacceptor used, which classifies biosensors into four main categories: enzyme biosensors (most popular class of biosensors), immunosensors (with high specificity and sensitivity—particularly useful in diagnosis), nucleic acid–based biosensors (high specificity for microbial strains and nucleic acid–containing analyte), and microbial or whole cell biosensors. The choice of bioacceptor-analyte combinations is limited, and hence, another classification scheme also involves combination of both. Some other

classification schemes exist based on the mode of use and its physical features such as in-line biosensors or off-line biosensors (described in Section 7). Additionally, some text also categorizes biosensors into four different classes based on operating principles (described in Section 7). Furthermore, biosensors can be classified as per the technology such as simple electrode based, flow cell based, or featured with multiplexing such as lab-on-chip formats (described in Section 8.2). (Fig. 6.6)

Indicators of Biosensor Quality

Quality of biosensor is detected on the basis of following parameters. Sensor specifications inform the user about deviations from the ideal behavior of the sensors. Anthony Turner reviewed in an excellent article entitled "Biosensors: Sense and Sensitivity" (Turner, 2013). Following are the various specifications of a sensor/transducer system.

Range

The range of a sensor indicates the limits between which the input can vary. The parameters of range may vary across different platforms, depending on the sensing technology. For example, in a thermosensor, a thermocouple for the measurement of temperature has a high range of 37–337°C. Similarly, for optical biosensors, range may be decided as per the wavelengths of radiations obtained. A slightly different term is span, which is defined as the difference between the maximum and minimum values of the input. Thus, the abovementioned thermocouple will have a span of 300°C.

Sensitivity

Sensitivity or limit of detection is the minimum amount of analyte that can be detected by a biosensor. In alternate terms, sensitivity of a sensor is defined as the ratio of change in output value of a sensor to the per unit change in input value that causes the output change. For example, a general-purpose thermocouple may have a sensitivity of 41 μV/°C. In animal biotechnology applications, a biosensor is required to detect analyte concentration of as low as ng/mL or even fg/mL to confirm the presence of traces of analyte in a sample. Hence, sensitivity is considered to be an important property of a biosensor.

Specificity

Specificity of a biosensor is defined as per the ability to detect a particular analyte in question rather than binding or producing a signal with any other analyte. Specificity depends particularly on the bioacceptor's

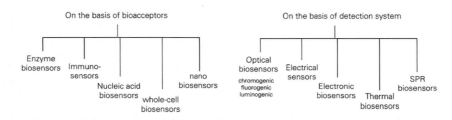

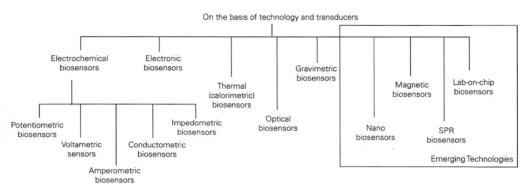

FIG. 6.6 Various scheme of biosensor classification.

ability to recognize the analyte. As antigen-antibody reactions are considered as most specific reactions, immunosensors are considered as one of the most specific biosensors.

Error and accuracy

Error is the difference between the result of the measurement and the true value of the quantity being measured. If a sensor gives a displacement reading of 29.8 mm, when the actual displacement had been 30 mm, then the error is −0.2 mm. The accuracy defines the closeness of the agreement between the actual measurement result and a true value of the measurement. It is often expressed as a percentage of the full-range output or full-scale deflection. A piezoelectric transducer used to evaluate dynamic pressure phenomena associated with explosions, pulsations, or dynamic pressure conditions in motors, rocket engines, compressors, and other pressurized devices is capable to detect pressures between 0.1 and 10,000 psig (0.7 KPa to 70 MPa). If it is specified with the accuracy of about ±1% full scale, then the reading given can be expected to be within ±0.7 MPa.

Factors Affecting the Biosensor Efficiency

Performance of biosensors depends on factors such as the chemical and physical conditions (pH, temperature, and contaminants), thickness, and stability of the materials (Kissinger, 2005). Sensitivity is the ability to detect lowest possible amount of analyte, and specificity indicates the ability of biosensor to be selective for the analyte, that is, it must not give the response for any other molecule than the target in a given mixture.

BIOSENSORS IN ANIMAL CELL CULTURE RESEARCH: CURRENT DEVELOPMENTS

Biosensors are of great need in animal biotechnology, particularly in processes involving large-scale use of animal cell cultures for extraction of a given metabolite or protein. This is so because in these processes a great deal of estimations at periodic intervals needs to be made. These include monitoring everything from culture conditions such as oxygen levels, presence of pyrogens, concentration of calcium and other ions, and nutrient levels. For example, during fermentation of food grains for alcoholic beverage manufacture, there has to be clear information that methanol levels are within permissible limits, the required taste, texture, and consistency and that there are no other unfavorable products being formed. In general, the metabolic state of the cells in the culture is monitored in all such processes. All of these requirements can be easily fulfilled by use of biosensors. For such processes, biosensors are classified as either invasive or noninvasive. Invasive sensors are placed in contact with the culture in a closed flow loop. Noninvasive sensors are usually placed on the exterior of the culture containing vessels. Sometimes a small sample of the culture may be taken out and analyzed using the noninvasive sensor. This is the preferred method in most of the industrial animal cell culture applications. This is the case as invasive biosensors, also called online biosensors, need to be able to withstand very inhospitable conditions such as high moisture (as fermenters are always humid), structural pressure (as biofilms may grow and encroach upon the sensor circuit), and regular commotion (in cases where impellers are used to aerate the culture). Thus, apart from sensitivity, they also have structural requirements that lead to higher cost. Hence, off-line biosensors (noninvasive biosensors) are the favored choice as they do not require high durability structural modifications. Also, they are not corroded or poisoned by other by-products such as peroxide generated during fermentation.

The efforts to produce biosensors that can estimate at least the basic physiological state of a cell, in terms of their energy metabolism rates, are one of the first steps toward the current trend of biosensor development and use. One of the first significant reviews on biosensor development, among many others, is by Owicki and Wallace Parce (1992). One finds the initial curiosity that stem from the desire to understand subcellular intricacies and quantify them. Although the review focuses on a fairly simple microphysiometer, it underscores the what to measure and why to measure. In the current world, biosensors are much more sophisticated. They have been divided into multiple classes, based on mechanism of action. Class I biosensors detect a change in chromophore spectral properties/quantum yield upon changes in ligand concentration. Class Ia biosensors are sensitive to H^+ concentrations, relating to ambient pH. Class Ib biosensors have a grafted sensing domain, which is dependent on binding of ions and heavy metals to induce a conformational change, which indirectly changes its spectral properties/quantum yield. Class II biosensors are Förster resonance energy transfer (FRET)—based biosensors, which are sensitive to change in distance and/or orientation between two fluorescent entities (e.g., GFP, YFP) covalently linked by a ligand binding/sensing domain. It is of two types: intermolecular (wherein the association/disassociation between a donor fluorescent protein and acceptor fluorescent protein causes change in FRET signal) and intramolecular (wherein a linker peptide connects the donor and acceptor fluorescent

proteins and this linker peptide changes conformation upon modification/ligand binding, leading to differential FRET signals). The intramolecular variant is more widely used, as the vagaries of transfection efficiency of the intermolecular class II biosensors are prevented and the stoichiometry of the fluorophores is near equal and constant for a single experiment. Class III biosensors are those which translocate/accumulate to particular cellular compartments. This class of biosensors basically takes advantage of intrinsic gradients, mostly electrochemical, present in membranes of the cellular compartments. Thus, they tend to accumulate in particular compartment in cells. Examples include NADH molecules traveling into mitochondria, DAG and PIP_3 accumulating in the plasma membrane, and weak bases within acidic organelles. Class III biosensors are mostly qualitative in nature. They tend not to be used for quantitative analysis. Class IV biosensors are nonfluorescent in nature. These constitute channel proteins that require ionic mediation for their activity. Also, certain sensors that change their NMR profiles (e.g., sensors of Ca^{2+}) are included in this class.

These classes of sensors are not mutually exclusive. Some sensors may fall into more than one category. An example would be use of photoswitchable fluorescent proteins (PSFPs) in both class I and class II biosensors. Also, use of such fluorescent proteins that serve as tags to certain cell compartment-specific proteins (GFP-tagged STAT3) (Bhattacharya and Schindler 2003) is another example of a single biosensor categorized in more than one class.

The modern-day requirements of biosensors are ever-increasing, as progress in science has led to not just more answers but also questions requiring deep investigations. Thus, from the time when simple knowledge of the mitochondrial ability to generate energy (in the form of ATP) was considered a scientific thunderbolt for molecular biologists, we now wish to understand and tinker with the basic molecular processes occurring within mitochondria to the point where tangible end results with great precision can be obtained, e.g., elimination of metabolic syndrome but without depressing regular systemic functioning and appetite. To this end, the first few steps and at least the last step would be monitoring the molecular processes that need to be tinkered with and also checking if these tinkered molecular processes are functioning as desired. As is evident, without biosensors, such endeavors would not be feasible. Energy metabolism, specific to humans/mammalians, requires oxygen. Thus, oxygen sensing is a very important aspect related to solving abovementioned problems. An important step in this direction is real time, direct, cost-effective, and chemical interference free methods to monitor oxygen uptake rates and oxygen consumption in cells. Biological oxygen demand (BOD), a key parameter for processes involving microbial energy generation, is particularly relevant in wastewater treatment and related fields. A major problem has been low efficiency of coulomb generation for a given BOD value and requirement of extensive maintenance. Kim et al. (2003) created a novel BOD sensor that overcame both these limitations by use of a microbial fuel cell. They used an electrochemically active bacterium, *S. putrefaciens*, to create the microbial fuel cell. Previously, the same authors using the same bacterium had created a microbial fuel cell that served as a lactate sensor (Kim et al., 1999). A microbial fuel cell type lactate biosensor uses a metal-reducing bacterium, *Shewanella putrefaciens* (Kim et al., 1999). Coming to oxygen sensing intracellularly, many significant findings have been made in this regard. Di Russo et al. (2015) demonstrated the applicability of fluorescence-based sensors to determining the O_2 kinetics in oxygenases. Super et al. (2016) have devised a microfluidics-based assay that is noninvasive and provides real-time monitoring of specific oxygen uptake rates of cells in a culture. Another assay by Deshpande and Heinzle (2004) used fluorescence ratios of an oxygen-sensitive and an oxygen-insensitive fluorophore. Very recently, Rivera and colleagues have developed an integrated phosphorescence-based photonic biosensor for controlling oxygen flow in a 3D cell culture setting. They coupled a photonic oxygen biosensor to a 3D tissue scaffold. This was used to control oxygen concentration in the 3D tissue. Quenching of a palladium-benzoporphyrin sensor by molecular oxygen transduced the local oxygen concentrations in the tissue scaffold (Rivera et al., 2019).

The second crucial requirement for metabolic systems after oxygen is glucose, which ultimately results in formation of the end-product lactate. An interesting method for online determination of glucose and lactate based on fiber optic detection of oxygen consumption was developed by Dremel et al. (1992). They utilized oxygen quenching the fluorescence of an indicator to estimate glucose and lactate levels via immobilized glucose oxidase and lactate oxidase, respectively. This system has been successfully applied to production of recombinant human antithrombin III. Another similar type of device for measuring glucose consumption/levels in a bioreactor has been developed by Tric et al. (2017), which consists of a commercially available oxygen sensor coated with cross-linked glucose oxidase.

They demonstrated in this study that biosensor stability is dependent not only on levels of hydrogen peroxide, which poisons the sensor but also on glucose oxidase turnover rate. This sensor displayed robustness by being in continuous operation for more than 52 days before being sterilized using UV irradiation. This is a great advance as reviews in the early 2000s stated that online monitoring and control of biological process has not yet been frequently used in research tools or industrial methods. Also, biosensors were only considered to be rudimentary variations of a glucose oxidase coupled to amperometric transducer (Schügerl, 2001).

A newer approach to estimate whether a particular process involves microbial production of certain products via normal metabolism and secretion or fermentation is to monitor the amino acid population of the cultures. Amino acid composition is also very important in animal diets, as improper composition of amino acids can lead to underweight animals. To this end, *Escherichia coli* has emerged as a valuable biosensor. Well-known bacterial genetics as well as similar patterns of uptake in the gut coupled with identical stereospecificity for both microbe and host have made *E. coli* a favored biosensor for monitoring amino acid composition and uptake. A simple strategy is to delete a particular gene for synthesis of an amino acid(s) in a microbe with GFP expression. This strategy has already been demonstrated to provide feasible results (Bertels et al., 2012).

Recent investigations have highlighted the influence of thiols/disulfides on various aspects of systems biology. A new domain has emerged, termed redox systems biology, which deals with thiol/disulfide effects on physiological function. The vastness of this domain can be imagined by the presence of thiols in posttranslational modifications affecting protein structure, distribution, and activity, and studies linking thiols to redox signaling, nitric oxide signaling, enzyme active sites, aging, neurodegeneration, hypoxia responses, etc. Studies have also implicated thiols in biologic functions as redox sensors (Go and Jones, 2013). Electrical biosensors based on gold electrodes have been constantly derided for poor long-term stability as the attachment of thiols is not stable. In a study by Kim et al., multithiol probes are shown to be a feasible solution to this problem. They have also demonstrated its applicability to label-free DNA microarrays, which are highly sensitive (Kim et al., 2017). A noteworthy approach is using "caged luciferin" strategy (i.e., luciferin is conjugated with another molecule, like boronic acid, which causes it to be selective for the molecule/radical that it will react with to provide bioluminescence) for detecting thiols in both glutathione and cysteine. Caged luciferin has already been shown to hold promise in the context of bioluminescent probes in an impressive review by Li et al. (2013). Realigning our attention toward biosensors using caged luciferin strategy, Hemmi et al. (2018) provide SEluc-1 (sulfonate ester luciferin), which is demonstrated to be chemoselective probe with ratiometric response, biologically relevant limit of detection, excellent selectivity, and high signal-to-noise ratio. Previous iterations of this strategy of using "caged luciferin" have also been used for thiol imaging in live cells using a variety of optical sensors (Li et al., 2010; Van de Bittner et al., 2010). A recent strategy for nitroxyl detection in vivo has been to target nitroxyl radicals using fluorophores that are thiol based. Pino et al. designed nitroxylfluor, which has excellent efficacy in presence of minute quantities of glutathione and is highly selective for nitroxyl radical over other reactive nitrogen, oxygen, and sulfur radicals. They reported its use for nitroxyl radical detection in live cells (Pino et al., 2017).

An important domain that requires biosensors so as to understand the energy dynamics of a living cell is the ATP status. Thus, to answer this fundamental question about energy dynamics, a fair amount of work has gone into developing a suitable ATP biosensor, mainly involving fluorescent detectors. Tantama et al. designed PercevalHR, a genetically encoded fluorescent biosensor, that is capable of detecting ATP:ADP ratio in mammalian cells. They reported its ability to monitor in real-time physiological changes in energy consumption and production in live neurons (Tantama et al., 2013). Another study by Tsuyama et al. (2013) used a previously available ATP biosensor (ATeam) optimizing it for low temperature and reported its utility as an effective FRET-based ATP biosensor in *Drosophila melanogaster* and *Caenorhabditis elegans* (Tsuyama et al., 2013). Very recently, De Col and colleagues have reported a map of MgATP^{2-} in *Arabidopsis* seedlings. The authors initially generated *Arabidopsis* sensor lines and used the sensor in vitro under conditions mimicking the plant cytosol. They also assayed ATP fluxes in isolated mitochondria to demonstrate the rapid and reliable nature of ATeam in sensing the ATP flux. They have also used ATeam ATP biosensor to demonstrate the possibility of ATP sensing in living plant cells (De Col et al., 2017). ATeam variants such as GO-ATeam and BTeam have also been developed. GO-ATeam utilizes redshifted ATP probe (pairing of green and orange fluorescent proteins as donor and acceptor, respectively) so that Ca^{2+} levels can also be simultaneously measured using fura-2 probe without

overlapping spectra (Nakano et al., 2011). ATeam and GO-ATeam are FRET-based biosensors. However, BTeam is a B(bioluminescence)RET-based biosensor wherein the sensor consists of the ATP-binding domain of bacterial ATP synthase subunit ε, yellow fluorescent protein, and an ATP-non-consuming luciferase. This BRET-based ATP biosensor was demonstrated to be capable of providing steady signal output and continuous monitoring of live cell populations (Yoshida et al., 2016).

The relation between calcium (Ca^{2+}) and ATP has been clearly demonstrated in multiple studies concerning energy metabolism and metabolic networks (Balaban, 2002; Brookes et al., 2004; Jouaville et al., 1999). Thus, monitoring intracellular Ca^{2+} levels is equally important in understanding cellular energy homeostasis and metabolism. Ca^{2+} biosensors have been developed, which help monitor its levels intracellularly. Most are biosynthetic and FRET based. Hara and colleagues developed a transgenic mammalian Ca^{2+} biosensor that was specific to mouse pancreas. The sensor was reported to remain stable in expression as the mice aged. It also did not impair function of the pancreas. The authors conclude that real-time optical imaging of Ca^{2+} dynamics is possible using this approach (Hara et al., 2004). Mank et al. developed a transgenic troponin-C and circular green fluorescent protein utilizing Ca^{2+} biosensor, named TN-XL. This biosensor was shown to have fastest rise and decay times of all known FRET-based Ca^{2+} biosensors when imaging presynaptic motoneuron terminals of transgenic *Drosophila* strains (Mank et al., 2006). Yoshikawa et al. (2016), interested in autoimmune disorder susceptibility prediction in prone mice, also developed a Ca^{2+} biosensor to assess role of Ca^{2+} flux in disease models using FRET-based Ca^{2+} indicator yellow cameleon 3.60 (YC3.60), based on the Cre/loxP system (YC3.60^{flox}).

Another important aspect related to energy metabolism is the study of redox molecules such as NADH, as they serve as a bridge between energy and redox status (Noctor et al., 2007). Biosensors for NADH, a jack of all trades in the cellular milieu, have also been developed and provide a very important insight into NADH flux in living systems. Hung et al. developed Peredox, a combination of circularly permuted GFP and Rex, a bacterial NADH-binding protein that was capable of sensing NADH-NAD(+) levels and could be calibrated with exogenous lactate and pyruvate. The authors used Peredox to measure NADH levels upon PI3K pathway inhibition in live cells (Hung et al., 2011). Zhao et al. (2011) developed an

NADH sensor named Frex that demonstrated high selectivity and responsiveness to NADH and allowed the authors to show spatiotemporal trends of NADH flux in various subcellular compartments. Recently, Mutyala et al. opened up a new paradigm in sensing of NADH by developing an NADH biosensor using nitrogen-doped graphene. The sensor was exceptionally stable and responsive over prolonged usage with linear detection range of 0.5–12 μM (Mutyala and Mathiyarasu, 2016).

RECENT DEVELOPMENTS IN BIOSENSORS FOR ANIMAL BIOTECHNOLOGY

A significant progress has been made in the development and advancement of the biosensor technology. Two main aspects have been greatly considered during the development of biosensors in the past one decade: (1) miniaturization for enhanced portability and (2) cost-effectiveness for enhanced commercial scale-up. Initially the first-generation biosensors were based on primarily clark's electrode; therefore, most biochemical reactions were coupled with oxygen-linked reaction, but with gradual improvement in the bioacceptors and improved transducers, it was possible to directly couple the reaction with secondary mediators such as metabolites or nanomaterials, rather than dependence on the oxygen (Fig. 6.7). Biosensors have also witnessed improvement in the bioacceptor types, where biomimetic nanomaterials have emerged as important molecules (discussed in the next section).

Similarly, an entirely new class of transducers called magnetic transducers has been added to the list. The use of magnetic sensing that relies on the use of micro/nanoparticles labeling has increased in recent years.

Biomolecules have virtually no magnetic properties; therefore, the addition of magnetic micro/nanobeads to biomolecular samples could be used to separate and quantify a known analyte within a given sample (Luka et al., 2015).

Emergence of microfluidic-based systems has contributed toward miniaturization to a great extent, as well as multiplexing of various detections has become possible, which proved to be highly useful in animal biotechnology. Integration of microfluidic system with the surface plasmon resonance has been widely used in various aspects of animal biotechnology, in particular study of biomolecular interactions kinetics, including protein-protein, drug-protein, nucleic acid and immunoglobulin interaction kinetics. In fact, SPR-based biosensors are one of the best and most sensitive sensors for quantitative measurements of bimolecular interactions.

Nanobiosensors

Nanomaterials are biomaterials with nanometer size in at least one dimension; they include some spherical particles such as quantum dots, linear structures such as nanowires, and tubular structures such as nanotubes. It has been demonstrated that nanomaterials are different in their physicochemical properties significantly from corresponding bulk material. This shift in the physical and chemical properties of material is an advantage for using them as potential materials in biosensors. Poncharal et al. made significant contribution toward exploration of biosensing abilities of nanomaterials. They observed electrostatic deflections and electromechanical resonances of carbon nanotubes, which proved to be useful in biosensing (Poncharal et al., 1999). Some of the nanosensors mimic biological

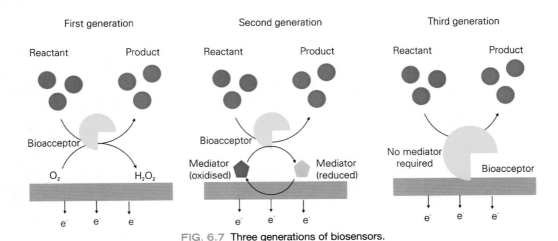

FIG. 6.7 Three generations of biosensors.

reactions such as phosphorylation and dephosphorylation that are known to play an important role in cellular regulation and signaling processes and apoptosis. Sun et al. demonstrated that the quenching efficiency of graphene oxide (GO) to the phosphorylated and dephosphorylated dye-labeled peptides could be distinguished by tuning the interaction between the phosphate group and the amino residue. Based on this hypothesis, an approach for fabricating GO-based fluorescent biosensors to survey the variation of phosphorylation state and phosphatase activity is presented (Sun et al., 2016). We (authors) have also developed a nanobiosensor based on biosensing abilities of cerium oxide using conventional cyclic voltametric detection. A detailed discussion on various types and applications of nanosensors is presented by Malik et al. (2013).

Lab-on-Chip biosensors

Microfluidics is considered as one of the most promising and futuristic technologies that involves multidisciplinary expertise, including chemistry, biochemistry, engineering, physics, microtechnology, nanotechnology, and biotechnology (Luka et al., 2015). These devices are fabricated with numerous microchannels embedded with antibodies, antigens, or oligonucleotides, enabling thousands of biochemical reactions from a single drop of blood. Commonly, polydimethylsiloxane (PDMS), thermoplastic polymers, glass, silicon, or paper-based technologies are used for fabrication of lab-on-a-chip. However, PDMS and paper-based lab-on-a-chip are more widely used because of their low cost and being easy to fabricate (Patel et al., 2016).

An important feature of the microfluids technology that differentiates it from classical biosensors is the large surface-to-volume ratio, which enables portability of microfluidic devices, which is important for on-site testing. Additionally, the advantages of lab-on-a-chip include its low cost, comparable sensitivity to conventional diagnostic methods, rapid testing time, ease of use, being handy to carry due to its compactness, low volume samples, and real-time monitoring, without much environmental interferences.

Microfluidic systems are mainly of three types: continuous-flow; droplet-based; and digital microfluidics. Continuous microfluidic devices consist of permanently etched microchannels and peripheral devices (such as micropumps and microvalves) used to manipulate a stream of fluid in these devices. Luka et al. (2015), in a review, provided a comprehensive description of pros and cons of various subtypes of microfluidic technology, their examples, and various printing technologies. In animal biotechnology, lab-on-chip technology has been widely accepted and holds a promise for rapid and cost-effective analysis of various parameters. Recently, Peter Ertl et al. have demonstrated that combination of microfabrication-based technologies can be used for the development of advanced in vitro diagnostic systems capable of analyzing cell cultures under physiologically relevant conditions. They developed a lab-on-a-chip-based device for stem cell analysis with tangible advantages of precise regulation of culturing conditions, while simultaneously monitoring relevant parameters using embedded sensory systems (Ertl et al., 2014).

CERIUM OXIDE AS POTENTIAL ELEMENT FOR BIOSENSORS

Cerium is the 57th element in periodic table, and its oxide has wide applications as antiabrasive in furnaces and exhaust pipes as well as glass cleaners (Karakoti et al., 2008). For over a decade, after the synthesis of nanosize particles of cerium oxide, its abilities have further been explored in biological sciences. Chen et al. made a pioneering discovery and unraveled the potential applications of nanoceria in preventing oxidative stress in eye, as it is one of the most vulnerable organs for oxidative damage. We and others have also explored several biological benefits of cerium oxide and learned that at optimal size and shape, nanoceria is potentially beneficial in a number of oxidative stress events caused by environmental stressors, especially hypoxia (Arya et al., 2013, 2016). These abilities of nanoceria are primarily attributed to its ability to cyclise between Ce^{3+} and Ce^{4+} oxidation states and mimicking superoxide dismutase and catalase activities in vivo. These abilities prove to be beneficial in biosensing abilities as well, especially sensing of hydrogen peroxides (Fig. 6.8).

Recently, we (authors) have designed a functional biosensor for hydrogen peroxide based on the redox cycling abilities of cerium oxide nanoparticles. In this biosensor, nanoceria was used as a bioacceptor and transducer to sense the hydrogen peroxide. Because of its stability and links with various diseases, sensing the level of H_2O_2 can be of great help in diagnosing these diseases, thereby easing disease management and amelioration. Nanoceria is a potent candidate in free radical scavenging as well as sensing because of its unique redox properties. These properties have been exploited, in the reported work, to sense and quantify peroxide levels. Nanoceria has been synthesized using different capping agents: hexamethylene-

(A) TEM of nanoceria

(B) Enzyme mimetic activities of nanoceria

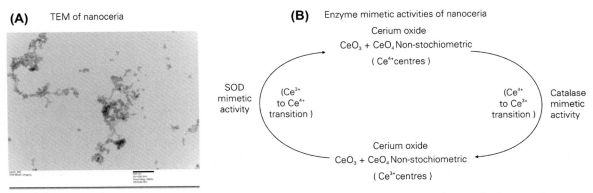

FIG. 6.8 TEM image of nanoceria and its redox cycling. **(A)** Transmission electron micrograph of nanoceria. **(B)** Enzyme mimetic activities of nanoceria.

tetra-amine (HMTA) and fructose. CeO_2-HMTA shows rhombohedral and cubic 6.4 nm particles, whereas CeO_2-fructose is found to be spherical with average particle diameter size 5.8 nm. CeO_2-HMTA, due to the better exposure of the active (200) and (220) planes relative to (111) plane, exhibits superior electrocatalytic activity toward H_2O_2 reduction. Amperometric responses were measured by increasing H_2O_2 concentration. The authors observed a sensitivity of 21.13 and 9.6 $\mu A/cm^2$ mM for CeO_2-HMTA and CeO_2-fructose, respectively. The response time of 4.8 and 6.5 s was observed for CeO_2-HMTA and CeO_2-fructose, respectively. The limit of detection is as low as 0.6 and 2.0 μM at S/N ratio 3 for CeO_2-HMTA and CeO_2-fructose, respectively. Ceria-HMTA was further tested for its antioxidant activity in an animal cell line in vitro, and the results confirmed its activity (Ujjain et al., 2014).

REFERENCES

Andreou, V.G., Clonis, Y.D., 2002. Novel fiber-optic biosensor based on immobilized glutathione S-transferase and sol—gel entrapped bromcresol green for the determination of atrazine. Analytica Chimica Acta 460, 151—161.

Arif, M., Setford, S.J., Burton, K.S., Tothill, I.E., 2002. L-Malic acid biosensor for field-based evaluation of apple, potato and tomato horticultural produce. Analyst 127, 104—108.

Arya, A., Sethy, N.K., Singh, S.K., Das, M., Bhargava, K., 2013. Cerium oxide nanoparticles protect rodent lungs from hypobaric hypoxia-induced oxidative stress and inflammation. International Journal of Nanomedicine 8, 4507—4520.

Arya, A., Gangwar, A., Singh, S.K., Roy, M., Das, M., Sethy, N.K., et al., 2016. Cerium oxide nanoparticles promote neurogenesis and abrogate hypoxia-induced memory impairment through AMPK-PKC-CBP signaling cascade. International Journal of Nanomedicine 11, 1159—1173.

Álvarez-González, M.I., Saidman, S.B., Lobo-Castañón, M.J., Miranda-Ordieres, A.J., Tuñón-Blanco, P., 2000. Electrocatalytic detection of NADH and glycerol by NAD+-modified carbon electrodes. Analytical Chemistry 72, 520—527.

Balaban, R.S., 2002. Cardiac energy metabolism homeostasis: role of cytosolic calcium. Journal of Molecular and Cellular Cardiology 34, 1259—1271.

Bertels, F., Merker, H., Kost, C., 2012. Design and characterization of auxotrophy-based amino acid biosensors. PLoS One 7, e41349.

Bhalla, N., Jolly, P., Formisano, N., Estrela, P., 2016. Introduction to biosensors. Essays in Biochemistry 60, 1—8.

Bhattacharya, S., Schindler, C., 2003. Regulation of Stat3 nuclear export. Journal of Clinical Investigation 111, 553—559.

Borgmann, S., Schulte, A., Neugebauer, S., Schuhmann, W., 2011. Advances in electrochemical science and engineering. In: Richard, C., Alkire, D.M.K., Lipkowski, J. (Eds.). WILEY-VCH Verlag GmbH & Co. KGaA, Weinheim, pp. 32885—32887.

Brookes, P.S., Yoon, Y., Robotham, J.L., Anders, M.W., Sheu, S.-S., 2004. Calcium, ATP, and ROS: a mitochondrial love-hate triangle. American Journal of Physiology — Cell Physiology **287**, C817—C833.

Chaplin, M.F., Bucke, C., 1990. Enzyme Technology. CUP Archive.

Choi, J.-W., Kim, Y.-K., Song, S.-Y., Lee, I-h, Lee, W.H., 2003. Optical biosensor consisting of glutathione-S-transferase for detection of captan. Biosensors and Bioelectronics 18, 1461—1466.

Corcoran, C.A., Rechnitz, G.A., 1985. Cell-based biosensors. Trends in Biotechnology 3, 92—96.

Cremer, M., 1906. Über die Ursache der elektromotorischen Eigenschaften der Gewebe, zugleich ein Beitrag zur Lehre von den polyphasischen Elektrolytketten. Zeitschrift für Biologie 47, 562—608.

D'Souza, S.F., 2001. Microbial biosensors. Biosensors and Bioelectronics 16, 337—353.

Danielsson, B., 1991. Calorimetric biosensors. Biochemical Society Transactions 19, 26.

De Col, V., Fuchs, P., Nietzel, T., Elsässer, M., Voon, C.P., Candeo, A., et al., 2017. ATP sensing in living plant cells reveals tissue gradients and stress dynamics of energy physiology. Elife.

Deshpande, R.R., Heinzle, E., 2004. On-line oxygen uptake rate and culture viability measurement of animal cell culture using microplates with integrated oxygen sensors. Biotechnology Letters 26, 763—767.

Di Russo, N.V., Bruner, S.D., Roitberg, A.E., 2015. Applicability of fluorescence-based sensors to the determination of kinetic parameters for O_2 in oxygenases. Analytical Biochemistry 475, 53—55.

dMonošík, R., Stredanský, M., Šturdík, E., 2012. Biosensors - classification, characterization and new trends 5, 109.

Dremel, B.A.A., Li, S.Y., Schmid, R.D., 1992. On-line determination of glucose and lactate concentrations in animal cell culture based on fibre optic detection of oxygen in flow-injection analysis. Biosensors and Bioelectronics 7, 133—139.

Ertl, P., Sticker, D., Charwat, V., Kasper, C., Lepperdinger, G., 2014. Lab-on-a-chip technologies for stem cell analysis. Trends in Biotechnology 32, 245—253.

Fogel, R., Limson, J., Seshia, A.A., 2016. Acoustic biosensors. Essays in Biochemistry 60, 101—110.

Go, Y.-M., Jones, D.P., 2013. Thiol/disulfide redox states in signaling and sensing. Critical Reviews in Biochemistry and Molecular Biology 48, 173—181.

Grime, J.K., 1985. Analytical Solution Calorimetry. J. Wiley.

Hara, M., Bindokas, V., Lopez, J.P., Kaihara, K., Luis, R., Landa, J., Harbeck, M., et al., 2004. Imaging endoplasmic reticulum calcium with a fluorescent biosensor in transgenic mice. American Journal of Physiology — Cell Physiology **287**, C932—C938.

Hemmi, M., Ikeda, Y., Shindo, Y., Nakajima, T., Nishiyama, S., Oka, K., et al., 2018. Highly sensitive bioluminescent probe for thiol detection in living cells. Chemistry - An Asian Journal 13, 648—655.

Holzinger, M., Le Goff, A., Cosnier, S., 2014. Nanomaterials for biosensing applications: a review. Frontiers in Chemistry 2, 63.

Huang, J., Li, J., Yang, Y., Wang, X., Wu, B., Anzai, J-i, et al., 2008. Development of an amperometric l-lactate biosensor based on l-lactate oxidase immobilized through silica sol—gel film on multi-walled carbon nanotubes/platinum nanoparticle modified glassy carbon electrode. Materials Science and Engineering: C 28, 1070—1075.

Hughes, W.S., 1922. The potential difference between glass and electrolytes in contact with the glass. Journal of the American Chemical Society 44, 2860—2867.

Hung, Y.P., Albeck, J.G., Tantama, M., Yellen, G., 2011. Imaging cytosolic NADH-NAD(+) redox state with a genetically encoded fluorescent biosensor. Cell Metabolism 14, 545—554.

Jouaville, L.S., Pinton, P., Bastianutto, C., Rutter, G.A., Rizzuto, R., 1999. Regulation of mitochondrial ATP synthesis by calcium: evidence for a long-term metabolic priming. Proceedings of the National Academy of Sciences 96, 13807.

Katrlík, J., Pizzariello, A., Mastihuba, Vr, Švorc, J., Stredanský, M., Miertuš, S., 1999. Biosensors for L-malate and L-lactate based on solid binding matrix. Analytica Chimica Acta 379, 193—200.

Karakoti, A.S., Monteiro-Riviere, N.A., Aggarwal, R., Davis, J.P., Narayan, R.J., Self, W.T., et al., 2008. Nanoceria as antioxidant: synthesis and biomedical applications. Journal of Occupational Medicine 60, 33—37.

Kim, B.H., Chang, I.S., Cheol Gil, G., Park, H.S., Kim, H.J., 2003. Novel BOD (biological oxygen demand) sensor using mediator-less microbial fuel cell. Biotechnology Letters 25, 541—545.

Kim, H.J., Hyun, M.S., Chang, I.S., Kim, B.H., 1999. A microbial fuel cell type lactate biosensor using a metal-reducing bacterium, Shewanella putrefaciens. Journal of Microbiology and Biotechnology 9, 365—367.

Kim, W.J., Shim, M.S., Chung, M., 2017. Immobilization and stability studies of multi-thiol probes for electrical biosensors by label-free analysis. Molecular Crystals and Liquid Crystals 654, 27—33.

Kissinger, P.T., 2005. Biosensors—a perspective. Biosensors and Bioelectronics 20, 2512—2516.

Labuda, J., Brett Ana Maria, O., Evtugyn, G., Fojta, M., Mascini, M., Ozsoz, M., et al., 2010. Electrochemical nucleic acid-based biosensors: concepts, terms, and methodology (IUPAC Technical Report). Pure and Applied Chemistry 1161.

Lakhin, A.V., Tarantul, V.Z., Gening, L.V., 2013. Aptamers: problems, solutions and prospects. Acta Naturae 5, 34—43.

Li, J.-P., Gu, H.-N., 2006. A selective cholesterol biosensor based on composite film modified electrode for amperometric detection. Journal of the Chinese Chemical Society 53, 575—582.

Li, J., Chen, L., Du, L., Li, M., 2013. Cage the firefly luciferin! — a strategy for developing bioluminescent probes. Chemical Society Reviews 42, 662—676.

Li, X., Qian, S., He, Q., Yang, B., Li, J., Hu, Y., 2010. Design and synthesis of a highly selective fluorescent turn-on probe for thiol bioimaging in living cells. Organic and Biomolecular Chemistry 8, 3627—3630.

Luka, G., Ahmadi, A., Najjaran, H., Alocilja, E., DeRosa, M., Wolthers, K., et al., 2015. Microfluidics integrated biosensors: a leading technology towards lab-on-a-chip and sensing applications. Sensors 15, 30011—30031.

Luqman, M. (Ed.), 2012. Recent Advances in Plasticizers. In tech open.

Malik, P., Katyal, V., Malik, V., Asatkar, A., Inwati, G., Mukherjee, T.K., 2013. Nanobiosensors: concepts and variations. ISRN Nanomaterials 2013, 9.

Maines, A., Prodromidis, M.I., Tzouwara-Karayanni, S.M., Karayannis, M.I., Ashworth, D., Vadgama, P., 2000. An enzyme electrode for extended linearity citrate measurements based on modified polymeric membranes. Electroanalysis 12, 1118—1123.

Mank, M., Reiff, D.F., Heim, N., Friedrich, M.W., Borst, A., Griesbeck, O., 2006. A FRET-based calcium biosensor with fast signal kinetics and high fluorescence change. Biophysical Journal 90, 1790—1796.

Moina, C., Ybarra, G., 2012. Fundamentals and Applications of Immunosensors. In: Advances in Immunoassay Technology. InTech.

Monošík, R., Stredansky, M., Tkac, J., Sturdik, E., 2012. Application of enzyme biosensors in analysis of food and beverages. Food Analytical Methods 5, 40–53.

Mieliauskiene, R., Nistor, M., Laurinavicius, V., Csöregi, E., 2006. Amperometric determination of acetate with a tri-enzyme based sensor. Sensors and Actuators B: Chemical 113, 671–676.

Mizutani, F., Hirata, Y., Yabuki, S., Iijima, S., 2003. Flow injection analysis of acetic acid in food samples by using trienzyme/poly(dimethylsiloxane)-bilayer membrane-based electrode as the detector. Sensors and Actuators B: Chemical 91, 195–198.

Mutyala, S., Mathiyarasu, J., 2016. A highly sensitive NADH biosensor using nitrogen doped graphene modified electrodes. Journal of Electroanalytical Chemistry 775, 329–336.

Nakano, M., Imamura, H., Nagai, T., Noji, H., 2011. Ca^{2+} regulation of mitochondrial ATP synthesis visualized at the single cell level. ACS Chemical Biology 6, 709–715.

Narang, U., Prasad, P.N., Bright, F.V., Ramanathan, K., Kumar, N.D., Malhotra, B.D., et al., 1994. Glucose biosensor based on a sol-gel-derived platform. Analytical Chemistry 66, 3139–3144.

Niculescu, M., Sigina, S., Csöregi, E., 2003. Glycerol dehydrogenase based amperometric biosensor for monitoring of glycerol in alcoholic beverages. Analytical Letters 36, 1721–1737.

Noctor, G., De Paepe, R., Foyer, C.H., 2007. Mitochondrial redox biology and homeostasis in plants. Trends in Plant Science 12, 125–134.

Owicki, J.C., Wallace Parce, J., 1992. Biosensors based on the energy metabolism of living cells: the physical chemistry and cell biology of extracellular acidification. Biosensors and Bioelectronics 7, 255–272.

Palchetti, I., Mascini, M., 2008. Nucleic acid biosensors for environmental pollution monitoring. Analyst 133, 846–854.

Pang, L., Li, J., Jiang, J., Shen, G., Yu, R., 2006. DNA point mutation detection based on DNA ligase reaction and nano-Au amplification: A piezoelectric approach. Analytical Biochemistry 358, 99–103.

Parra-Alfambra, A.M., Casero, E., Petit-Domínguez, M.D., Barbadillo, M., Pariente, F., Vázquez, L., et al., 2011. New nanostructured electrochemical biosensors based on three-dimensional (3-mercaptopropyl)-trimethoxysilane network. Analyst 136, 340–347.

Patel, S., Nanda, R., Sahoo, S., Mohapatra, E., 2016. Biosensors in health care: the milestones achieved in their development towards lab-on-chip-analysis. Biochem Res Int, 3130469.

Peña, N., Tárrega, R., Reviejo, A.J., Pingarrón, J.M., 2002. Reticulated vitreous carbon-based composite bienzyme electrodes for the determination of alcohols in beer samples. Analytical Letters 35, 1931–1944.

Pereira, A.C., Aguiar, M.R., Kisner, A., Macedo, D.V., Kubota, L.T., 2007. Amperometric biosensor for lactate based on lactate dehydrogenase and Meldola Blue coimmobilized on multi-wall carbon-nanotube. Sensors and Actuators B: Chemical 124, 269–276.

Pino, N.W., Davis, J., Yu, Z., Chan, J., 2017. NitroxylFluor: a thiol-based fluorescent probe for live-cell imaging of nitroxyl. Journal of the American Chemical Society 139, 18476–18479.

Poncharal, P., Wang, Z.L., Ugarte, D., de Heer, W.A., 1999. Electrostatic deflections and electromechanical resonances of carbon nanotubes. Science 283, 1513.

Pollegioni, L., Piubelli, L., Sacchi, S., et al., 2007. Cell. Mol. Life Sci. 64, 1373. https://doi.org/10.1007/s00018-007-6558-4.

Prodromidis, M.I., Tzouwara-Karayanni, S.M., Karayannis, M.I., Vadgama, P.M., 1997. Bioelectrochemical determination of citric acid in real samples using a fully automated flow injection manifold. Analyst 122, 1101–1106.

Rathee, K., Dhull, V., Dhull, R., Singh, S., 2016. Biosensors based on electrochemical lactate detection: a comprehensive review. Biochemistry and Biophysics Reports 5, 35–54.

Rivera, K.R., Pozdin, V.A., Young, A.T., Erb, P.D., Wisniewski, N.A., Magness, S.T., et al., 2019. Integrated phosphorescence-based photonic biosensor (iPOB) for monitoring oxygen levels in 3D cell culture systems. Biosensors and Bioelectronics 123, 131–140.

Rocchitta, G., Spanu, A., Babudieri, S., Latte, G., Madeddu, G., Galleri, G., et al., 2016. Enzyme biosensors for biomedical applications: strategies for safeguarding analytical performances in biological fluids. Sensors 16, 780.

Roederer, J.E., Bastiaans, G.J., 1983. Microgravimetric immunoassay with piezoelectric crystals. Analytical Chemistry 55, 2333–2336.

Romero, M.R., Ahumada, F., Garay, F., Baruzzi, A.M., 2010. Amperometric biosensor for direct blood lactate detection. Analytical Chemistry 82, 5568–5572.

Sacchi, S., Pollegioni, L., Pilone, M.S., Rossetti, C., 1998. Determination of D-amino acids using a D-amino acid oxidase biosensor with spectro-metric and potentiometric detection. Biotechnol. Tech. 12, 149–153.

Sasso, S.V., Pierce, R.J., Walla, R., Yacynych, A.M., 1990. Electropolymerized 1,2-diaminobenzene as a means to prevent interferences and fouling and to stabilize immobilized enzyme in electrochemical biosensors. Analytical Chemistry 62, 1111–1117.

Schügerl, K., 2001. Progress in monitoring, modeling and control of bioprocesses during the last 20 years. Journal of Biotechnology 85, 149–173.

Sheu, J.T., Chen, C.C., Chang, K.S., Li, Y.K., 2008. A possibility of detection of the non-charge based analytes using ultrathin body field-effect transistors. Biosensors and Bioelectronics 23, 1883–1886.

Smutok, O., Ngounou, B., Pavlishko, H., Gayda, G., Gonchar, M., Schuhmann, W., 2006. A reagentless bienzyme amperometric biosensor based on alcohol oxidase/

peroxidase and an Os-complex modified electrodeposition paint. Sens. Actuators B: Chem. 113, 590–598.

Soldatkin, O.O., Peshkova, V.M., Dzyadevych, S.V., Soldatkin, A.P., Jaffrezic-Renault, N., El'skaya, A.V., 2008. Novel sucrose three-enzyme conductometric biosensor. Materials Science and Engineering: C 28, 959–964.

Sharma, S., Byrne, H., O'Kennedy, R.J., 2016. Antibodies and antibody-derived analytical biosensors. Essays in Biochemistry 60, 9–18.

Surareungchai, W., Worasing, S., Sritongkum, P., Tantichareon, M., Kirtikara, K., 1999. Dual electrode signal-subtracted biosensor for simultaneous flow injection determination of sucrose and glucose. Analytica Chimica Acta 380, 7–15.

Sun, T., Xia, N., Liu, L., 2016. A graphene oxide-based fluorescent platform for probing of phosphatase activity. Nanomaterials 6, 20.

Super, A., Jaccard, N., Marques, M.P.C., Macown, R.J., Griffin, L.D., Veraitch, F.S., et al., 2016. Real-time monitoring of specific oxygen uptake rates of embryonic stem cells in a microfluidic cell culture device. Biotechnology Journal 11, 1179–1189.

Tan, W., Wang, H., Chen, Y., Zhang, X., Zhu, H., Yang, C., et al., 2011. Molecular aptamers for drug delivery. Trends in Biotechnology 29, 634–640.

Tantama, M., Martínez-François, J.R., Mongeon, R., Yellen, G., 2013. Imaging energy status in live cells with a fluorescent biosensor of the intracellular ATP-to-ADP ratio. Nature Communications 4, 2550.

Tkáč, J., Voštiar, I., Sturdík, E., Gemeiner, P., Mastihuba, Vr, Annus, J., 2001. Fructose biosensor based on d-fructose dehydrogenase immobilised on a ferrocene-embedded cellulose acetate membrane. Analytica Chimica Acta 439, 39–46.

Tkáč, J., Vostiar, I., Gemeiner, P., Sturdík, E., 2002. Stabilization of ferrocene leakage by physical retention in a cellulose acetate membrane. The fructose biosensor. Bioelectrochemistry (Amsterdam, Netherlands) 55, 149–151.

Tric, M., Lederle, M., Neuner, L., Dolgowjasow, I., Wiedemann, P., Wölfl, S., et al., 2017. Optical biosensor optimized for continuous in-line glucose monitoring in animal cell culture. Analytical and Bioanalytical Chemistry 409, 5711–5721.

Tsuyama, T., Kishikawa, J-i, Han, Y.-W., Harada, Y., Tsubouchi, A., Noji, H., et al., 2013. In vivo fluorescent adenosine 5′-Triphosphate (ATP) imaging of *Drosophila melanogaster* and *Caenorhabditis elegans* by using a genetically encoded fluorescent ATP biosensor optimized for low temperatures. Analytical Chemistry 85, 7889–7896.

Turner, A.P.F., 2013. Biosensors: sense and sensibility. Chemical Society Reviews 42, 3184–3196.

Umar, A., Rahman, M.M., Vaseem, M., Hahn, Y.-B., 2009. Ultra-sensitive cholesterol biosensor based on low-temperature grown ZnO nanoparticles. Electrochemistry Communications 11, 118–121.

Ujjain, S.K., Das, A., Srivastava, G., Ahuja, P., Roy, M., Arya, A., et al., 2014. Nanoceria based electrochemical sensor for hydrogen peroxide detection. Biointerphases 9, 031011.

Van de Bittner, G.C., Dubikovskaya, E.A., Bertozzi, C.R., Chang, C.J., 2010. In vivo imaging of hydrogen peroxide production in a murine tumor model with a chemoselective bioluminescent reporter. Proceedings of the National Academy of Sciences of the United States of America **107**, 21316–21321.

Vo-Dinh, T., Cullum, B., 2000. Biosensors and biochips: advances in biological and medical diagnostics. Fresenius' Journal of Analytical Chemistry 366, 540–551.

Vidal, J.-C., Espuelas, J., Garcia-Ruiz, E., Castillo, J.-R., 2004. Amperometric cholesterol biosensors based on the electropolymerization of pyrrole and the electrocatalytic effect of Prussian-Blue layers helped with self-assembled monolayers. Talanta 64, 655–664.

Yoshida, T., Kakizuka, A., Imamura, H., 2016. BTeam, a novel BRET-based biosensor for the accurate quantification of ATP concentration within living cells. Scientific Reports 6, 39618.

Yoshikawa, S., Usami, T., Kikuta, J., Ishii, M., Sasano, T., Sugiyama, K., et al., 2016. Intravital imaging of Ca2+ signals in lymphocytes of Ca2+ biosensor transgenic mice: indication of autoimmune diseases before the pathological onset. Scientific Reports 6, 18738.

Wang, Y., Lin, H., Shao, J., Cai, Z.-S., Lin, H.-K., 2008. A phenylhydrazone-based indole receptor for sensing acetate. Talanta 74, 1122–1125.

Zhang, D., Lu, Y., Zhang, Q., Yao, Y., Li, S., Li, H., et al., 2015. Nanoplasmonic monitoring of odorants binding to olfactory proteins from honeybee as biosensor for chemical detection. Sensors and Actuators B: Chemical 221, 341–349.

Zhao, Y., Jin, J., Hu, Q., Zhou, H.-M., Yi, J., Yu, Z., et al., 2011. Genetically encoded fluorescent sensors for intracellular NADH detection. Cell Metabolism 14, 555–566.

CHAPTER 7

Nanoparticle-Mediated Oxidative Stress Monitoring and Role of Nanoparticle for Treatment of Inflammatory Diseases

VIKRAM DALAL • SAGARIKA BISWAS, PHD

LIST OF ABBREVIATIONS

5-LO	5-Lipoxygenase
8-oxo-dG	8-Hydroxy-20-deoxyguanosine
AgNPs	Silver nanoparticles
ATM	Ataxia telangiectasia mutated
ATR	ATM- and Rad3-related proteins
AuNP	Gold nanoparticle
BMPO	5-Tert butoxycarbonyl-5-methyl-1-pyrroline N-oxide
BSP	Betamethasone sodium phosphate
CACs	Circulating angiogenic cells
CCP	Cyclic citrullinated peptide
CIA	Collagen-induced arthritis
Co_3O_4	Tricobalt tetraoxide
CoONP	Cobalt oxide nanoparticle
CuNP	Copper nanoparticle
DCFDA	Dichlorofluorescein diacetate
DCFH	Dichlorofluorescein
DECPO	5,5-Diethylcarbonyl-1-pyrroline N-oxide
DEPMPO	5-Diethoxyphosphoryl-5-methyl-1-pyrroline N-oxide
DHR	Dihydrorhodamine 123
DMARD	Disease-modifying antirheumatic drug
DMPO	5,5-Dimethyl-1-pyrroline N-oxide
DUOX1	Dual oxidases
ELISA	Enzyme-linked immunosorbent assay
EPPN	N-2-(2-Ethoxycarbonyl-propyl)-a-phenylnitrone
ERK	Extracellular signal—regulated kinase
ESR	Electron spin resonance
Fpg	Formamidopyrimidine glycosylase
GC-MS	Gas chromatography-mass spectroscopy
GPx1	Glutathione peroxidase
HA-AuNP/TCZ	Hyaluronate-gold nanoparticle/tocilizumab
HEKs	Human epidermal keratinocytes
HPLC	High-performance liquid chromatography
HUVEC	Human umbilical vein endothelial cell
IBD	Inflammatory bowel diseases
IL	Interleukin
JAK-STAT	Janus kinase signal transducer and activator of transcription
JNK	c-Jun N-terminal kinase
LC-MS	Liquid chromatography-mass spectroscopy
LDH	Lactate dehydrogenase
MAPK/SAPK	Mitogen-activated protein kinase/stress-activated protein kinase
MAP-LC3	Microtubule-associated protein 1 light chain 3
MBL	Mannose-binding lectin
MCP-1	Monocyte chemoattractant protein-1
MDA	Malondialdehyde
MTX	Methotrexate
NADPH	Nicotinamide adenine dinucleotide phosphate
NFκB	Nuclear factor κB
NiO	Nickel oxide
NOX	Catalytic subunit of NADPH oxidases
NP	Nanoparticle
NSAID	Nonsteroidal antiinflammatory drug
nZVI; Fe_0	Nanoparticulate zero-valence iron

Nanotechnology in Modern Animal Biotechnology. https://doi.org/10.1016/B978-0-12-818823-1.00007-7

ONOO⁻	Peroxynitrite
•O₂⁻	Superoxide anion
PBN	N-Tertbutyl-p-phenylnitrone
PCOOH	Phosphatidylcholine
PDIA3	Protein disulfide isomerase associated 3
PEGylated AuNP	Polyethylene glycol–functionalized AuNP
PEOOH	Phosphatidylethanolamine
PLGA	Poly(lactic-co-glycolic acid)
PVP-AgNP	Polyvinyl pyrrolidone–coated silver nanoparticle
RANKL	Receptor activator of nuclear factor-κB ligand
RNS	Reactive nitrogen species
ROS	Reactive oxygen species
RS	Reactive species
SiNPs	Silica nanoparticles
siRNA	Small interfering RNA
SLN	Solid lipid nanoparticle
SOD	Superoxide dismutase
TAC	Total antioxidant capacity
TCZ	Tocilizumab
TiO₂	Titanium oxide
TLR	Toll-like receptor
TNF-α	Tumor necrosis factor-α
TXNL1	Thioredoxin-like protein 1
YFP	Yellow fluorescent protein
ZnONPs	Zinc oxide nanoparticles

NANOPARTICLES AND OXIDATIVE STRESS
Introduction
Metallic and metallic oxide nanoparticles (NPs) have a wide range of applications in areas of bioimaging, biosensors, and drug delivery applications (Arvizo et al., 2012; Doane and Burda, 2012; Sasidharan and Monteiro–Riviere, 2015; Thurn et al., 2007). Nanoparticles are used in several therapeutic applications like the treatment of infectious and inflammatory diseases, construction of membranes, mats, and hydrogels (Gaharwar et al., 2014; Ikoba et al., 2015; Zhan et al., 2015). They exhibit remarkable properties such as optical sensitivity, conductivity, and reactivity because of their unique physicochemical properties such as controllable shape and size, high surface area to volume ratio, and superparamagnetism (Alam et al., 2009; Ramesh et al., 2014). Metallic NPs can generate the toxicity such as oxidative stress and inflammation, which can further lead to cell injury and death (Giovanni et al., 2015; Nel et al., 2006; Wang et al., 2015). Metallic NPs can form the radicals, which can

participate in redox reactions and affect the cellular uptake. Nanoparticles can directly interact with DNA, proteins, and lipids, which can cause mutagenesis, protein degradation, and membrane damage (Nel et al., 2006).

The imbalance between the levels of production of antioxidants and oxidants within the cell refers to cellular oxidative stress. The disturbance of generation of oxidants such as reactive nitrogen species (RNS), reactive oxygen species (ROS), peroxynitrite (ONOO⁻), hydroxyl radicals (•OH), and superoxide anion (•O₂⁻) and antioxidants such as dismutase, scavengers, and catalase is called as oxidative stress. Production of ROS like hydroxyl radical (•OH), hydrogen peroxide (H₂O₂), ONOO⁻, nitric oxide (NO), and O₂•⁻ is one of the main causes of oxidative stress. ROS are generated as by-products of biochemical reactions such as cytochrome P450, neutrophil-mediated phagocytosis, enzymatic metabolism, and mitochondrial respiration. Oxidative stress can result in several diseases such as rheumatoid arthritis, Alzheimer's diseases, Parkinson's diseases, systemic lupus erythematosus, myocardial infractions, and chronic fatigue (Valko et al., 2007). Oxidative stress arises when oxidants or prooxidants are produced more as compared with antioxidants. Oxidative stress plays a key role in inflammatory processes, breaks down immunological tolerance, and can cause the apoptosis and even cell death (Dalal et al., 2017; Messner and Imlay, 2002; Rice-Evans and Gopinathan, 1995). Several redox potential flavoproteins play a pivotal role in the generation of oxidants (Messner and Imlay, 2002). Oxidative stress can oxidize the low-density lipoproteins, which can further result in plaque formation and is linked to cardiovascular diseases. Antioxidants exhibit the antimutagenic, antiinflammatory, anticarcinogenic, antibacterial, and antiviral properties (Owen et al., 2000). Report shows that ROS generated by mitochondria can induce inflammation in phagocytic cells and oxidative damage of telomeric DNA is less well repaired (Kepp et al., 2011; Sorbara and Girardin, 2011; von Zglinicki, 2002). Oxidative stress induced the production of 8-oxo-7,8-dihydro-20-deoxyguanosine (8-oxodG) at the GGG triplet of the telomere of the chromosome that results into the loss of telomere (Kawanishi and Oikawa, 2004). Metallic NPs induce biological responses such as disruption of the cell cycle, DNA damage, and lysosomal dysfunction in the cells (Bhabra et al., 2009; Mahmoudi et al., 2011a,b; Stern et al., 2012). Nanoparticles can activate the inflammatory cells such as neutrophils and macrophages, which can further enhance the production of ROS (Bancos et al., 2015; Goncalves et al., 2010).

Oxidative Stress and Antioxidants Mechanism

Exogenous or endogenous sources can induce the oxidative stress in the human. These sources generate various oxidants. Antioxidants are necessary to counter the oxidants, which further leads to a decrease in oxidative stress. So, the balance between antioxidants and oxidants can be used to determine the oxidative stress in a human.

Endogenous sources

Intracellular enzymes such as oxidases, peroxisomes, NADPH oxidases, lipoxygenases, and cytochromes (CYPs) can produce the oxidants within the cells. Nicotinamide adenine dinucleotide phosphate oxidase is the main source of ROS in vasculatures that exist in different isoforms such as catalytic subunit of NADPH oxidases (NOX1), NOX2, NOX3, and NOX4 (Ushio-Fukai, 2006). In tumor cells, the NADPH oxidase plays a pivotal role in the production of ROS (Landry and Cotter, 2014). It has been found that nuclear factor erythroid 2–related factor 2–Kelch-like ECH–associated protein 1 (Nrf2–Keap1) pathway plays an important role in the generation of cytosolic and mitochondrial ROS by superoxide dismutase (SOD1), NADPH oxidase, thioredoxin, and glutathione (Kensler et al., 2007; Kovac et al., 2015). CYP450 is a heme-containing superfamily of protein, which can degrade the xenobiotic compounds. In the CYP450 catalytic cycle, dioxygen is activated to O_2 by single-electron reduction reaction (Lewis, 2002). CYP2E1 is also known as "leaky" enzyme, which can produce ROS in the absence of substrate (Robertson et al., 2001). Catalytic residues of CYP are loose coupled, which can cause the generation of oxy-radicals and further initiate the microsomal NADPH-independent lipid peroxidation.

Lipoxygenase is the iron-containing enzyme, which can metabolize the eicosanoids such as prostaglandins and leukotrienes. Arachidonic acid is metabolized into 5-lipoxygenase (5-LO), which is essential for the production of leukotriene, and it can cause several inflammatory responses (Dixon et al., 1990). It has been reported that leukotriene and 5-LO play a pivotal role in the regulation of inflammation and oxidation in diseases such as arthritis, asthma, and several neurodegenerative conditions (Joshi and Praticò, 2015). Peroxisomes are the multifunctional enzymes, which play a vital role in catalysis of oxidation reactions. H_2O_2 is the by-product, and furthermore, it is degraded by catalases (Nordgren and Fransen, 2014). It has been found that excessive production of ROS by peroxisome can cause

mitochondrial-mediated cell deaths in human (Wang et al., 2013). In mitochondria, four multiprotein complexes (complex I–IV) of electron transport chain regulate the oxidative phosphorylation and the electrochemical proton gradient (Liu et al., 2002). Complex I can couple with molecular oxygen and release the O_2^-, whereas complex III releases O_2^- into the cytoplasm (St-Pierre et al., 2002). It has been reported that perturbation of electron transport chain (ETC) can raise the level of oxidative stress, which can further damage the mitochondrial DNA (Indo et al., 2007).

Exogenous source

External environmental factors such as NPs, ionization radiation (IR), inflammatory cytokines, and bacterial and fungi toxins can also regulate the production of oxidants, which can alter the intracellular oxidative stress. Exposure of these external factors to a specific cell type may produce ROS, which can affect the adjacent cells. Dendritic cells and macrophages along with cell surface expression of Toll-like receptors (TLRs) play a significant role in the innate immune response. Coexpression of multiple TLRs can trigger the oxidative stress mediated by disturbance of production of antiinflammatory and proinflammatory cytokine (Lavieri et al., 2014).

It has been found that longer exposure of thyroid cells to IR can produce the ROS (Ameziane-El-Hassani et al., 2015). IR is the main cause of breakage of bonds and production of the free radicals. It can trigger the formation of NADPH oxidase dual oxidases (DUOX1)-dependent H_2O_2, which can further result in oxidative stress. Report shows that *Streptococcus pneumonia* can cause the oxidative stress in epithelial cells of lungs (Zahlten et al., 2014). Induction of oxidative stress mediated by *Streptococcus pneumonia* is dependent on pneumococcal autolysin LytA. In human lymphocytes, mycotoxins can induce oxidative stress and genotoxicity. *Fusarium* produces the deoxynivalenol (DON), which can damage the membrane and reduce the cell viability (Yang et al., 2014). Lymphocytes treated with DON raise the levels of ROS, lipid peroxidation, and 8-hydroxy-2'-deoxyguanosine.

Oxidative Stress Induced by Metallic and Metallic Oxide Nanoparticle

Nanoparticles play a crucial role in oxidative stress because of their physical size, high biopersistence, and surface area to volume ratio. Exposure of human bronchial epithelial cells to nanoparticulate zero-valence iron (nZVI; Fe_0) produces the ROS (Keenan et al., 2009). Entry of NPs via the respiratory system to the epithelial lining of the lung can result in deposition of inflammatory phagocytes, which can further produce RNS and ROS that lead to lipids, proteins, and DNA damage (Bonner, 2007). In rats, injection of nickel oxide (NiO) and titanium dioxide (TiO$_2$) NPs can raise the level of 8-OHdG (Kawai et al., 2014). Exposure of human umbilical vein endothelial cells (HUVECs) to silica nanoparticles (SiNPs) up to 200 μg/mL enhances the production of ROS, mitochondrial depolarization, and cell death. Furthermore, the rise in the levels of chemokine monocyte chemoattractant protein-1 (MCP-1), cytokines interleukin-8 (IL-8) and interleukin-6 (IL-6), and lactate dehydrogenase also observed after treatment of HUVECs with SiNPs (Liu and Sun, 2010). Cellular uptake of copper nanoparticles (CuNPs) can raise the level of generation of ROS in the cells (Bulcke et al., 2014).

Molecular level

Interaction of metal NPs may damage the cellular biomolecules such as DNA, protein, and lipids. ROS can induce DNA strand breakage, genetic mutation, and cross-linking of DNA (Golbamaki et al., 2015). Exposure of human liver HepG2 cells with zinc oxide nanoparticles (ZnONPs) can cause the oxidative stress, which further leads to DNA damage in these cells by increasing the production of formamidopyrimidine glycosylase (Fpg)—sensitive sites (Sharma et al., 2012a,b). Fpg-modified single cell gel electrophoresis (comet assay) was utilized to detect the oxidative genomic damage in kidney and liver of mice (Sharma et al., 2012a,b). Oral administration of polyvinylpyrrolidone-coated silver nanoparticles (PVP-AgNPs) to mice can cause permanent genetic alterations and DNA damage in different tissues. PVP-AgNPs can trigger DNA deletion in embryos and permanent chromosomal aberration in bone marrow that downregulate the base pair excision (BER) genes (Kovvuru et al., 2015). Exposure of lung fibroblasts with gold nanoparticles (AuNPs) can induce genotoxicity and lipid peroxidation, which can be determined by fluorescence in situ hybridization and comet assay (Li et al., 2011; Ng et al., 2013).

Treatment of HUVECs with SiNPs enhances the production of ROS that results into lipid peroxidation (Guo et al., 2015). The level of malondialdehyde (MDA) increased after 24 h exposure of HUVECs to SiNPs. It has been reported that fibrinogen unfolds in plasma after binding with negatively charged poly(acrylic acid)—conjugated AuNPs (PAA-AuNPs) (Deng et al., 2011). The interactions of superparamagnetic iron oxide nanoparticles (SPIONs) with iron-saturated transferrin can release the iron and alter its main function (Mahmoudi et al., 2011a,b). Silver nanoparticles (AgNPs) treatment of human monocytic THP-1 cells causes NP-induced oxidative stress by increasing the levels of heme oxygenase-1 and protein carbonylation (Haase et al., 2011).

Biochemical level

Oxidative stress is related to activation of certain signaling pathways at the biochemical level. It can dysregulate various molecules such as DNA repair molecules p66, p44, and p53 and signaling pathways such as nuclear factor κB (NFκB), Janus kinase signal transducer and activator of transcription (JAK-STAT), phosphoinositide phospholipase C (PLCγ), and mitogen-activated protein kinase/stress-activated protein kinase (MAPK/SAPK) (Martindale and Holbrook, 2002). DNA damage plays a major role in cellular aging and carcinogenesis. Various enzymes such as DNA damage response serine/threonine kinases Ataxia telangiectasia mutated (ATM), catalytic subunit (DNA-PKcs), ATM- and Rad3-related proteins (ATR), and DNA-dependent protein kinase play a vital role in cellular DNA repair mechanism (Chen et al., 2012). Exposure of lung fibroblasts with 20 mm size AuNPs can reduce the expression of DNA damage response genes such as ATLD/HNGS1, BRCA1, AT-V1/AT-V2, and HUS1 (Li et al., 2008). Comparative proteomic analysis by using two-dimensional gel electrophoresis shows that treatment of MRC-5 lung fibroblasts with AuNPs can cause differential expression of oxidative stress—related proteins such as heterogeneous nuclear ribonucleic protein C1/C2 (hnRNP C1/C2), NADH ubiquinone oxidoreductase (NDUFS1), thioredoxin-like protein 1 (TXNL1), and protein disulfide isomerase associated 3 (PDIA3) (Li et al., 2011). Oxidative stress affects the expression of protein kinases such as members of MAPK family. Exposure of human epithelial embryonic cells to silver nanoparticles (AgNPs) can increase the generation of ROS and extracellular signal—regulated kinase (ERK) and phosphorylation of c-Jun N-terminal kinase (JNK) (Rinna et al., 2014).

Oxidative stress can induce fibroblasts to activate specific p53 transcriptional responses (Gambino et al., 2013). Study shows that p53 plays a vital role in zinc oxide nanoparticles (ZnONPs)—induced oxidative stress using BJ skin fibroblast p53 knockdown system (Setyawati et al., 2013). Treatment of cells with low levels of ZnoNPs reveals that p53 is activated after the expression of antioxidant genes such as glutathione peroxidase (GPx1) and SOD2. Exposure to higher levels of ZnONPs showed a rise in levels of ROS, which can trigger the apoptosis. These results confirm that p53 plays a crucial role in NP-induced oxidative stress. In human leukemia cells, cobalt oxide nanoparticles (CoONPs) can increase the level of production of ROS (Chattopadhyay et al., 2015). CoONPs can enhance the levels of proinflammatory cytokine tumor necrosis factor-α (TNF-α), which can further induce the apoptosis via triggering of p38, caspase-3, and caspase-8. Treatment of AgNPs to HUVECs can inhibit the proliferation of HUVECs. Moreover, the expression levels of chemokines, adhesion molecules, and inflammatory cytokines also increased in the endothelial cells (Shi et al., 2014). AgNPs can dysfunction the HUVECs, which can further activate the inhibitors of κ B kinase (IKK)/NFκB pathway. Different sizes of SPIONs treatment with normal human-murine epidermal cells [JB6 P(+)] and human epidermal keratinocytes (HEKs) showed an activation of activator protein 1 (AP-1) and rise in levels of proinflammatory mediators IL-8 and IL-6 in JB6 cells and HEKs cells, respectively (Murray et al., 2013). Treatment of JB6 cells with ultraviolet radiation B (UVB) can trigger the NFκB, which further activates the IL-6 (Libermann and Baltimore, 1990). Exposure of ultraviolet B (UVB) and NPs to JB6 released the lactate dehydrogenase (LDH), which can cause the release of cytokines, depletion of glutathione, and cellular damage.

Cellular level

Oxidative stress can affect at cellular levels such as proliferation, autophagy, cell senescence, growth arrests, and cell death also. Exposure of increasing concentrations (up to 100 μg/mL) of tricobalt tetraoxide (Co$_3$O$_4$) and TiO$_2$ to human circulating angiogenic cells (CACs) can decrease the cell viability and raise the levels of lipid peroxidation and caspase activity (Spigoni et al., 2015). Zinc oxide induces apoptosis and decreases the cell viability in primary astrocytes and activates the caspase-3, which can further increase the release of LDH and raise the levels of ROS (Wang et al., 2014). Exposure of ZnO can induce apoptotic events such as cleave polymerase-1 (PARP) and nuclear condensation. c-Jun N-terminal kinase inhibitor (SP600125) can decrease the cleavage of caspase-3 expression and PARP. Human epithelial cells HK-2 were treated with soluble ZnONPs and cadmium sulfide (CdS) (Pujalté et al., 2015). Concentration-dependent exposure of ZnONPs and cadmium sulfide (CdS) cause the release of Zn^{2+} and Cd^{2+}, which can further result in cell death. The rise in the level of ROS production and lipid peroxidation causes the destabilization of lysosomal membrane and nuclear condensation of cells. Exposure of ZnONPs can induce tumor suppressor p53 by triggering oxidative stress, which

can further result in apoptosis in BJ fibroblasts (Ng et al., 2011). Exposure of AgNPs can affect the stability of lysosomes, which can result in cellular toxicity (Setyawati et al., 2014). Endocytosis of AuNPs by normal rat kidney cells can trigger the accumulation of autophagosome (Ma et al., 2011). Silver nanoparticles can also induce the formation of autophagosome in lung fibroblasts along with biochemical changes such as upregulation of autophagy gene 7 (ATG 7) and microtubule-associated protein 1 light chain 3 (MAP-LC3) (Li et al., 2010). Polyethylene glycol—functionalized AuNPs (PEGylated AuNPs) can decrease the oxygen-carrying ability of erythrocytes, which further results in loss of CD47 from erythrocyte membranes (He et al., 2014).

Measurement of Oxidative Stress

Various techniques have been utilized to determine the principal markers of DNA, protein, and lipids oxidative damage.

Reactive species measuring in vivo

L band electron spin resonance (ESR) with nitroxy probes can be utilized to detect the reactive species (RS) directly (Berliner et al., 2001). RS cannot determine in vivo because of their short time of persistence in cells. Some RS such as NO. and H$_2$O$_2$ can be detected in vivo. Transient RS can be detected by determining the level of trapped RS species or measurement of damaged level because of RS during oxidative stress. Oxidative stress can be determined by a reduction in total antioxidant capacity (TAC) of body fluids and erythrocytes. The serum or plasma TAC is due to the presence of urate, albumin —SH groups, and ascorbate (Prior and Cao, 2001). Level of plasma urate or ascorbate may be varied by the consumption of certain ingredients.

Trapping of reactive species

Free radicals generated during oxidative stress can be measured by ESR only. However, limitation of ESR is that it cannot determine the level of reactive radicals. This limitation can be sorted by using some trap reagents or probes, which can interact with unreactive radicals and produce stable radicals; these stable radicals can be measured by ESR. Different traps such as 5-tert butoxy carbonyl-5-methyl-1-pyrroline N-oxide (BMPO), 5,5-diethyl carbonyl-1-pyrroline N-oxide (DECPO), N-tertbutyl-p-phenylnitrone (PBN), 5-diethoxy phosphoryl-5-methyl-1-pyrroline N-oxide (DEPMPO), N-2-(2-ethoxy carbonyl-propyl)- a-phenylnitrone (EPPN), and 5,5-dimethyl-1-pyrroline N-oxide (DMPO) have been tested. The major limitation of ESR is that main product of reaction, which gives ESR signal, can be removed by enzymatic metabolism or electrochemical reaction. For example, 5,5-dimethyl-1-pyrroline N-oxide (DMPO-OH), which traps the OH·, can be reduced by ascorbate and results in no signal in ESR.

Ex vivo measurement

Spin traps. Spin traps can only be utilized for body fluids or tissue samples because of the unknown toxicity of trapping the endogenous free radicals. Hydroxylamine and DMPO probes can be utilized to determine the concentration of free radicals in liver biopsies and skin, respectively (Haywood et al., 1999). ESR can measure the secondary radicals such as lipid-derived radicals (peroxyl, alkoxy, etc.) and protein radicals generated during oxidative stress. ESR can measure the semidehydroascorbate radical, which is produced after reaction of ascorbic acid (vitamin C) with different RS. The level of concentration of generation of free radicals in organs, skin, and plasma can be determined by detection of semidehydroascorbate (Haywood et al., 1999).

Aromatic traps. Aromatic traps are more widely used as compared with spin traps, as they can be utilized for human consumption, including phenylalanine and salicylate. Phenylalanine and salicylate are utilized to determine the ex vivo radical production in human blood of RA patient (Liu et al., 1997). Production of OH• in saliva can be determined by phenylalanine. Salicylate can be utilized to determine the OH• in a myocardial infarction, diabetes, and alcoholism.

Measurement of blood pressure. The rise in the level of oxidative stress can reduce the activity of NO• (Kojda and Harrison, 1999). In blood, O_2• can be produced by fibroblasts, phagocytes, lymphocytes, enzyme xanthine oxidase, and vascular endothelium. The rise in the level of O_2 can produce NAD(P)H, which plays an important role in vascular tension and hypertension. The increment in the level of antioxidant can eliminate ROS, which can restore NO• that results into the dysfunction of human endothelial. Thus, short-term vascular effects can be determined by blood pressure to check the level of oxidative stress.

Reactive species fingerprinting. The level of RS can be determined by unique chemical fingerprint given by the combination of RS with biological molecules (Halliwell and Gutteridge, 2015). Direct methods of detection of RS cannot be utilized for humans, as they can cause toxicity in cells, for example, reactive radicals can break the strand or mutate the DNA. Lipids can be oxidized or chlorinated by RS (Halliwell and Gutteridge, 2015). Modification of sugars and nitrogenous bases by oxidative damage can be determined by gas chromatography-mass spectroscopy (GC-MS), high-performance liquid chromatography (HPLC), liquid chromatography (LC-MS), and antibody-based assays. Rates of oxidative stress in the body can be measured by detection of 8-hydroxy-20-deoxyguanosine (8-oxo-dG) by HPLC, mass spectroscopy, and enzyme-linked immunosorbent assay (ELISA). Oxidative stress can alter the structure or functions of proteins, and it can produce new antigens, which result into immune responses (Halliwell, 1978). All amino acids can be attacked by the different mechanism by RS (Headlam and Davies, 2003). The reaction of free radicals with protein can generate the amino acid radicals. Cross-linking of amino acid radicals can produce the peroxyl radicals, which can further induce the production of more radicals (Headlam and Davies, 2003).

Reactive species measurement in cells

Spin traps and aromatic compounds can be utilized to determine the concentration of ROS production in cells. Oxidative stress damage can be detected by assays of different end products such as oxidized protein carbonyls, 3-nitrotyrosine, and DNA bases.

Dichlorofluorescein diacetate. Dichlorofluorescein diacetate (DCFDA) utilized to determine the concentration of cellular peroxidase (Ischiropoulos et al., 1999). Esterases convert DCFDA into dichlorofluorescein (DCFH), and further RS can convert it into dichlorofluorescein (DCF). DCF can excite at 488 nm and further be visualized at 525 nm.

Dihydrorhodamine 123. DHR can be utilized to determine the NO_2•, $ONOO^-$, and OH• (Ischiropoulos et al., 1999). It is more sensitive and can be used for the measurement of HOCl than DCFDA (Ischiropoulos et al., 1999). DHR can oxidize into rhodamine 123, which is a positive charge molecule and fluorescent at 536 nm. Rhodamine can accumulate in mitochondria and can measure the free single O_2 in the mitochondria.

Luminol. Luminol can determine the generation of RS by activated phagocytosis (Faulkner and Fridovich, 1993). O_2•$^-$ can oxidize luminol, and luminal can react with O_2•$^-$ to produce the light-emitting products. Luminol cannot be used for human cells, as it can reduce O_2 also to generate $O_2^\bullet$ (Faulkner and Fridovich, 1993).

Diphenyl-1-pyrenylphosphine. Reactions of peroxidation with diphenyl-1-pyrenylphosphine produced the fluorescent product, which can be measured at 380 nm after excitation at 351 nm (Takahashi et al., 2001). The fluorescent products can stay up to 2 days in living cells.

Cis-parinaric acid. Incorporation of cis-parinaric acid into peroxidized lipid can oxidize the cis-parinaric and lose its fluorescence at 413 nm (Ritov et al., 1996). It is a polyunsaturated structure, and it is highly susceptible to nonspecific oxidation, so it should be kept under dark in N_2 atmosphere.

Measurement of the output of probes
Fluorescence microplate reader. It is an application which can be used to detect the increase or decrease in relative fluorescence. But the quality and sensitivity of results depend on the machines and the addition of extra equipment such as emission filters and excitation, which increase the cost. Cells are preserved in suspension form for the determination of top reading fluorescence. Bottom reading measurement is advantageous as it does not require the extra supplement of trypsin (Halliwell, 2003).

Flow cytometry. Intracellular fluorescence of cells in culture can be measured by flow cytometer. It has the disadvantage that it required the extra addition of trypsin, which can generate the oxidative stress.

Confocal microscopy. Here, cells are loaded along with fluorescent dyes, and results can be visualized in real-time in situ in a culture chamber at 37°C. Counterstains such as lysotracker dyes, ER tracker, or milto tracker can be utilized to view the role of lysosomal, endoplasmic, and mitochondria roles in oxidative stress.

Other techniques
High-performance liquid chromatography. The HPLC has been used to measure the lipid peroxidation by measuring the phosphatidylethanolamine (PEOOH) and phosphatidylcholine (PCOOH) levels in the liver of mouse (Miyazawa et al., 1987). Lipid peroxidation generates endoperoxides, hydroperoxides, pentane, ethane, and end products of MDA, so detection of PEOOH and PCOOH is the most reliable

technique for measurement of oxidative stress (Miyazawa et al., 1987).

Amplex red and multiphoton microscopy. Amplex red and multiphoton microscopy can be utilized to determine the roles of antioxidant in oxidative stress. It has found that antioxidant treatment can reduce the level of pathological neuronal alterations in Alzheimer mouse models (Garcia-Alloza et al., 2006). But all antioxidants are not effective, so antioxidants that can reduce the generation of ROS in animals have been identified.

Biosensors. Here, combination of yellow fluorescent protein (YFP) along with regulatory domain of OxyR called hyper is used. It has been used to measure the biological roles of hydrogen peroxide (Belousov et al., 2006). H_2O_2 oxidizes the cysteine residues of the regulatory domain of OxyR, which results in fluorescence at 516 nm. Hyper sensor is the most effective sensor, as it can measure even submicromolar concentrations of H_2O_2 also, so it has been utilized to determine the role of H_2O_2 in wound responses of zebrafish (Niethammer et al., 2009).

NANOPARTICLE ROLES IN THE TREATMENT OF INFLAMMATORY DISEASES
Introduction
Local delivery of antiinflammatory drugs is necessary to avoid the side effects of the drug. Several nanoparticles such as polylactic-co-glycolic acid (PLGA) NPs, gold half-shell multifunctional NPs, and solid lipid nanoparticles (SLPs) have been used to increase drug localization at the target site (Gref et al., 1994; Lee et al., 2012; Ye et al., 2008). Nanoparticles such as AuNPs, PLGA, infliximab and methotrexate, actarit-loaded SLPs, and gold half-shell NPs have been used for the treatment of inflammatory diseases. Gold NPs exhibit various properties such as facile surface modification, biocompatibility, and versatile conjugation so they can be utilized for various biomedical applications (Ghosh et al., 2008; Giljohann et al., 2010; Lee et al., 2012; Thakor et al., 2011). It has an antiangiogenic property, which can bind to a vascular endothelial factor, which plays a pivotal role in the pathogenesis of rheumatoid arthritis (Bhattacharya et al., 2004; Firestein, 2003; Tsai et al., 2007).

Nanoparticles targeted to macrophages can be used for the treatment of autoimmune blood disorders such as rheumatoid arthritis (RA) and diabetes (Barrera et al., 2000; Chellat et al., 2005; Moghimi et al., 2005). Intravenously injected NPs have the ability to accumulate in the organs such as liver and spleen so they can be utilized for treatment of RA to remove the particulates from the blood (Moghimi et al., 2001). Spleen produced the macrophages, which further cause the inflammatory responses. The spleen can be targeted to enhance the therapeutic efficacy and to decline the side effects of disease-modifying antirheumatic drugs (DMARDs). In mice, it has been found that actarit plays an important role in an inhibitory effect on progression lesions (Yoshida, 1987). Actarit can improve the plasma albumin/globulin ratio of adjuvant RA rats (Fujisawa et al., 1990, 1994). However, oral administration of actarit leads to higher accumulation of actarit in gastrointestinal tract and kidney as compared with spleen (Sugihara et al., 1990). Most adverse effects of actarit are related to gastrointestinal disorders (Matsubara, 1999). DMARDs can induce the renal dysfunction and urinary abnormalities in RA (Makino et al., 2002). Biodegradation of polymers into glycolic acid, metabolite monomers, and lactic acid can be obtained by using PLGA. It can be easily metabolized in Krebs cycle and hence highly used for drug delivery and biomaterial applications (Kumari et al., 2010). PLGA NPs are internalized in the body by clathrin-mediated endocytosis and pinocytosis.

Inflammatory Bowel Diseases

Crohn's diseases and ulcerative colitis possessed different pathogenesis, but both required corticosteroids and 5-aminosalicylic acid to trigger the remission of the treatment. However, the significant increment of uptake is required to equilibrate the pharmacological effects, but it can result in adverse side effects (Meissner and Lamprecht, 2008). PLGA-based NPs are a promising candidate for the delivery of drugs to the colon in inflammatory bowel diseases (IBDs). Smaller NPs can be taken up easily in the areas of inflammations (Lamprecht et al., 2001a,b). High negative charged PLGA can be used for drug delivery in ulcerated tissues. Confocal laser endomicroscopy showed that PLGA NPs accumulate in ulcerous lesions of inflammatory bowel disease patients and further nanoparticles lead to the release of antiinflammatory drugs (Schmidt et al., 2010). PLGA NPs can accumulate in inflamed regions and can release the entrapped drugs in inflammation regions (Lamprecht et al., 2001a,b). Tacrolimus-engulfed PLGA NPs can release the tacrolimus in the inflamed regions and protect the tacrolimus from P-gp efflux and mucosal metabolism (Lamprecht et al., 2005).

Inflammatory Lung Diseases

PLGA encapsulated PS-341, which is a chymotryptic threonine protease inhibitor that can decrease the inflammation of cystic fibrosis. *Pseudomonas aeruginosa*—induced inflammation can be controlled by PLGA-encapsulated PS-341, whereas direct administration of PS-341 did not show any effect on inflammatory neutrophils and macrophages (Vij et al., 2010). Oral administration of PLGA NPs encapsulating curcumin increases the treatment in cystic fibrosis mice as compared with direct delivery of curcumin (Cartiera et al., 2009).

Ophthalmic Inflammatory Diseases

Direct administration of free drugs in eyes is not preferable because of poor bioavailability in tissues, impermeability of drugs, and rapid turnover in the corneal epithelial membrane. So, only 5% of free drugs reach up to intraocular tissues after cornea penetration (Zhang et al., 2004). PLGA NPs have a suitable size for drug delivery in ophthalmic inflammatory diseases. PLGA and its metabolites are nontoxic and biodegradable (Garinot et al., 2007). Nonsteroidal antiinflammatory drugs are used to cure seasonal allergic conjunctivitis, postoperative inflammation, and reduction of bacterial colonization and inhibit bacterial adhesion to human corneal cells (Schalnus, 2003). PLGA encapsulated with poloxamer 188 (P188) can be used to graft cationic polymer on the anionic surface of NPs. The NPs reveal a better antiinflammatory efficacy and no toxicity as compared with eye drops. It has been found that PLGA-encapsulated flurbiprofen shows high precorneal residence time at the inflammatory site (Araújo et al., 2009; Vega et al., 2008). PLGA can be used in combination with other nanoparticles to modify NPs' properties to target it to the more specific site in the eye. PLGA/Eudragit @RL (75:25) encapsulated with ciprofloxacin reveals a higher drug concentration in tear film (Dillen et al., 2006).

Rheumatoid Arthritis

Rheumatoid arthritis is a chronic systemic disorder, which can cause joint and bone destruction. RA involved environmental factors, genetic risk factors, and activation of immune responses. RA is characterized by inflammation such as heat, redness, pain, and swelling and loss of function of connection. RA can be diagnosed by clinical parameter, the presence of anticyclic citrullinated peptide (CCP)/rheumatoid factor/anti—mannose-binding lectin (MBL), c reactive protein, radiographic evidence of joint destruction, and complete blood count. Several antiinflammatory drugs such as glucocorticoids are used for the treatment of RA, but they can cause serious systemic side effects (Lee and Kavanaugh, 2003). It has been found that macrophage can phagocytose the PLGA nanoparticles, but PLGA microsphere stayed in the synovium of rat (Horisawa et al., 2002a,b). Betamethasone sodium phosphate (BSP), a steroid encapsulated in PLGA, has been used to release the drug in animal arthritis model (Higaki et al., 2005). Direct intraarticular injection of BSP encapsulated with PLGA nanoparticles enhances the pharmacological efficacy (Horisawa et al., 2002a,b).

Monocyte-derived myeloid cells comprise potential effectors and scavenger cells during inflammatory processes (Iwamoto et al., 2007; Jongstra-Bilen et al., 2006; Taylor et al., 2003). Uncontrolled activation of myeloid cells can result in chronic inflammations. Myeloid cells—mediated chronic inflammation plays a crucial role in chronic illnesses such as cardiovascular, neurodegenerative, autoimmune, and pulmonary metabolic disorders (Boyd et al., 2008; Croce et al., 2009; Deng et al., 2009; Qu et al., 2009; Soehnlein and Weber, 2009). Curcuma longa (turmeric) has antineoplastic, antiinflammatory, chemopreventive, and antioxidant properties (Anand et al., 2008a,b; Anand et al., 2008a,b; Dhillon et al., 2006; Ravindran et al., 2009). Thus, curcumin has been encapsulated in biodegradable microspheres, liposomes, hydrogels, biodegradable nanoparticles, and cyclodextrin (Choi et al., 2009; Cui et al., 2009; Kurzrock and Li, 2005; Mulik et al., 2009; Narayanan et al., 2009), and exosomes can deliver the curcumin to activate the myeloid cells (Sun et al., 2010). This technology can be used to treat inflammatory cells by releasing the curcumin at the site of inflammation.

Tumor necrosis factor-α (TNF-α) secretion from macrophages plays an important role in progression and development of RA. Double-stranded RNAs cleaved into very short 21-22 mers fragments, which can enter into ribonuclear protein complex called RNA-induced silencing complex. siRNA antisense strand can guide the RNA-induced silence complex for the degradation of specific mRNA (Elbashir et al., 2001). Report shows that direct articular injection of TNF-α-specific siRNA can decrease the joint inflammation in murine collagen-induced arthritis (CIA) (Schiffelers et al., 2005). In mice, TNF-α-specific siRNA encapsulated with cationic lipid-based NPs shows antiinflammatory effects (Khoury et al., 2006). Knockdown of TNF-α chitosan/small interfering RNA (siRNA) NPs can control the inflammation in RA. Chitosan NPs-engulfed anti-TNF-α dicer substrate (DsiRNA) can result in

knockdown of TNF-α in primary peritoneal macrophages (Howard et al., 2009). It has been reported that NPs-mediated TNF-α knockdown in peritoneal macrophage can be used to decline the local and systemic inflammation (Howard et al., 2009).

Gold nanoparticles can inhibit the receptor activator of NFκB ligand (RANKL)—induced synthesis of osteoclasts by quenching ROS, which can cause the destruction of cartilage and bones (Sul et al., 2010). Gold nanoparticles can be utilized as a nanoprobe for diagnosis and detection of RA. Tocilizumab (TCZ) humanized antibody against interleukin-6 (IL-6) can be used to inhibit the pathogenesis of RA. It has been observed that hyaluronate-AuNPs/tocilizumab (HA-AuNP/TCZ) complex can be used for the treatment of RA (Lee et al., 2014). Lipophilic anti-RA drug incorporated into solid NPs. SLN-encapsulated actarit can be used as a passive targeting agent in RA and may reduce dosing frequency, doses, and toxicity and improve compliance (Ye et al., 2008). Methotrexate (MTX) is a DMARD, which is used for the treatment of RA (Saag et al., 2008). Methotrexate exhibits a good therapeutic efficacy, but the long-term use of MTX can cause serious systemic infections such as bone marrow suppression, infections, and hepatitis (Schnabel and Gross, 1994; Van Ede et al., 1998). Increase in dose concentration of MTX in RA patients can cause several disorders such as bone marrow dysfunction and intestinal lung diseases. Alternatively, for better treatment, MTX containing (Arg-Gly-Asp) RGD-attached gold (Au) half-shell NPs can be used for delivery of MTX at inflammation site, which further can be used for the treatment of RA (Lee et al., 2012).

Nonsteroidal antiinflammatory drugs (NSAIDs) can inhibit cyclooxygenase, which converts arachidonic acid into prostaglandins, so NSAIDs can be utilized for the treatment of inflammatory disorders. NSAIDs decrease the antiinflammation and antiproliferation, which further can suppress the expression of NFκB (Kusunoki et al., 2008). siRNA and glucocorticoids can be codelivered by encapsulating into PLGA to suppress the expression of unnecessary proteins and genes of RA (Park et al., 2012).

CONCLUSION

Metallic NPs exhibit high surface area to volume ratio, optical sensitivity, and superparamagnetism. Nanoparticles are widely used in biomedical applications such as drug delivery, biosensors, and bioimaging and therapeutic applications such as treatment of inflammatory diseases, construction of hydrogels, mats, and membranes. Nanoparticles produce the radicals, which can lead to oxidative stress, which further can result in cell injury and cell death. NPs can interact with lipids, proteins, and DNA and degrade protein and membranes.

Oxidative stress is the disturbance of production of prooxidants such as superoxide anion ($\cdot O_2^-$), ROS, $ONOO^-$, and RNS and antioxidants such as catalase and dismutase. The endogenous and exogenous source is the main cause of oxidative stress in a human. Intracellular enzymes such as NADPH oxidases, CYPs, peroxisomes, and oxidases can generate the oxidants within the cells. Lipoxygenase can metabolize the prostaglandins and leukotrienes, which further play a crucial role in the regulation of inflammation and oxidation in several diseases such as asthma, arthritis, and neurodegenerative diseases. IR, NPs, and bacterial and fungi toxins are the exogenous sources of oxidative stress.

Nanoparticles can affect the cells and organs at biochemical and molecular levels. Exposure of NPs such as nZVI, silica NPs, and copper NPs can produce the RNS and ROS, which can damage the lipids, protein, and DNA. Human liver cells HepG2 exposure to zinc oxide NPs (ZnONPs) can cause oxidative stress, which results in DNA damage. Silica nanoparticles exposure to HUVEC increases the generation of RS, which can result in lipids peroxidation. Oxidative stress can dysregulate several molecules such as DNA repair molecules p66, p44, p53, and signaling pathways such as NFκB, an activator of transcription (JAK-STAT), mitogen-activated protein kinase/stress-activated protein kinase (MAPK/SAPK), and Janus kinase signal transducer. DNA damage can result in carcinogenesis and cellular aging. Oxidative stress can alter the various cellular level processes such as autophagy, proliferation, cell senescence, and growth arrests. TiO_2 and tricobalt tetraoxide (Co_3O_4) to human CACs can increase lipid peroxidation and caspase activity.

Oxidative stress can be measured in vivo by L band ESR with nitroxy probes and ESR with traps such as BMPO, DECPO, PBN, 5DEPMPO, EPPN, and DMPO. Oxidative stress can determine ex vivo by measurement of free radicals by using spin traps and aromatic traps. In spin traps, DMPO probes and hydroxylamine are used to measure the concentration of free radicals. Salicylate and phenylalanine are used to detect the generation of free radicals in the blood of RA patients. End products, such as oxidized protein carbonyls, 3-nitrotyrosine, and DNA bases, of oxidative stress can be measured by dichlorofluorescein diacetate (DCFDA), luminol, cis-parinaric acid, DHR, and diphenyl-1-pyrenyl phosphine. Several techniques such as fluorescence microplate reader, flow cytometry, and confocal microscopy can be utilized to measure the oxidative stress. In oxidative stress, lipid peroxidation can be determined by measuring the level of PEOOH and PCOOH. Biosensors like hyper can measure even submicromolar concentrations of H_2O_2 in oxidative stress.

Various NPs such as gold half-shell multifunctional NPs, PLGA NPs, and SLN have been utilized for the delivery of drug at the inflammation site in inflammatory diseases. Nanoparticles such as AuNPs, PLGA, infliximab and MTX, actarit-loaded SLN, gold half-shell NPs have been used for the treatment of inflammatory diseases. PLGA-based NPs have been used for delivery of drugs to the colon in inflammatory bowel diseases. Poloxamer encapsulated in PLGA can be utilized to cure sodium arachidonate-induced inflammation in the rabbit. Curcumin encapsulated in biodegradable microspheres, liposomes, hydrogels, biodegradable NPs, and cyclodextrin can be used for delivery of drug at inflammation site for the treatment of RA.

In the past decade, scientific research has been focused on the determination of NPs-mediated oxidative stress at the cellular, biochemical, and molecular level. Roles of several new techniques such as ESR with traps such as PBN and 5-diethoxy phosphoryl-5-methyl-1-pyrroline N-oxide (DEPMPO), fluorescence plate reader, flow cytometry, HPLC, and biosensor such as hyper have been determined for the measurement of the concentration of RS in oxidative stress. There is a need for newly developed techniques, which can measure the generation of RS or level of oxidative stress inside the cell. New biodegradable NPs are required for the biomedical applications and therapeutic approaches. Advanced technology has to be utilized for the encapsulation of drugs in NPs so that they can release the drugs at the infection site.

REFERENCES

Alam, S., Anand, C., Ariga, K., Mori, T., Vinu, A., 2009. Unusual magnetic properties of size-controlled iron oxide nanoparticles grown in a nanoporous matrix with tunable pores. Angewandte Chemie International Edition 48, 7358–7361.

Ameziane-El-Hassani, R., Talbot, M., Dos Santos, M.C.d.S., Al Ghuzlan, A., Hartl, D., Bidart, J.-M., De Deken, X., Miot, F., Diallo, I., de Vathaire, F., 2015. NADPH oxidase DUOX1 promotes long-term persistence of oxidative stress after an exposure to irradiation. Proceedings of the National Academy of Sciences 112, 5051–5056.

Anand, P., Sundaram, C., Jhurani, S., Kunnumakkara, A.B., Aggarwal, B.B., 2008a. Curcumin and cancer: an "old-age" disease with an "age-old" solution. Cancer Letters 267, 133–164.

Anand, P., Thomas, S.G., Kunnumakkara, A.B., Sundaram, C., Harikumar, K.B., Sung, B., Tharakan, S.T., Misra, K., Priyadarsini, I.K., Rajasekharan, K.N., 2008b. Biological activities of curcumin and its analogues (Congeners) made by man and Mother Nature. Biochemical Pharmacology 76, 1590–1611.

Araújo, J., Vega, E., Lopes, C., Egea, M., Garcia, M., Souto, E., 2009. Effect of polymer viscosity on physicochemical properties and ocular tolerance of FB-loaded PLGA nanospheres. Colloids and Surfaces B: Biointerfaces 72, 48–56.

Arvizo, R.R., Bhattacharyya, S., Kudgus, R.A., Giri, K., Bhattacharya, R., Mukherjee, P., 2012. Intrinsic therapeutic applications of noble metal nanoparticles: past, present and future. Chemical Society Reviews 41, 2943–2970.

Bancos, S., Stevens, D.L., Tyner, K.M., 2015. Effect of silica and gold nanoparticles on macrophage proliferation, activation markers, cytokine production, and phagocytosis in vitro. International Journal of Nanomedicine 10, 183.

Barrera, P., Blom, A., Van Lent, P.L., Van Bloois, L., Beijnen, J.H., Van Rooijen, N., De Waal Malefijt, M.C., Van De Putte, L., Storm, G., Van Den Berg, W.B., 2000. Synovial macrophage depletion with clodronate-containing liposomes in rheumatoid arthritis. Arthritis & Rheumatology 43, 1951–1959.

Belousov, V.V., Fradkov, A.F., Lukyanov, K.A., Staroverov, D.B., Shakhbazov, K.S., Terskikh, A.V., Lukyanov, S., 2006. Genetically encoded fluorescent indicator for intracellular hydrogen peroxide. Nature Methods 3, 281.

Berliner, L.J., Khramtsov, V., FLUII, H., Clanton, T.L., 2001. Unique in vivo applications of spin traps. In: Bio-Assays for Oxidative Stress Status. Elsevier, pp. 262–272.

Bhabra, G., Sood, A., Fisher, B., Cartwright, L., Saunders, M., Evans, W.H., Surprenant, A., Lopez-Castejon, G., Mann, S., Davis, S.A., 2009. Nanoparticles can cause DNA damage across a cellular barrier. Nature Nanotechnology 4, 876.

Bhattacharya, R., Mukherjee, P., Xiong, Z., Atala, A., Soker, S., Mukhopadhyay, D., 2004. Gold nanoparticles inhibit VEGF165-induced proliferation of HUVEC cells. Nano Letters 4, 2479–2481.

Bonner, J.C., 2007. Lung fibrotic responses to particle exposure. Toxicologic Pathology 35, 148–153.

Boyd, J.H., Kan, B., Roberts, H., Wang, Y., Walley, K.R., 2008. S100A8 and S100A9 mediate endotoxin-induced cardiomyocyte dysfunction via the receptor for advanced glycation end products. Circulation Research 102, 1239–1246.

Bulcke, F., Thiel, K., Dringen, R., 2014. Uptake and toxicity of copper oxide nanoparticles in cultured primary brain astrocytes. Nanotoxicology 8, 775–785.

Cartiera, M.S., Ferreira, E.C., Caputo, C., Egan, M.E., Caplan, M.J., Saltzman, W.M., 2009. Partial correction of cystic fibrosis defects with PLGA nanoparticles encapsulating curcumin. Molecular Pharmaceutics 7, 86–93.

Chattopadhyay, S., Dash, S.K., Tripathy, S., Das, B., Kar Mahapatra, S., Pramanik, P., Roy, S., 2015. Cobalt oxide nanoparticles induced oxidative stress linked to activation of TNF-α/caspase-8/p38-MAPK signaling in human leukemia cells. Journal of Applied Toxicology 35, 603–613.

Chellat, F., Merhi, Y., Moreau, A., Yahia, L.H., 2005. Therapeutic potential of nanoparticulate systems for macrophage targeting. Biomaterials 26, 7260–7275.

Chen, B.P., Li, M., Asaithamby, A., 2012. New insights into the roles of ATM and DNA-PKcs in the cellular response to oxidative stress. Cancer Letters 327, 103–110.

Choi, H.S., Ipe, B.I., Misra, P., Lee, J.H., Bawendi, M.G., Frangioni, J.V., 2009. Tissue-and organ-selective biodistribution of NIR fluorescent quantum dots. Nano Letters 9, 2354–2359.

Croce, K., Gao, H., Wang, Y., Mooroka, T., Sakuma, M., Shi, C., Sukhova, G.K., Packard, R.R., Hogg, N., Libby, P., 2009. Myeloid-related protein-8/14 is critical for the biological response to vascular injury. Circulation 120, 427–436.

Cui, J., Yu, B., Zhao, Y., Zhu, W., Li, H., Lou, H., Zhai, G., 2009. Enhancement of oral absorption of curcumin by self-microemulsifying drug delivery systems. International Journal of Pharmaceutics 371, 148–155.

Dalal, V., Sharma, N.K., Biswas, S., 2017. Oxidative stress: diagnostic methods and application in medical science. In: Oxidative Stress: Diagnostic Methods and Applications in Medical Science. Springer, pp. 23–45.

Deng, Z.b., Liu, Y., Liu, C., Xiang, X., Wang, J., Cheng, Z., Shah, S.V., Zhang, S., Zhang, L., Zhuang, X., 2009. Immature myeloid cells induced by a high-fat diet contribute to liver inflammation. Hepatology 50, 1412–1420.

Deng, Z.J., Liang, M., Monteiro, M., Toth, I., Minchin, R.F., 2011. Nanoparticle-induced unfolding of fibrinogen promotes Mac-1 receptor activation and inflammation. Nature Nanotechnology 6, 39.

Dhillon, N., Wolff, R., Abbruzzese, J., Hong, D., Camacho, L., Li, L., Braiteh, F., Kurzrock, R., 2006. Phase II clinical trial of curcumin in patients with advanced pancreatic cancer. Journal of Clinical Oncology 24, 14151-14151.

Dillen, K., Vandervoort, J., Van den Mooter, G., Ludwig, A., 2006. Evaluation of ciprofloxacin-loaded Eudragit® RS100 or RL100/PLGA nanoparticles. International Journal of Pharmaceutics 314, 72–82.

Dixon, R., Diehl, R., Opas, E., Rands, E., Vickers, P., Evans, J., Gillard, J., Miller, D., 1990. Requirement of a 5-lipoxygenase-activating protein for leukotriene synthesis. Nature 343, 282.

Doane, T.L., Burda, C., 2012. The unique role of nanoparticles in nanomedicine: imaging, drug delivery and therapy. Chemical Society Reviews 41, 2885–2911.

Elbashir, S.M., Harborth, J., Lendeckel, W., Yalcin, A., Weber, K., Tuschl, T., 2001. Duplexes of 21-nucleotide RNAs mediate RNA interference in cultured mammalian cells. Nature 411, 494.

Faulkner, K., Fridovich, I., 1993. Luminol and lucigenin as detectors for $O_2^{\bullet-}$. Free Radical Biology and Medicine 15, 447–451.

Firestein, G.S., 2003. Evolving concepts of rheumatoid arthritis. Nature 423, 356.

Fujisawa, H., Nishimura, T., Inoue, Y., Ogaya, S., Shibata, Y., Nakagawa, Y., Sato, S., Kimura, K., 1990. Antiinflammatory properties of the new antirheumatic agent 4-acetylaminophenylacetic acid. Arzneimittel Forschung 40, 693–697.

Fujisawa, H., Nishimura, T., Motonaga, A., Inoue, Y., Inoue, K., Suzuka, H., Yoshifusa, H., Kimura, K., Muramatsu, M., 1994. Effect of actarit on type II collagen-induced arthritis in mice. Arzneimittel Forschung 44, 64–68.

Gaharwar, A.K., Peppas, N.A., Khademhosseini, A., 2014. Nanocomposite hydrogels for biomedical applications. Biotechnology and Bioengineering 111, 441–453.

Gambino, V., De Michele, G., Venezia, O., Migliaccio, P., Dall'Olio, V., Bernard, L., Minardi, S.P., Fazia, M.A.D., Bartoli, D., Servillo, G., 2013. Oxidative stress activates a specific p53 transcriptional response that regulates cellular senescence and aging. Aging Cell 12, 435–445.

Garcia-Alloza, M., Dodwell, S.A., Meyer-Luehmann, M., Hyman, B.T., Bacskai, B.J., 2006. Plaque-derived oxidative stress mediates distorted neurite trajectories in the Alzheimer mouse model. Journal of Neuropathology & Experimental Neurology 65, 1082–1089.

Garinot, M., Fiévez, V., Pourcelle, V., Stoffelbach, F., des Rieux, A., Plapied, L., Theate, I., Freichels, H., Jérôme, C., Marchand-Brynaert, J., 2007. PEGylated PLGA-based nanoparticles targeting M cells for oral vaccination. Journal of Controlled Release 120, 195–204.

Ghosh, P., Han, G., De, M., Kim, C.K., Rotello, V.M., 2008. Gold nanoparticles in delivery applications. Advanced Drug Delivery Reviews 60, 1307–1315.

Giljohann, D.A., Seferos, D.S., Daniel, W.L., Massich, M.D., Patel, P.C., Mirkin, C.A., 2010. Gold nanoparticles for biology and medicine. Angewandte Chemie International Edition 49, 3280–3294.

Giovanni, M., Yue, J., Zhang, L., Xie, J., Ong, C.N., Leong, D.T., 2015. Pro-inflammatory responses of RAW264. 7 macrophages when treated with ultralow concentrations of silver, titanium dioxide, and zinc oxide nanoparticles. Journal of Hazardous Materials 297, 146–152.

Golbamaki, N., Rasulev, B., Cassano, A., Robinson, R.L.M., Benfenati, E., Leszczynski, J., Cronin, M.T., 2015. Genotoxicity of metal oxide nanomaterials: review of recent data and discussion of possible mechanisms. Nanoscale 7, 2154–2198.

Goncalves, D., Chiasson, S., Girard, D., 2010. Activation of human neutrophils by titanium dioxide (TiO2) nanoparticles. Toxicology in Vitro 24, 1002–1008.

Gref, R., Minamitake, Y., Peracchia, M.T., Trubetskoy, V., Torchilin, V., Langer, R., 1994. Biodegradable long-circulating polymeric nanospheres. Science 263, 1600–1603.

Guo, C., Xia, Y., Niu, P., Jiang, L., Duan, J., Yu, Y., Zhou, X., Li, Y., Sun, Z., 2015. Silica nanoparticles induce oxidative stress, inflammation, and endothelial dysfunction in vitro via activation of the MAPK/Nrf2 pathway and nuclear factor-κB signaling. International Journal of Nanomedicine 10, 1463.

Haase, A., Arlinghaus, H.F., Tentschert, J., Jungnickel, H., Graf, P., Mantion, A., Draude, F., Galla, S., Plendl, J., Goetz, M.E., 2011. Application of laser postionization secondary neutral mass spectrometry/time-of-flight secondary ion mass spectrometry in nanotoxicology: visualization of nanosilver in human macrophages and cellular responses. ACS Nano 5, 3059–3068.

Halliwell, B., Gutteridge, J.M., 2015. Free Radicals in Biology and Medicine. Oxford University Press, USA.

Halliwell, B., 1978. Biochemical mechanisms accounting for the toxic action of oxygen on living organisms: the key role of superoxide dismutase. Cell Biology International Reports 2, 113–128.

Halliwell, B., 2003. Oxidative stress in cell culture: an under-appreciated problem? FEBS Letters 540, 3–6.

Haywood, R.M., Wardman, P., Gault, D.T., Linge, C., 1999. Ruby laser irradiation (694 nm) of human skin biopsies: assessment by electron spin resonance spectroscopy of free radical production and oxidative stress during laser depilation. Photochemistry and Photobiology 70, 348–352.

He, Z., Liu, J., Du, L., 2014. The unexpected effect of PEGylated gold nanoparticles on the primary function of erythrocytes. Nanoscale 6, 9017–9024.

Headlam, H.A., Davies, M.J., 2003. Cell-mediated reduction of protein and peptide hydroperoxides to reactive free radicals. Free Radical Biology and Medicine 34, 44–55.

Higaki, M., Ishihara, T., Izumo, N., Takatsu, M., Mizushima, Y., 2005. Treatment of experimental arthritis with poly (D, L-lactic/glycolic acid) nanoparticles encapsulating betamethasone sodium phosphate. Annals of the Rheumatic Diseases 64, 1132–1136.

Horisawa, E., Kubota, K., Tuboi, I., Sato, K., Yamamoto, H., Takeuchi, H., Kawashima, Y., 2002a. Size-dependency of DL-lactide/glycolide copolymer particulates for intra-articular delivery system on phagocytosis in rat synovium. Pharmaceutical Research 19, 132–139.

Horisawa, E., Hirota, T., Kawazoe, S., Yamada, J., Yamamoto, H., Takeuchi, H., Kawashima, Y., 2002b. Prolonged anti-inflammatory action of DL-lactide/glycolide copolymer nanospheres containing betamethasone sodium phosphate for an intra-articular delivery system in antigen-induced arthritic rabbit. Pharmaceutical Research 19, 403–410.

Howard, K.A., Paludan, S.R., Behlke, M.A., Besenbacher, F., Deleuran, B., Kjems, J., 2009. Chitosan/siRNA nanoparticle–mediated TNF-α knockdown in peritoneal macrophages for anti-inflammatory treatment in a murine arthritis model. Molecular Therapy 17, 162–168.

Ikoba, U., Peng, H., Li, H., Miller, C., Yu, C., Wang, Q., 2015. Nanocarriers in therapy of infectious and inflammatory diseases. Nanoscale 7, 4291–4305.

Indo, H.P., Davidson, M., Yen, H.-C., Suenaga, S., Tomita, K., Nishii, T., Higuchi, M., Koga, Y., Ozawa, T., Majima, H.J., 2007. Evidence of ROS generation by mitochondria in cells with impaired electron transport chain and mitochondrial DNA damage. Mitochondrion 7, 106–118.

Ischiropoulos, H., Gow, A., Thom, S.R., Kooy, N.W., Royall, J.A., Crow, J.P., 1999. [38] Detection of reactive nitrogen species using 2, 7-dichlorodihydrofluorescein and dihydrorhodamine 123. Methods in Enzymology 367−373. Elsevier.

Iwamoto, S., Iwai, S.-i., Tsujiyama, K., Kurahashi, C., Takeshita, K., Naoe, M., Masunaga, A., Ogawa, Y., Oguchi, K., Miyazaki, A., 2007. TNF-α drives human CD14+ monocytes to differentiate into CD70+ dendritic cells evoking Th1 and Th17 responses. The Journal of Immunology 179, 1449−1457.

Jongstra-Bilen, J., Haidari, M., Zhu, S.-N., Chen, M., Guha, D., Cybulsky, M.I., 2006. Low-grade chronic inflammation in regions of the normal mouse arterial intima predisposed to atherosclerosis. Journal of Experimental Medicine 203, 2073−2083.

Joshi, Y.B., Praticò, D., 2015. The 5-lipoxygenase pathway: oxidative and inflammatory contributions to the Alzheimer's disease phenotype. Frontiers in Cellular Neuroscience 8, 436.

Kawai, K., Li, Y.-S., Kawasaki, Y., Morimoto, Y., 2014. Oxidative stress in rat lung after exposure to titanium dioxide and nickel oxide nanoparticles. European Respiratory Journal 44, P4785.

Kawanishi, S., Oikawa, S., 2004. Mechanism of telomere shortening by oxidative stress. Annals of the New York Academy of Sciences 1019, 278−284.

Keenan, C.R., Goth-Goldstein, R., Lucas, D., Sedlak, D.L., 2009. Oxidative stress induced by zero-valent iron nanoparticles and Fe (II) in human bronchial epithelial cells. Environmental Science and Technology 43, 4555−4560.

Kensler, T.W., Wakabayashi, N., Biswal, S., 2007. Cell survival responses to environmental stresses via the Keap1-Nrf2-ARE pathway. Annual Review of Pharmacology and Toxicology 47, 89−116.

Kepp, O., Galluzzi, L., Kroemer, G., 2011. Mitochondrial control of the NLRP3 inflammasome. Nature Immunology 12, 199.

Khoury, M., Louis-Plence, P., Escriou, V., Noel, D., Largeau, C., Cantos, C., Scherman, D., Jorgensen, C., Apparailly, F., 2006. Efficient new cationic liposome formulation for systemic delivery of small interfering RNA silencing tumor necrosis factor α in experimental arthritis. Arthritis & Rheumatology 54, 1867−1877.

Kojda, G., Harrison, D., 1999. Interactions between NO and reactive oxygen species: pathophysiological importance in atherosclerosis, hypertension, diabetes and heart failure. Cardiovascular Research 43, 652−671.

Kovac, S., Angelova, P.R., Holmström, K.M., Zhang, Y., Dinkova-Kostova, A.T., Abramov, A.Y., 2015. Nrf2 regulates ROS production by mitochondria and NADPH oxidase. Biochimica et Biophysica Acta (BBA) − General Subjects 1850, 794−801.

Kovvuru, P., Mancilla, P.E., Shirode, A.B., Murray, T.M., Begley, T.J., Reliene, R., 2015. Oral ingestion of silver nanoparticles induces genomic instability and DNA damage in multiple tissues. Nanotoxicology 9, 162−171.

Kumari, A., Yadav, S.K., Yadav, S.C., 2010. Biodegradable polymeric nanoparticles based drug delivery systems. Colloids and Surfaces B: Biointerfaces 75, 1−18.

Kurzrock, R., Li, L., 2005. Liposome-encapsulated curcumin: in vitro and in vivo effects on proliferation, apoptosis, signaling, and angiogenesis. Journal of Clinical Oncology 23, 4091-4091.

Kusunoki, N., Yamazaki, R., Kawai, S., 2008. Pro-apoptotic effect of nonsteroidal anti-inflammatory drugs on synovial fibroblasts. Modern Rheumatology 18, 542−551.

Lamprecht, A., Schäfer, U., Lehr, C.-M., 2001a. Size-dependent bioadhesion of micro-and nanoparticulate carriers to the inflamed colonic mucosa. Pharmaceutical Research 18, 788−793.

Lamprecht, A., Ubrich, N., Yamamoto, H., Schäfer, U., Takeuchi, H., Maincent, P., Kawashima, Y., Lehr, C.-M., 2001b. Biodegradable nanoparticles for targeted drug delivery in treatment of inflammatory bowel disease. Journal of Pharmacology and Experimental Therapeutics 299, 775−781.

Lamprecht, A., Yamamoto, H., Takeuchi, H., Kawashima, Y., 2005. Nanoparticles enhance therapeutic efficiency by selectively increased local drug dose in experimental colitis in rats. Journal of Pharmacology and Experimental Therapeutics 315, 196−202.

Landry, W.D., Cotter, T.G., 2014. ROS Signalling, NADPH Oxidases and Cancer. Portland Press Limited.

Lavieri, R., Piccioli, P., Carta, S., Delfino, L., Castellani, P., Rubartelli, A., 2014. TLR costimulation causes oxidative stress with unbalance of proinflammatory and anti-inflammatory cytokine production. The Journal of Immunology 192, 5373−5381.

Lee, S.J.-A., Kavanaugh, A., 2003. Pharmacological treatment of established rheumatoid arthritis. Best Practice & Research Clinical Rheumatology 17, 811−829.

Lee, S.-M., Kim, H.J., Ha, Y.-J., Park, Y.N., Lee, S.-K., Park, Y.-B., Yoo, K.-H., 2012. Targeted chemo-photothermal treatments of rheumatoid arthritis using gold half-shell multifunctional nanoparticles. ACS Nano 7, 50−57.

Lee, M.-Y., Yang, J.-A., Jung, H.S., Beack, S., Choi, J.E., Hur, W., Koo, H., Kim, K., Yoon, S.K., Hahn, S.K., 2012. Hyaluronic acid−gold nanoparticle/interferon α complex for targeted treatment of hepatitis C virus infection. ACS Nano 6, 9522−9531.

Lee, H., Lee, M.-Y., Bhang, S.H., Kim, B.-S., Kim, Y.S., Ju, J.H., Kim, K.S., Hahn, S.K., 2014. Hyaluronate−gold nanoparticle/tocilizumab complex for the treatment of rheumatoid arthritis. ACS Nano 8, 4790−4798.

Lewis, D.F.V., 2002. Oxidative stress: the role of cytochromes P450 in oxygen activation. Journal of Chemical Technology and Biotechnology 77, 1095−1100.

Li, J.J., Zou, L., Hartono, D., Ong, C.N., Bay, B.H., Lanry Yung, L.Y., 2008. Gold nanoparticles induce oxidative damage in lung fibroblasts in vitro. Advanced Materials 20, 138−142.

Li, J.J., Hartono, D., Ong, C.-N., Bay, B.-H., Yung, L.-Y.L., 2010. Autophagy and oxidative stress associated with gold nanoparticles. Biomaterials 31, 5996−6003.

Li, J.J., Lo, S.-L., Ng, C.-T., Gurung, R.L., Hartono, D., Hande, M.P., Ong, C.-N., Bay, B.-H., Yung, L.-Y.L., 2011. Genomic instability of gold nanoparticle treated human lung fibroblast cells. Biomaterials 32, 5515–5523.

Libermann, T.A., Baltimore, D., 1990. Activation of interleukin-6 gene expression through the NF-kappa B transcription factor. Molecular and Cellular Biology 10, 2327–2334.

Liu, X., Sun, J., 2010. Endothelial cells dysfunction induced by silica nanoparticles through oxidative stress via JNK/P53 and NF-κB pathways. Biomaterials 31, 8198–8209.

Liu, L., Leech, J.A., Urch, R.B., Silverman, F.S., 1997. In vivo salicylate hydroxylation: a potential biomarker for assessing acute ozone exposure and effects in humans. American Journal of Respiratory and Critical Care Medicine 156, 1405–1412.

Liu, Y., Fiskum, G., Schubert, D., 2002. Generation of reactive oxygen species by the mitochondrial electron transport chain. Journal of Neurochemistry 80, 780–787.

Ma, X., Wu, Y., Jin, S., Tian, Y., Zhang, X., Zhao, Y., Yu, L., Liang, X.-J., 2011. Gold nanoparticles induce autophagosome accumulation through size-dependent nanoparticle uptake and lysosome impairment. ACS Nano 5, 8629–8639.

Mahmoudi, M., Azadmanesh, K., Shokrgozar, M.A., Journeay, W.S., Laurent, S., 2011a. Effect of nanoparticles on the cell life cycle. Chemical Reviews 111, 3407–3432.

Mahmoudi, M., Shokrgozar, M.A., Sardari, S., Moghadam, M.K., Vali, H., Laurent, S., Stroeve, P., 2011b. Irreversible changes in protein conformation due to interaction with superparamagnetic iron oxide nanoparticles. Nanoscale 3, 1127–1138.

Makino, H., Yoshinaga, Y., Yamasaki, Y., Morita, Y., Hashimoto, H., Yamamura, M., 2002. Renal involvement in rheumatoid arthritis: analysis of renal biopsy specimens from 100 patients. Modern Rheumatology 12, 148–154.

Martindale, J.L., Holbrook, N.J., 2002. Cellular response to oxidative stress: signaling for suicide and survival. Journal of Cellular Physiology 192, 1–15.

Matsubara, T., 1999. The basic and clinical use of DMARDs. Rheumatology 22, 81–97.

Meissner, Y., Lamprecht, A., 2008. Alternative drug delivery approaches for the therapy of inflammatory bowel disease. Journal of Pharmaceutical Sciences 97, 2878–2891.

Messner, K.R., Imlay, J.A., 2002. Mechanism of superoxide and hydrogen peroxide formation by fumarate reductase, succinate dehydrogenase, and aspartate oxidase. Journal of Biological Chemistry 277, 42563–42571.

Miyazawa, T., Yasuda, K., Fujimoto, K., 1987. Chemiluminescence-high performance liquid chromatography of phosphatidylcholine hydroperoxide. Analytical Letters 20, 915–925.

Moghimi, S.M., Hunter, A.C., Murray, J.C., 2001. Long-circulating and target-specific nanoparticles: theory to practice. Pharmacological Reviews 53, 283–318.

Moghimi, S.M., Hunter, A.C., Murray, J.C., 2005. Nanomedicine: current status and future prospects. The FASEB Journal 19, 311–330.

Mulik, R., Mahadik, K., Paradkar, A., 2009. Development of curcuminoids loaded poly (butyl) cyanoacrylate nanoparticles: physicochemical characterization and stability study. European Journal of Pharmaceutical Sciences 37, 395–404.

Murray, A.R., Kisin, E., Inman, A., Young, S.-H., Muhammed, M., Burks, T., Uheida, A., Tkach, A., Waltz, M., Castranova, V., 2013. Oxidative stress and dermal toxicity of iron oxide nanoparticles in vitro. Cell Biochemistry and Biophysics 67, 461–476.

Narayanan, N.K., Nargi, D., Randolph, C., Narayanan, B.A., 2009. Liposome encapsulation of curcumin and resveratrol in combination reduces prostate cancer incidence in PTEN knockout mice. International Journal of Cancer 125, 1–8.

Nel, A., Xia, T., Mädler, L., Li, N., 2006. Toxic potential of materials at the nanolevel. Science 311, 622–627.

Ng, K.W., Khoo, S.P., Heng, B.C., Setyawati, M.I., Tan, E.C., Zhao, X., Xiong, S., Fang, W., Leong, D.T., Loo, J.S., 2011. The role of the tumor suppressor p53 pathway in the cellular DNA damage response to zinc oxide nanoparticles. Biomaterials 32, 8218–8225.

Ng, C.-T., Li, J.J.E., Gurung, R.L., Hande, M.P., Ong, C.-N., Bay, B.-H., Yung, L.-Y.L., 2013. Toxicological profile of small airway epithelial cells exposed to gold nanoparticles. Experimental Biology and Medicine 238, 1355–1361.

Niethammer, P., Grabher, C., Look, A.T., Mitchison, T.J., 2009. A tissue-scale gradient of hydrogen peroxide mediates rapid wound detection in zebrafish. Nature 459, 996.

Nordgren, M., Fransen, M., 2014. Peroxisomal metabolism and oxidative stress. Biochimie 98, 56–62.

Owen, R., Giacosa, A., Hull, W., Haubner, R., Spiegelhalder, B., Bartsch, H., 2000. The antioxidant/anticancer potential of phenolic compounds isolated from olive oil. European Journal of Cancer 36, 1235–1247.

Park, J.S., Yang, H.N., Jeon, S.Y., Woo, D.G., Kim, M.S., Park, K.-H., 2012. The use of anti-COX2 siRNA coated onto PLGA nanoparticles loading dexamethasone in the treatment of rheumatoid arthritis. Biomaterials 33, 8600–8612.

Prior, R.L., Cao, G., 2001. In vivo total antioxidant capacity: comparison of different analytical methods. In: Bio-Assays for Oxidative Stress Status. Elsevier, pp. 39–47.

Pujalté, I., Passagne, I., Daculsi, R., de Portal, C., Ohayon-Courtès, C., L'Azou, B., 2015. Cytotoxic effects and cellular oxidative mechanisms of metallic nanoparticles on renal tubular cells: impact of particle solubility. Toxicology Research 4, 409–422.

Qu, P., Du, H., Li, Y., Yan, C., 2009. Myeloid-specific expression of Api6/AIM/Spα induces systemic inflammation and adenocarcinoma in the lung. The Journal of Immunology 182, 1648–1659.

Ramesh, G.V., Kodiyath, R., Tanabe, T., Manikandan, M., Fujita, T., Matsumoto, F., Ishihara, S., Ueda, S., Yamashita, Y., Ariga, K., 2014. NbPt3 intermetallic nanoparticles: highly stable and CO-tolerant electrocatalyst for fuel oxidation. ChemElectroChem 1, 728–732.

Ravindran, J., Prasad, S., Aggarwal, B.B., 2009. Curcumin and cancer cells: how many ways can curry kill tumor cells selectively? The AAPS Journal 11, 495—510.

Rice-Evans, C.A., Gopinathan, V., 1995. Oxygen toxicity, free radicals and antioxidants in human disease: biochemical implications in atherosclerosis and the problems of premature neonates. Essays in Biochemistry 29, 39.

Rinna, A., Magdolenova, Z., Hudecova, A., Kruszewski, M., Refsnes, M., Dusinska, M., 2014. Effect of silver nanoparticles on mitogen-activated protein kinases activation: role of reactive oxygen species and implication in DNA damage. Mutagenesis 30, 59—66.

Ritov, V.B., Banni, S., Yalowich, J.C., Day, B.W., Claycamp, H.G., Corongiu, F.P., Kagan, V.E., 1996. Nonrandom peroxidation of different classes of membrane phospholipids in live cells detected by metabolically integrated cis-parinaric acid. Biochimica et Biophysica Acta (BBA) — Biomembranes 1283, 127—140.

Robertson, G., Leclercq, I., Farrell, G.C., 2001. II. Cytochrome P-450 enzymes and oxidative stress. American Journal of Physiology — Gastrointestinal and Liver Physiology 281, G1135—G1139.

Saag, K.G., Teng, G.G., Patkar, N.M., Anuntiyo, J., Finney, C., Curtis, J.R., Paulus, H.E., Mudano, A., Pisu, M., Elkins-Melton, M., 2008. American College of Rheumatology 2008 recommendations for the use of nonbiologic and biologic disease-modifying antirheumatic drugs in rheumatoid arthritis. Arthritis Care & Research 59, 762—784.

Sasidharan, A., Monteiro-Riviere, N.A., 2015. Biomedical applications of gold nanomaterials: opportunities and challenges. Wiley Interdisciplinary Reviews: Nanomedicine and Nanobiotechnology 7, 779—796.

Schalnus, R., 2003. Topical nonsteroidal anti-inflammatory therapy in ophthalmology. Ophthalmologica 217, 89—98.

Schiffelers, R.M., Xu, J., Storm, G., Woodle, M.C., Scaria, P.V., 2005. Effects of treatment with small interfering RNA on joint inflammation in mice with collagen-induced arthritis. Arthritis & Rheumatology 52, 1314—1318.

Schmidt, C., Collnot, E.M., Bojarski, C., Schumann, M., Schulzke, J.D., Lehr, C.M., Stallmach, A., 2010. W1266 confocal laser endomicroscopy (CLE) reveals mucosal accumulation of PLGA-nanoparticles in ulcerous lesions of patients with inflammatory bowel diseases. Gastroenterology 138, S—687.

Schnabel, A., Gross, W.L., 1994. Low-dose methotrexate in rheumatic diseases—efficacy, side effects, and risk factors for side effects. Seminars in Arthritis and Rheumatism 310—327.

Setyawati, M.I., Tay, C.Y., Leong, D.T., 2013. Effect of zinc oxide nanomaterials-induced oxidative stress on the p53 pathway. Biomaterials 34, 10133—10142.

Setyawati, M.I., Yuan, X., Xie, J., Leong, D.T., 2014. The influence of lysosomal stability of silver nanomaterials on their toxicity to human cells. Biomaterials 35, 6707—6715.

Sharma, V., Anderson, D., Dhawan, A., 2012a. Zinc oxide nanoparticles induce oxidative DNA damage and ROS-triggered mitochondria mediated apoptosis in human liver cells (HepG2). Apoptosis 17, 852—870.

Sharma, V., Singh, P., Pandey, A.K., Dhawan, A., 2012b. Induction of oxidative stress, DNA damage and apoptosis in mouse liver after sub-acute oral exposure to zinc oxide nanoparticles. Mutation Research: Genetic Toxicology and Environmental Mutagenesis 745, 84—91.

Shi, J., Sun, X., Lin, Y., Zou, X., Li, Z., Liao, Y., Du, M., Zhang, H., 2014. Endothelial cell injury and dysfunction induced by silver nanoparticles through oxidative stress via IKK/NF-κB pathways. Biomaterials 35, 6657—6666.

Soehnlein, O., Weber, C., 2009. Myeloid cells in atherosclerosis: initiators and decision shapers. Seminars in Immunopathology 35—47.

Sorbara, M.T., Girardin, S.E., 2011. Mitochondrial ROS fuel the inflammasome. Cell Research 21, 558.

Spigoni, V., Cito, M., Alinovi, R., Pinelli, S., Passeri, G., Zavaroni, I., Goldoni, M., Campanini, M., Aliatis, I., Mutti, A., 2015. Effects of TiO_2 and Co_3O_4 nanoparticles on circulating angiogenic cells. PLoS One 10 e0119310.

Stern, S.T., Adiseshaiah, P.P., Crist, R.M., 2012. Autophagy and lysosomal dysfunction as emerging mechanisms of nanomaterial toxicity. Particle and Fibre Toxicology 9, 20.

St-Pierre, J., Buckingham, J.A., Roebuck, S.J., Brand, M.D., 2002. Topology of superoxide production from different sites in the mitochondrial electron transport chain. Journal of Biological Chemistry 277, 44784—44790.

Sugihara, K., Morino, A., Nomura, A., Iida, S., Sugiyama, M., 1990. Pharmacokinetics of 4-acetylaminophenylacetic acid. 1st communication: absorption, distribution, metabolism and excretion in mice, rats, dogs and monkeys after single administration of 14C-labeled compound. Arzneimittel Forschung 40, 800—805.

Sul, O.-J., Kim, J.-C., Kyung, T.-W., Kim, H.-J., Kim, Y.-Y., Kim, S.-H., Kim, J.-S., Choi, H.-S., 2010. Gold nanoparticles inhibited the receptor activator of nuclear factor-κb ligand (RANKL)-induced osteoclast formation by acting as an antioxidant. Bioscience Biotechnology and Biochemistry 74, 2209—2213.

Sun, D., Zhuang, X., Xiang, X., Liu, Y., Zhang, S., Liu, C., Barnes, S., Grizzle, W., Miller, D., Zhang, H.-G., 2010. A novel nanoparticle drug delivery system: the anti-inflammatory activity of curcumin is enhanced when encapsulated in exosomes. Molecular Therapy 18, 1606—1614.

Takahashi, M., Shibata, M., Niki, E., 2001. Estimation of lipid peroxidation of live cells using a fluorescent probe, diphenyl-1-pyrenylphosphine. Free Radical Biology and Medicine 31, 164—174.

Taylor, E., Megson, I., Haslett, C., Rossi, A., 2003. Nitric oxide: a key regulator of myeloid inflammatory cell apoptosis. Cell Death and Differentiation 10, 418.

Thakor, A., Jokerst, J., Zavaleta, C., Massoud, T., Gambhir, S., 2011. Gold nanoparticles: a revival in precious metal administration to patients. Nano Letters 11, 4029—4036.

Thurn, K.T., Brown, E., Wu, A., Vogt, S., Lai, B., Maser, J., Paunesku, T., Woloschak, G.E., 2007. Nanoparticles for applications in cellular imaging. Nanoscale Research Letters 2, 430.

Tsai, C.Y., Shiau, A.L., Chen, S.Y., Chen, Y.H., Cheng, P.C., Chang, M.Y., Chen, D.H., Chou, C.H., Wang, C.R., Wu, C.L., 2007. Amelioration of collagen-induced arthritis in rats by nanogold. Arthritis & Rheumatology 56, 544–554.

Ushio-Fukai, M., 2006. Localizing NADPH oxidase–derived ROS. Science's STKE 2006 re8-re8.

Valko, M., Leibfritz, D., Moncol, J., Cronin, M.T., Mazur, M., Telser, J., 2007. Free radicals and antioxidants in normal physiological functions and human disease. The International Journal of Biochemistry & Cell Biology 39, 44–84.

Van Ede, A.E., Laan, R.F., Blom, H.J., De Abreu, R.A., van de Putte, L.B., 1998. Methotrexate in rheumatoid arthritis: an update with focus on mechanisms involved in toxicity. Seminars in Arthritis and Rheumatism 277–292.

Vega, E., Gamisans, F., Garcia, M., Chauvet, A., Lacoulonche, F., Egea, M., 2008. PLGA nanospheres for the ocular delivery of flurbiprofen: drug release and interactions. Journal of Pharmaceutical Sciences 97, 5306–5317.

Vij, N., Min, T., Marasigan, R., Belcher, C.N., Mazur, S., Ding, H., Yong, K.-T., Roy, I., 2010. Development of PEGylated PLGA nanoparticle for controlled and sustained drug delivery in cystic fibrosis. Journal of Nanobiotechnology 8, 22.

von Zglinicki, T., 2002. Oxidative stress shortens telomeres. Trends in Biochemical Sciences 27, 339–344.

Wang, B., Van Veldhoven, P.P., Brees, C., Rubio, N., Nordgren, M., Apanasets, O., Kunze, M., Baes, M., Agostinis, P., Fransen, M., 2013. Mitochondria are targets for peroxisome-derived oxidative stress in cultured mammalian cells. Free Radical Biology and Medicine 65, 882–894.

Wang, J., Deng, X., Zhang, F., Chen, D., Ding, W., 2014. ZnO nanoparticle-induced oxidative stress triggers apoptosis by activating JNK signaling pathway in cultured primary astrocytes. Nanoscale research letters 9, 117.

Wang, P., Wang, X., Wang, L., Hou, X., Liu, W., Chen, C., 2015. Interaction of gold nanoparticles with proteins and cells. Science and Technology of Advanced Materials 16, 034610.

Yang, W., Yu, M., Fu, J., Bao, W., Wang, D., Hao, L., Yao, P., Nüssler, A.K., Yan, H., Liu, L., 2014. Deoxynivalenol induced oxidative stress and genotoxity in human peripheral blood lymphocytes. Food and Chemical Toxicology 64, 383–396.

Ye, J., Wang, Q., Zhou, X., Zhang, N., 2008. Injectable actarit-loaded solid lipid nanoparticles as passive targeting therapeutic agents for rheumatoid arthritis. International Journal of Pharmaceutics 352, 273–279.

Yoshida, H., 1987. Effect of MS-932 (4-acetylaminophenylacetic acid) on articular lesions in MRL/1 mice. International Journal of Immunotherapy 261–264.

Zahlten, J., Kim, Y.-J., Doehn, J.-M., Pribyl, T., Hocke, A.C., García, P., Hammerschmidt, S., Suttorp, N., Hippenstiel, S., Hübner, R.-H., 2014. Streptococcus pneumoniae–induced oxidative stress in lung epithelial cells depends on pneumococcal autolysis and is reversible by resveratrol. The Journal of Infectious Diseases 211, 1822–1830.

Zhan, Y., Zeng, W., Jiang, G., Wang, Q., Shi, X., Zhou, Z., Deng, H., Du, Y., 2015. Construction of lysozyme exfoliated rectorite-based electrospun nanofibrous membranes for bacterial inhibition. Journal of Applied Polymer Science 132.

Zhang, W., Prausnitz, M.R., Edwards, A., 2004. Model of transient drug diffusion across cornea. Journal of Controlled Release 99, 241–258.

CHAPTER 8

Biomedical Applications of Nanoparticles

SUBHASHINI BHARATHALA • PANKAJ SHARMA, PHD

INTRODUCTION

Nanomaterials have been used since the advent of human civilization. Simply put, nanotechnology encompasses creation, design, and study of functional materials and devices with a size range of 1–100 nm (nanoscale). Because of their small size and high surface area, nanoparticles (NPs) offer unique properties with a wide range of applications. There is evidence that nanotechnology was used by humans as early as 500 BC in a variety of applications, including pottery coatings (Hunault et al., 2017), stained glass (Fermo and Padeletti, 2012), and medicinal formulations (Krukemeyer et al., 2015). In 1857, Michael Faraday noted changes in electronic and optical properties of materials based on their small size (Sattler, 2011). However, it is generally accepted that the seeds for modern nanotechnology and nanomedicine were sown by Dr. Richard Freynman, who delivered the famous lecture "There's Plenty of Room at the Bottom" during the Annual Meeting of the American Physical Society in 1959 (Junk and Riess, 2006). He conceptualized methods to manipulate individual atoms and molecules that would result in change of their physical properties, paving the way for molecular nanotechnology. The term "nanotechnology" was first used by Norio Taniguchi in 1974 who defined it as the "processing of, separation, consolidation, and deformation of materials by one atom or one molecule" (Taniguchi, 1974). The term "nanotechnology" was used again by Eric Drexler who was unaware of the work of Taniguchi and popularized the concept of nanoscale assemblers initially proposed by Freynman (Drexler, 1981).

The progress in the field of nanoscience was facilitated by development of powerful imaging tools that allowed the visualization and characterization of NPs. These included the development of immersion ultramicroscope by Richard Zsigmondy and Henry Siedentopf in 1912 (Mappes et al., 2012), the transmission electron microscope by Max Knoll and Ernst Ruska in 1931 (Ruska, 1987), the field electron microscope by Erwin Muller in 1936 and its improvement to the field ion microscope in 1951 (Muller et al., 1968). Furthermore, development of the scanning tunneling microscope by Gerd Binnig and Heinrich Rohrer in 1981 (Binnig and Rohrer, 1982) and the atomic force microscope in 1986 (Binnig et al., 1986) allowed scientists to characterize and manipulate nanosize particles. Presently, nanoscience and nanotechnology is truly a multidisciplinary that needs expertise from diverse domains, including, but not limited to, physics, chemistry, biology, medicine, electronics, engineering, computers, and information technology. Nanomaterials share unique properties very different from macromaterials. They offer a much a larger surface area that is excellent for dissolution and serve as catalysts. The surface area offered by them limits optical distortion, and the property of reduced electrical resistance offers excellent transport properties (Kuna et al., 2009). Nanotechnology has found diverse applications in cosmetics, building materials, foods, agriculture, and medicine.

In 1990s, the first nanotechnology-based companies started emerging and were involved in bulk use of passive nanomaterials in sunscreen lotions and other cosmetic products, packaging materials, paints and coatings, household appliances, clothing materials, and disinfectants. Rapid progress in the past two decades has seen an explosion in consumer products available in the market that contain (or at least claim to contain) one or more nanomaterials. These include home appliances, electronics and computers, telecommunication, construction materials, paints and coatings, cosmetics, food and beverages, dietary supplements, clothing materials, etc. Nanomaterials are also being used extensively in the automobile and aerospace industries (Asmatulu, 2016). In 2013, there were more than 1800 products available in the market, which contain nanomaterials

Nanotechnology in Modern Animal Biotechnology. https://doi.org/10.1016/B978-0-12-818823-1.00008-9

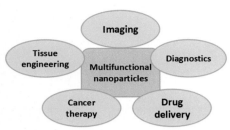

FIG. 8.1 **Biomedical Applications of Nanoparticles.**
Nanoparticles are used for various biomedical applications
and have multifunctional role in therapeutics, diagnostics,
imaging, and drug delivery. Nanomaterials are also used as
scaffolds in tissue engineering and as biomaterials for
medical applications.

(Vance et al., 2015). With rapid progress in nanoscience
and nanotechnology, these numbers should only be
taken as a rough estimate, and the actual numbers
may be significantly more.

The unique physicochemical properties attained by
nanosizing of materials attracted the attention of
biomedical scientists, and the concept of nanopharmaceuticals was developed. Nanopharmaceuticals have
been defined as "Pharmaceuticals engineered on the
nanoscale, i.e., pharmaceuticals where the nanomaterial
plays the pivotal therapeutic role or adds additional
functionality to the previous compound" (Rivera Gil
et al., 2010). The first "nanodrug," Doxil, was approved
in 1995 (Barenholz, 2012). It is a liposomal encapsulated form of doxorubicin that belongs to the anthracy-
cline group and is used for cancer chemotherapy.
Initially approved for Kaposi's sarcoma, it is also being
used for ovarian cancer and multiple myeloma. It is less
cytotoxic than the nonencapsulated form but tends to
cause hand-foot syndrome, limiting its use. NPs have
advantages over traditional therapies, as they are modified to perform altered physicochemical properties such
as solubility, size, shape, and biocompatibility, which
made them as target molecules in different fields of
biomedicine (Szefler, 2018). Nanotechnology has great
potential in almost all domains of biomedicine,
including imaging, diagnostics, novel drug formulations for targeted drug delivery, cancer treatment, and
tissue engineering (Fig. 8.1).

CLASSIFICATION OF NANOMATERIALS

Nanomaterials are defined as a "material with any
external dimension in the nanoscale or having internal
structure or surface structure in the nanoscale" with
nanoscale defined as the "length range approximately

from 1 to 100 nm" Nanostructured materials possess
eminent magnetic, optical, electrical, and mechanical
properties, compared with their bulk materials; henceforth the controlled synthesis of nanostructured materials with defined morphology has recently received
much attention. According to their size, nanomaterials
are classified into zero, single, two, and three dimensions. NPs, quantum dots, and nanocrystals were
considered as zero-dimensional structures. Nanofibers,
nanorods, nanoribbons, and nanowires are considered
as one dimensional; thin films and nanorods are two
dimensional. Three-dimensional nanostructures are
the colloids and fullerenes with complexity in their
structure (Rosenman et al., 2011).

The nanomaterials can also be distinguished into
three types such as metallic, nonmetallic, and biodegradable according to their source of origin (Fig. 8.2).
These materials are synthesized from their bulk materials, modified with various functional groups, which
allow them to be conjugated with antibodies, ligands,
and drugs of interest, thus opening a wide range of potential applications in biotechnology. Nanomaterials
are synthesized from bulk materials using either a top-
down approach (by methods such as ball milling,
plasma arching, and vapor destination) or a bottom-
up approach using sol-gel and electrodisposition
methods; only the former has been used at a commercial scale(Ghadi et al., 2014).

Metallic Nanoparticles

Metal NPs are those of metallic origin, which exhibits
strong optical absorption because of their localized surface effect called plasmon resonance, which is mostly
explored for novel applications in biomedicine. Metal
NPs such as gold, silver, and copper have a broad absorption band in the visible region of electromagnetic
spectrum, which is the basis for a wide variety of applications (Giner-Casares et al., 2016). Among the metallic
NPs, gold, silver, and iron have been extensively used
for medical applications. Iron oxide NPs are used as
MRI contrast agents, hyperthermia treatment, drug delivery, gene delivery, and bioimaging (Vallabani and
Singh, 2018). Gold (AU) NPs are used in bioimaging,
photothermal therapy, and drug delivery (Elahi et al.,
2018), whereas silver (Ag) NPs are used for drug delivery, wound dressing, antimicrobial agents, and cancer
therapy (Long et al., 2015). Zinc oxide NPs, although
used predominantly in the paint and cosmetic industries, have recently found antimicrobial and anticancer
applications because of their ability to generate reactive
oxygen species (Long et al., 2015), and NPs synthesized
using cadmium and copper have also been used for a

Classification of nanomaterials

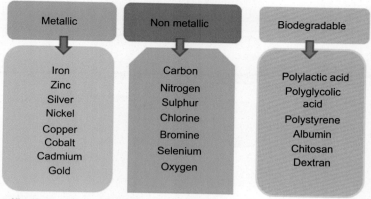

FIG. 8.2 **Classification of Nanomaterials.** Nanomaterials are classified as metallic, nonmetallic, and biodegradable, based on their source of origin and chemical structure.

variety of biomedical applications (Rubilar et al., 2013; Wang et al., 2016).

Nonmetallic/Inorganic Nanoparticles

Apart from metals, nonmetallic and inorganic NPs have also been used extensively for various biomedical applications (Conde et al., 2014) for their better imaging and drug delivery characteristics (Colombeau et al., 2016). Among the various nonmetals, carbon has been most extensively used for biomedical applications such as drug delivery, bioimaging, tissue engineering, and biosensing (Mohajeri et al., 2018). Single-walled and nonmetallic in origin multiwalled carbon nanotubes (CNTs) were used for drug delivery and tissue engineering and as biosensors (Dineshkumar et al., 2015). Fullerenes, which are spherical in shape and made up of 60 carbon atoms, were also used for drug delivery. CNTs doped with sulfur, boron, nitrogen, and fluorine can significantly alter the properties in delivery of therapeutic biomolecules and bioimaging (Lamanna et al., 2012). Mesoporous silica NPs are relatively easier to synthesize, chemically stable, and biocompatible; these properties along with their high surface area and drug carrying capacities make them attractive carriers for drug delivery (Li et al., 2017). Calcium phosphate nanomaterials are extensively used as prosthetics and scaffolds in tissue engineering applications (Li et al., 2017).

Biodegradable Nanoparticles

Biodegradable nanomaterials are widely used in the nanodrug delivery field. The polymeric materials break down to natural metabolites and are eliminated through natural pathways and offer less toxicity compared with other materials. A wide collection of natural and synthetic nanomaterials are used for drug delivery, vaccine delivery, cancer therapy, and diagnostics because of their versatility in nature and function. The eminent biocompatibility, encapsulation, and drug release profile of biodegradable NPs justified them to use in a wide variety of applications in drug targeting, medicine, and other biological purpose. In the recent years, they are widely used in medical applications because of their low toxicity, eminent biocompatibility, and eliminated by normal metabolic pathways. The materials on hydrolysis form the degradation compounds, which can be eliminated as CO_2 and water. The biodegradable NPs are classified into natural and synthetic in nature based on their source of origin, which have a variety of applications (Kumari et al., 2010).

Biodegradable polymers can be mainly classified as agropolymers, polymers synthesized by chemical methods and also by microbial production (Fig. 8.3). Natural polymers are formed naturally or synthesized during the growth cycles of organisms. Among the biodegradable polymers, synthetic polymers constitute the largest group, which can be produced under controlled conditions, which were extensively used as scaffolds for tissue engineering and carriers for drug delivery. Synthetic polymers have a wide range of applications such as scaffolds, wound dressing stents, and drug delivery (Asti and Gioglio, 2014). Natural biodegradable polymers such as chitosan, albumin, dextran, and lectin are also being used as drug delivery vehicles in the recent years. Chitosan is a polyaminosaccharide,

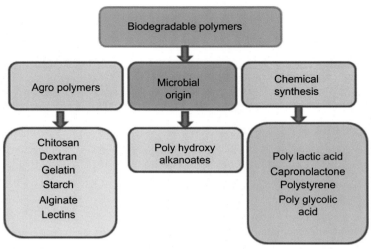

FIG. 8.3 **Classification of Biodegradable Polymers.** Biodegradable polymers are classified into various types based on the origin.

which is obtained by deacetylation of chitin, which is obtained from shells of crustaceans, is the second polymer, which is largely used after cellulose. Albumin is a protein present in the egg white, the ovalbumin, and bovine serum albumin, whereas human serum albumin is the major soluble protein of the circulating system involved in the maintenance of osmotic balance and also plays a significant role in the transport of nutrients to the cells (An and Zhang, 2017). Different forms are extensively used as nanocarriers for drug and antibody delivery (Bronze-Uhle et al., 2017; Yu et al., 2014; Fadaeian et al., 2015). Dextran, a natural polysaccharide synthesized by the bacteria, is extensively used in the medical field, as nanocarrier for drug delivery, and as an alternative for PEGylation to avoid the interaction between the chemical surfaces. Dextran polymers are extensively used as micelles in nanodrug delivery (Banerjee and Bandopadhyay, 2016). Gelatin, a biodegradable polymer obtained from collagen, also used in drug delivery because of its biocompatibility and reduced toxicity (Azimi et al., 2014). Gelatin is also used to generate scaffolds in tissue engineering applications, such as cardiac and vascular tissues and also in biosensing, drug, and gene delivery (Yue et al., 2015). Lectins are carbohydrates obtained from plants, viruses, and other microorganisms and are also recently being tested for experimental treatments of cancer, wounds, and certain immunological diseases. They are capable of inducing mitosis and immune responses, which heal the infections and inflammations (Coelho et al., 2017). Synthetic biodegradable polymers are more

advantageous because of their uniformity and versatile functions. Polymers such as polylactic acid and polyglycolic acid are widely administered in biomedical applications because of their carbon-based chemistry, which is closer to the biological tissue, which can be used for targeted interaction between the material and the system (Maitz, 2015). They are widely used as prosthetics, implants, and biomaterials also used as carriers for nanodrug delivery (Kumbar et al., 2014). Another source of NP synthesis is through microbial enzymes and microorganisms to develop nontoxic, clean, green, and eco-friendly NPs (Iravani, 2014).

BIOMEDICAL APPLICATIONS OF NANOMATERIALS

Nanomaterials of different varieties have significant applications in biomedical field, especially in diagnosis, imaging, drug delivery, and biomaterials as prostheses and implants, as they are in the size of biological structures and are well integrated into the system (Ramos et al., 2017). The nanomaterials are available from various sources such as metals, nonmetals, chemicals, microbes, and plants (Makarov et al., 2014). Although metallic and nonmetallic NPs have been most extensively used in biomedicine, nanomaterials made up of lipids and proteins, namely, liposomes and micelles are increasingly being preferred as nanocarriers for drug delivery (Allen and Cullis, 2013). At present, there are approximately 175 unique nanomedicinal products of which a majority of them (more than 100) are at various stages of clinical trials, most of them being developed for treatment of cancer or infectious diseases (Weissig et al., 2014).

Diagnosis

Visualization of anatomical structures by using contrasting agents has been used for a long time to identify diseased tissues and help clinicians to decide the appropriate mode of treatment. The possibilities of attaining unique physicochemical properties based on material type, size, shape, and manufacturing approaches make nanotechnology applications in anatomical and functional imaging very attractive (Pelaz et al., 2017). Nanomaterials may be engineered to develop contrast agents with higher specificity and sensitivity while minimizing their toxicity in different imaging modalities.

Despite the use of X-rays, computed tomography (CT) is the most applied technique in clinical diagnosis because of its high signal-to-noise ratio. However, its low sensitivity is primarily due to weak interaction between soft tissues and incident X-rays. The development of NP-based contrast agents that bind or accumulate in specific cells using materials with high atomic number and high electron density such as gold has been demonstrated to increase the sensitivity of CT (Bao et al., 2014). The properties of these NPs may be changed by changing their size and shape; furthermore, tissue/cell specificity may be obtained by conjugating them with specific targeting molecules such as folic acid for targeting cancer cells (Zhang et al., 2015a). Combined with emerging CT technologies such as phase-contrast CT and those that depend on phase shift or scattering rather than X-ray attenuation (Mohammadi et al., 2014), using NPs with distinct X-ray diffraction properties will allow reduced X-ray dosage, making CT safer and more sensitive (Krenkel et al., 2015; Zhou and Brahme, 2008).

Unlike CT, magnetic resonance imaging (MRI) does not use ionizing radiations and is based on differences in the time taken for the magnetically aligned protons to return to equilibrium after exposure to a radio frequency pulse. Compared with CT, MRI has better spatial resolution and has the ability to record both anatomical and functional information. It is different for different tissues, allowing soft tissue contrast with excellent spatial resolution (James and Gambhir, 2012). To improve MRI resolution, paramagnetic chelates of lanthanide metals such as gadolinium (Xu et al., 2012) for enhancing the positive contrast and superparamagnetic iron oxide nanoparticles (SPIONs) for enhancing the negative contrast; their use is limited due to their toxicity (Khawaja et al., 2015; Singh et al.,

2010). Dual-mode contrast agents with a superparamagnetic core and a paramagnetic shell have been tested (Ali et al., 2011) but suffer from signal interference; it may be overcome by introducing a separating layer between the core and the shell (Shin et al., 2014). Alternatively the two contrast agents may be separated by using a "bridge" in a dumbbell design to minimize magnetic interference while allowing their surfaces to be modified for further targeting (Cheng et al., 2014). Exploiting minor differences between normal tissues and tumors, smart NP-based probes for MRI have been developed that are able to sense minor changes in temperature (Li et al., 2017) or cellular redox status (Zhu et al., 2015).

A number of NP-based radiolabels have been synthesized for use in PET/SPECT imaging; most of these are hybrid structures with a radiolabel core and an organic coating for colloidal stability and conjugation with various ligands for specific tissue/cell targeting (Polyak and Ross, 2017). Different radiolabels with distinct energies may also be conjugated to different parts of NPs to study biodistribution and elimination of distinct parts of the NPs; this will be greatly beneficial for development of efficient NP-mediated targeted drug delivery systems (Ulbrich et al., 2016). Traditional fluorophores used for fluorescence imaging suffer from aggregation-caused quenching and proximity-caused quenching and hence are rapidly photobleached. Ultra-bright NPs consisting of a silica or polymeric matrix exploiting aggregation-induced emission (AIE) have recently been developed (Li et al., 2017).

Inorganic fluorescent quantum dots (QDs) are resistant to photobleaching and have very narrow emission bandwidth, allowing for simultaneous visualization of multiple biological entities using multiplexing (Xu et al., 2017). The use of QDs is limited because of their cytotoxicity that may be partially overcome by using QDs of less than 10 nm diameter. In contrast, the newly developed AIE fluorochromes increase in fluorescence intensity following aggregation and are resistant to photobleaching, allowing long-term, real-time imaging (Ding et al., 2013). By adding different functional groups, ligands, and receptors on the surface, the versatility and targeting AIEs may be enhanced, permitting their use for monitoring targeted drug delivery and drug release profiles (Gu et al., 2017; Gao and Tang, 2017). Other nanostructures are being tested for fluorescence imaging, including carbon dots and nanodiamonds that have several advantages including small size, high photostability, bright fluorescence with tunable surface chemistry, environmentally safe, and nontoxic (Montalti et al., 2015). "Smart fluorophores" that change emission characteristics depending on the microenvironment such as temperature and pH are also being tested to "sense" changes in the cellular milieu for monitoring disease status (Freidus et al., 2018; Luby et al., 2017).

Apart from disease detection, localization, and monitoring inside the body, NPs can also be used for similar applications outside the body in body fluids such as blood (or blood components), urine, saliva, and sweat (Shinde et al., 2012; Huang et al., 2018). Typically, NPs are coated with ligands for specific biological molecules, which has led to significant improvements in sensitivity and specificity of detection, such as for hormones (Leuvering et al., 1981) and specific DNA moieties (Elghanian et al., 1997). Recently, they have also been used for diagnostic purposes from breath by metabolite and/or other biochemical analyses (Vishinkin and Haick, 2015; Fernandes et al., 2015).

Because of their increased stability along with precise and tunable emission characteristics, fluorescent QDs have been used to develop high-throughput, rapid, and multiplexed detection of multiple molecules (Elghanian et al., 1997). Another strategy for high-throughput multiplexing involves multiple "noses" with each "nose" coated with a separate agent for detection of a specific analyte and has been used for rapid detection of bacteria (Elghanian et al., 1997) and cancer cells in the serum (Bajaj et al., 2009), or for rapid characterization of cancer drug mechanisms by sensing cellular physicochemical changes (Rana et al., 2015). Microfluidic-based platforms have been used for simultaneous, more sensitive detection of up to 10 different individual proteins (Pan et al., 2010) or nucleic acids (Zhang et al., 2014a). Attempts are being made to develop microfluidic chips that can simultaneously detect multiple proteins and nucleic acids for robust, point-of-care diagnostics.

Although the use of fluorescent NPs for passive sensing has been quite successful, active sensing using functional NPs that respond to specific intracellular changes has been more challenging. Nevertheless, CdSe/ZnS QDs functionalized with dopamine or oxidized cytochrome C have been shown to sense intracellular changes in pH (Medintz et al., 2010) or ROS (Li et al., 2011) levels, respectively. Carbon-based QDs have also been functionalized for intracellular sensing of pH (Du et al., 2013) and metal ions (Jiang et al., 2016). However, these QDs are limited in their applications because of their inherent toxicity and limited capability to enter the cells. Inorganic and organic NPs have also been tested for intracellular

sensing. Rather than surface functionalization of metallic and carbon QDs, both functional and nonfunctional fluorophores are embedded in the silica NPs and have been used for sensing intracellular changes in pH (Wu et al., 2011), oxygen (Koo et al., 2004), and metal ion (Park et al., 2003) levels by ratiometric measurements. Organic polymer NPs are most flexible for intracellular sensing because of availability of a large variety of nontoxic starting materials and flexibility to render either hydrophobic or hydrophilic properties. They have been used for fluorescent sensing of intracellular changes in levels of metal ions (Park et al., 2003; Sumner et al., 2002) and glucose (Xu et al., 2002).

Apart from fluorescent detection, plasmonic NPs have also been used for detection of biomolecules; in this case, the shift in surface plasma resonance of the NPs following binding to the target molecule is detected. This method has been used for detection of intracellular changes in enzyme levels (Aili et al., 2011) and specific cancer-related point mutations (Valentini et al., 2013). Plasmonic NPs have also been used for antibody-based immunoassays (Endo et al., 2006) with significantly increased sensitivity (Polo et al., 2013). Another NP-based method for molecular detection is based on surface-enhanced Raman scattering (SERS). Although limited by the ability to detect individual analytes in complex biological samples, this has been, to some extent, overcome by using chemoreceptors to the target molecules as intermediaries. Using this strategy, SERS has been used as a biosensor for detection of drugs (Sanles-Sobrido et al., 2009), metabolites associated with disease conditions (Guerrini et al., 2015), and toxic ions (Guerrini et al., 2014), as also for intracellular detection of changes in pH (Kneipp et al., 2006) and nitric oxide levels (Rivera Gil et al., 2013). Semiconductor nanowire field-effect transistors (FETs) are able to detect analytes in biological fluids such as blood (Leonardi et al., 2018), urine (Chen et al., 2015), saliva (Zhang et al., 2015b), and sweat (Liu et al., 2018) and convert it into electrical signals, the possible downside being its relative insensitivity to detect disease markers in buffers with high ionic strength (Liu et al., 2018), possibly due to ionic interference. Specific ion channels (Oblatt-Montal et al., 1994) and even intact primary cells (Milligan et al., 2009) and stem cells (Haythornthwaite et al., 2012) that are excitable have been used as transducers; it has been made relatively easy with the development of planar patch clamp devices that allow for high-throughput screening (Bosca et al., 2014).

Drug Delivery

Another field of application for nanostructures is the drug delivery on cellular level. This method of active substance administration improves the pharmacokinetic properties of the drug, which translates into better dissolution kinetics, quicker absorption, and achieving the therapeutic concentration in the target tissue. Nanodrug delivery offers several pharmacokinetic advantages such as high stability, improved bioavailability, and high membrane permeability (Onoue et al., 2014). Nanodrugs possess altered physicochemical properties such as surface charge, size, and solubility, allowing them to penetrate targeted sites by various endocytotic processes. Functionalization of NPs with surface ligands enhances the intracellular transport of entrapped drugs (Hu et al., 2016). Carbohydrate-binding ligands on the surface of biodegradable nanospheres increase cellular binding (Weissenböck et al., 2004), and attachment of specific proteins such as antibodies to the NP surface may enable a more specific immunologically directed targeting of the particles (Arruebo et al., 2009). The aim of NP entrapment of drugs is to enhance delivery to the target cells and reduction in the toxicity of the free drug to nontarget organs (De Jong and Borm, 2008), which result in an increase of therapeutic index and reduced toxicity to other organ systems.

Although NPs have been used for drug delivery for a number of diseases, the prime focus has been to improve cancer treatment. However, despite sustained efforts, the results have been rather disappointing. Only marginal improvements in drug efficacy have been achieved (Stirland et al., 2013), and efforts are ongoing to achieve clinically relevant improvements in efficacy of cancer drugs (Cheng et al., 2012). The efficacy of oncodrugs is determined by circulation in blood, accumulation and penetration into the tumor, cellular internalization, and intracellular drug release, the so-called CAPIR cascade (Sun et al., 2014). An ideal nanocarrier for cancer drugs should be "stealthy" while in circulation and hold the drug tightly during transport to the tumor tissues, where it should accumulate and "stick" to the tumor cells and dump its cargo that is subsequently taken up efficiently by the tumor cells (Sun et al., 2012). Furthermore, the nanoformulation should be able to access all tumor cells, even at locations distant from the blood vessels.

Although the main focus has been on cancer, nanotechnology is also being tested for the treatment of infectious diseases that cause maximum mortality, cardiovascular disease that are ranked at the top in terms of mortality for noncommunicable diseases, and also for neurodegenerative diseases that currently have very poor treatment options (Lepeltier et al., 2015). Unlike anticancer drugs that are small and can be easily packaged within the NPs, the large macromolecules that are effective against the other human diseases are large in size, are metabolized rapidly, and are unable to penetrate various biological barriers such as tissues and cellular membranes.

The different types of nanodrug delivery systems are (1) liposomes, (2) polymeric NPs, (3) dendrimers, (4) carbon-based nanocarriers, and (5) micelles (Fig. 8.4).

Liposomes

Liposomes were first described by Bangham in 1961 (Bangham et al., 1965) as artificial lipid vesicles and have been extensively emerged as drug carriers, as they mimic the biophysical properties of the cell (Zylberberg and Matosevic, 2016). Liposomes are amphiphilic vesicles with an outer lipid bilayer and an inner aqueous compartment. Water-soluble drugs are entrapped in the aqueous compartment of liposomes, whereas hydrophobic drugs are incorporated into the lipid bilayer (Babu et al., 2014). Liposomes organized with DNA called lipoplexes are designed for gene therapy and form significant lipid-based gene delivery vehicles (Kono et al., 2001); siRNA-incorporated liposomes have been used for gene silencing in various diseases (Lee et al., 2012). Liposomes are classified in terms of their intracellular delivery and composition into five types such as conventional, pH sensitive, long circulating, immunoliposomes, and cationic liposomes (Maherani et al., 2011; Akbarzadeh et al., 2013). Nanoliposomes with site-directing ligands attached to the membrane surface have been developed to enhance the selectivity of liposomal drug delivery (Allen and Cullis, 2013) to the targeted areas. The advancements in liposomal drug delivery utilized the combination of monoclonal antibodies surface stabilized with polyethylene glycols for longer circulation to enhance the pharmacokinetic properties of the formulation (Sercombe et al., 2015). The rapid advancement in liposomal drug delivery made many liposomal formulations available clinically for treatment. The first nanosized liposomal product was Doxil used for ovarian cancer in Israel and USA (Barenholz, 2012); subsequently, many products are available in the management of various cancers. Liposomes have also become important carrier systems in vaccine development because of the development of products Epaxal and Inflexal V by Crucell, Berna Biotech for vaccination against hepatitis and influenza, respectively (Schwendener, 2014). Several liposomal formulations were successfully translated into the clinical use and

Nano carriers

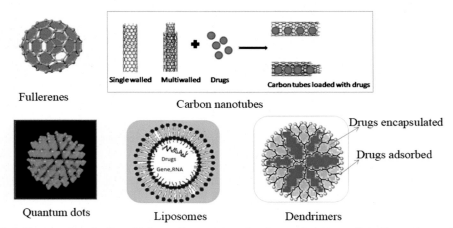

Fullerenes   Carbon nanotubes

Quantum dots   Liposomes   Dendrimers

FIG. 8.4 **Nanocarriers for Drug Delivery.** Nanocarriers of various materials are efficiently used as carriers for drug delivery. Liposomes and dendrimers are multifunctional in nature, which deliver multiple drugs, gene, and RNA. Carbon-based nanocarriers such as fullerenes and nanotubes are used for drug and siRNA delivery. Quantum dots are involved in drug delivery and diagnostics and also as sensing agents.

in different phases of clinical investigation. Liposomes coupled with folate have been developed for the delivery of various drugs, including platinum-based carboplatin in ovarian cancer model (Gabizon et al., 2010); liposomal systems of folic acid-based were coupled with the cell-penetrating peptide (CPP) TAT using paclitaxel as the chemotherapeutic agent (Niu et al., 2011). Liposomes prepared using nonionic surfactants such as alkyl ethers or alkyl esters are called as niosomes. Niosomes are biodegradable, biocompatible, and nonimmunogenic in nature with long shelf life, exhibit high stability, and ensure targeted drug delivery in the sustained or rapid release. The composition, surface charge, and various nonionic surfactants have been reported in niosome development entrapped with a wide variety of drugs (Ag Seleci et al., 2016). In summary, the liposome-based nanodrug delivery may provide benefits to the diversified patient population for various therapeutic applications (Bulbake et al., 2017).

Polymeric nanoparticles
Polymeric nanoparticles (PNPs) synthesized using either synthetic or natural polymers offer a wide scope in drug delivery, as they are more biocompatible and biodegradable in nature (Singh and Lillard, 2009). Although NPs synthesized using any kind of polymer may technically be called as PNPs, the term is specifically used for nanospheres and nanocapsules (Rao

and Geckeler, 2011). The ability of PNPs for controlled drug delivery to combine both therapy and imaging offers targeted drug delivery and protection of drugs, which facilitates the improvement of therapeutic index (Krasia-Christoforou and Georgiou, 2013). Nonbiodegradable polymers are also approved by the FDA for the drug delivery, which release the drugs by diffusion control unlike the burst release by biodegradable polymers. Majority of the polymers are silicone, polyvinyl acetate, and ethyl vinyl acetate, which have been developed by retinitis, uveitis, and diabetic retinopathy treatment (Yang and Pierstorff, 2012). Polymers are typically not present as bulk materials and can be applied as coatings on biomedical devices or can be encapsulated with drugs and were used as nanodrug delivery systems. The coatings may be homogeneous, cross-linked coatings, or layer-by-layer deposited films. The polymeric NPs can be formulated as dendrimers, micelles, nanogels, and nanocapsules (Stuart et al., 2010; Mohamed et al., 2014). The chemistry of polymeric NPs affects their stability, biodegradability, biocompatibility, biodistribution, and cellular and subcellular fate. Incorporation of functional groups modulates the reaction of NPs in biological environments under variable pH, enzymatic, redox conditions, or in response to external stimuli, such as temperature variation, radiation and activation with magnetic fields (Elsabahy and Wooley, 2012). Recent advancements in drug delivery are the amphiphilic biodegradable polymeric NPs for diagnosis and cancer therapy (Zhang et al., 2014b). Among the polymeric NPs, synthetic polyesters such as polycyanoacrylate and related polymers poly(lactide-co-glycolide) PLA or poly(lactic acid) were used in targeted drug delivery, gene silencing, and siRNA delivery (Tong et al., 2012). Natural polymer—based NPs are more beneficial over traditional methods of drug delivery system in terms of efficiency and targeted drug delivery. The various natural polymers such as dextran, gelatin, albumin, chitosan, and alginate were used to prepare the NPs. However, they have some disadvantages such as poor batch-to batch reproducibility and prone to easy degradation (Marin et al., 2013).

Dendrimers

Dendrimers are nanocarriers, which are highly branched with well-defined structural architecture developed in 1978 for the drug delivery (Buhleier et al., 1978). They are the organized nanocarriers with molecular weight ranging from 5000 to 50,000 g/mol with a low polydispersity index characterized by layers known as "generations." They differentiated into the inner core followed by branches attached radially, which are associated with various functional groups. Chemotherapeutic drug delivery to the targeted site is essential to reduce the toxic effects of the drugs, which were achieved by the surface modifications of the dendrimers with attached moieties such as folic acid, antibodies, and peptides. The encapsulation of the drug inside the dendrimers resulted in enhanced solubility and in vitro release of the drug (Madaan et al., 2014). The most used dendrimers in drug delivery are the polyamidoamines (PAMAMs), which are biocompatible, non-immunogenic, and hydrophilic in nature. The core of this dendrimers was made up of hydrophobic molecules with branching made of ethylenediamine and methylacrylate terminated with carboxyl and amine groups. PAMAM dendrimers are used as carriers in drug delivery (Chanphai et al., 2017), gene transfection (Shcharbin et al., 2009), diagnostic agents (Kojima et al., 2011), and boron-neutron capture therapy for metastatic brain tumors (Sun et al., 2016). Tertial amines in PAMAM dendrimers are available for hydrogen bonding and other noncovalent interactions with binding molecules, making them effective for solubilizing hydrophobic drugs. Polypropylene imine (PPI) dendrimers were also widely used in drug delivery with presence of alkyl chains as side branches, as they are more hydrophobic than PAMAM dendrimers (Shao et al., 2011) Poly-L-lysine (PLL) dendrimers

were used as gene carriers because of their excellent good biocompatibility, biodegradability, and hydrophilic nature, with their core and branching units made with amino acid lysine (Maiellaro and Taylor, 2007). The characteristics of dendrimers, with high branching, and multiple drug encapsulating ability, with variable chemical composition and high biological compatibility, make them idea carriers in biomedical applications.

Carbon-based nanocarriers

CNTs were prepared by using benzene and hydrogen (Oberlin et al., 1976) by vapor growth method in 1976, and the advancements in nanotubes are the multiwalled (SWNTS) described by Iijima and Ichihashi (1993). CNTs later developed to multiwalled (MWCNTS) also offer many benefits in targeted drug delivery, imaging, and diagnosis (Dineshkumar et al., 2015). CNTs are widely used in different fields of science and biomedicine because of their novel properties. Such carbon molecules have novel properties that made them useful in a wide variety of applications in mechanical, structural, thermal, electronics, optical, biomedical, and other fields of science and medicine. Surface-engineered CNTs used as nanocarriers for the drug delivery have shown enormous potential in targeting specific cells without affecting the normal cells with less amount of dose than the native drug (Nirosa and Roy, 2013). The surface modifications of CNTs with ligands, which allow the formation of complexes with DNA, made them suitable carriers for gene delivery (Ramos-Perez et al., 2013) and association through lipopolymers with covalent bonding for siRNA delivery (Siu et al., 2014). CNTs were also used for delivering small proteins such as recombined ricin A chain protein toxin (RAT) for targeting the breast cancer cells (Weng et al., 2009). Drug targeting to the lymphatic system will be more effective to treat metastatic tumors such as gemcitabine loaded with CNTs effectively accumulated in the lymph nodes to deliver the drug through the lymphatic circulation (Yang et al., 2009). Buckminsterfullerene (C60), popularly known as the buckyball, is a spherical cage—like closed structure, which was discovered in 1985; it is highly promising because of its antihuman immunodeficiency virus (HIV) activity, carrier for drug delivery, gene delivery, and diagnostic applications (Bakry et al., 2007). A variety of carbon-based nanocapsules and nano rods were also demonstrated as drug carriers. Even though CNTs are shown as promising drug carriers for drug delivery, their reproducibility in drug delivery is quite challenged, which impedes their commercialization as drug delivery vehicles (Zhang et al., 2011).

Polymeric micelles

Polymeric micelles are spherical, colloidal, and supra-molecular nanoconstructs and synthesized from nano-sized amphiphilic diblock or triblock copolymers for drug and gene delivery (Wakaskar, 2017; Kedar et al., 2010). The hydrophobic drugs are encapsulated in the micellar core by chemical or physical conjugation, whereas the hydrophilic portion forms the shell or the corona. At low critical micellar concentration (CMC), they exhibit greater stability, allow prolonged circula-tion time in blood, and avoid reticuloendothelial uptake with their smaller size not able to get degraded by the blood components (Kulthe et al., 2012). Multi-functional micelles with attached ligands are the ideal carriers for drug delivery, which are generally more sta-ble with a low CMC, significantly achieving a higher drug concentration at target site. Micelles have demon-strated a variety of shapes, which also influence the effectiveness of the formulation (Ahmad et al., 2014). The incorporation of hydrophobic drugs in micellar structure improves the bioavailability and therapeutic efficiency of poorly soluble drugs (Kalepu and Nek-kanti, 2015). For pharmaceutical applications, micelles were prepared using two strategies such as active or pas-sive targeting/EPR effect for targeted drug delivery. Active targeting includes carbohydrates, folate mole-cules, antibodies, aptamers, protein, and peptide-based ligands. Multifunctional micelles with two or three attached ligands are the recent advancements (Movassaghian et al., 2015), which combine many functions within a single carrier, with each individual component functioning in accurate coordination with the other components.

NANOPARTICLES FOR TISSUE ENGINEERING AND REGENERATIVE MEDICINE

Tissue engineering (TE) is an interdisciplinary field incorporating biology and engineering to synthesize the substitute or enhance the organ function (Langer and Vacanti, 1993). The basic principle of TE is the isolation of cells from the tissues of interest, seeded on a scaffold, which provides the surface for adhesion, mimicking the cellular internal structures on which cells get deposited and get functionalized. Nanotechnology in TE has a wide range of applications in generation of neural cell tissue, cartilage and bone, vascular cells, corneas, heart valves, and hepatic cells (Danie Kingsley et al., 2013). Nanomaterials provide primary mechani-cal support and serve as a sink to recruit cells, facilitating

their proliferation and differentiation to replace the damaged tissue (Pina et al., 2015). Working as a surro-gate for the extracellular matrix (ECM), these biomate-rials also modulate appropriate function of the recruited cells (Dutta and Dutta, 2009). Rather than providing mere mechanical support, the scaffolds used in TE mimic the ECM in providing the physical, chemical, and biological support in appropriate regen-eration of damaged tissues in vivo (Gu et al., 2014). Ideally these "scaffolds" should degenerate over a period of time once the repair of the issues is complete (Mano et al., 2007).

The application of nanotechnology in TE such as nanofibers and nanopatterns is a breakthrough in the preparation of more compatible biomimicking scaf-folds, as most of the biomolecules are in the micro- and nanosizes, which provide the best internal microenvi-ronment (Rivron et al., 2008; Nakanishi et al., 2008). Techniques such as electrospinning and self-assembly are used to fabricate the scaffolds with fibers to form an interconnected, porous architecture and surface modified by attaching peptides and other molecules, which enhance cell growth and differentiation (Reneker and Yarin, 2008). Nanomaterials from natural sources are now being preferred, as they are biocompatible and closely resemble the endogenous biological mole-cules present in the microenvironment where TE is be-ing attempted, besides facilitating cellular recruitment to the damaged tissues (Mano et al., 2007; Renth and Detamore, 2012). Among the various materials used as scaffolds, the widely accepted are the natural poly-mers because of their unique properties such as high surface-to-volume ratio, biocompatibility, and biode-gradable in nature (Guo and Ma, 2014). Nanobiomate-rials offer various advantages such as improved electrical, mechanical, magnetic, and optical properties with increased surface area for protein adsorption, which govern the cell attachment and allow the signals to internal proteins for cell migration, proliferation, and tissue formation (Walmsley et al., 2015). Because of these inherent advantages, natural polymers have been explored for tissue regeneration (Suarato et al., 2018) and even for organ transplantation (Suarato et al., 2018). The natural biomaterials being explored for TE may be classified as either proteins, such as collagen, gelatin and fibronectin (DeFrates et al., 2018), or polysaccharides such as hyaluronan, alginate, and chitosan (Pellá et al., 2018). However, natural poly-mers inherently have weak mechanical properties; because of their source of origin, there are often issues of inconsistency in terms of physical and chemical properties (Suarato et al., 2018). In some cases, these naturally sourced nanomaterials may also induce both innate and adaptive immune responses in the host (Atala et al., 2012).

To overcome the inherent limitations of naturally occurring polymers, it is being explored if biomimetics of these natural compounds can be used as alternatives for TE (Reddy and Reddy, 2018). The use of synthetic polymers for TE is another attractive option, as their size, shape, tensile strength, biodegradability, and other properties can be controlled and modified as per specific requirements (Ye et al., 2018). Poly(α-hydroxy acids) (PHAs) are the most widely used synthetic class of polymers used and include polylactic acid (PLA), polyglycolic acid (PGA), and their copolymer, poly(lactic-co-glycolic acid) (PLGA) (Koller, 2018). Other popular synthetic polymers for TE are polyethylene glycol (PEG) (Cruz-Acuña et al., 2018) and poly(ε-caprolactone) (PCL) (Siddiqui et al., 2018).

The natural process of wound healing is a multistep process involving multiple cellular and extracellular components over time and includes the participation of multiple signaling pathways. This creates unique challenges for TE and means that using only a single material is unable to mimic the natural process that involves complex cell-to-cell interactions as also the interactions of different cells with the extracellular matrix. To address these issues, hybrid nanomaterials combining the natural and synthetic polymers are being tested (Li et al., 2018). The complex architecture of the tissues being regenerated poses another significant challenge, which is being addressed by 3D printing methods that allow precise arrangement of different kinds of cells within a defined extracellular environment (Guerra et al., 2018). However, the process is too time-consuming to be upscaled for clinical applications. Other strategies such as use of polymer brushes (Kim and Jung, 2016) and fiber-assisted molding (FAM) (Hosseini et al., 2014) are also being explored. Apart from cells and ECM, wound healing also involves production and release of biomolecules such as growth factors and cytokines, which presents another kind of challenge. This is being addressed using a variety of sophisticated drug delivery systems (Hosseini et al., 2014; Raza et al., 2018; Ferracini et al., 2018; De Witte et al., 2018; Tsai et al., 2018; Guo et al., 2018; Dong et al., 2018; Du et al., 2018; Sun et al., 2018; Kim et al., 2018). Nevertheless, significant challenges remain but are being addressed as we develop a better spatiotemporal understanding of the natural process of wound healing, as also the organization of different tissues and organs in utero.

FACTORS INFLUENCING UTILITY OF NANOPARTICLES IN BIOMEDICINE

The quick adaptative nature in size, shape, and composition of NPs is highly utilized in biomedical applications. The factors that influence the nanoparticle utility are the size, pH composition, solubility, stability, reduced toxicity, and biodegradability of the nanomaterial (Patra and Baek, 2014). For the successful application of NPs in drug delivery, the NPs that respond to heat and pH were developed, which are also conjugated with antibodies and DNA, which were successfully developed for the treatment of various diseases. The other significant factor is the solubility of NPs for the effective drug and gene delivery to the target molecules, which will be achieved by their nanosize and increased surface area (Sutradhar and Amin, 2014).

Materials Used for Nanoparticle Synthesis

The materials of metallic and inorganic NPs were extensively used because of their adaptability in biomedical applications. Metal oxide NPs such as iron oxide and cadmium were extensively used for diagnosis. Inorganic NPs are also effectively used in drug delivery (Corr, 2014). They are less toxic compared with metals and metal oxides, which are also biocompatible and hydrophilic in nature. However, they are mainly used in drug delivery and tissue engineering applications (Arvizo et al., 2012). The recent update in nanobiotechnology is the use of biodegradable materials whose applications are very well studied, which is an alternative to reduce the toxicity induced by the nonbiodegradable NPs. The excellent biocompatibility of biodegradable NPs rationalized them to use in diversification of applications in drug targeting, medicine, and other biological purpose, which are less toxic in nature and eliminated by the body (Luckachan and Pillai, 2011).

Nanoparticle Characteristics

The characteristics of NPs play a significant role in their diverse applications. The main parameters that influence the use of NPs are the physicochemical characteristics such as size, shape, surface area, topography, solubility, and topography. The chemical properties that influence the NPs are the structure, composition of nanomaterial surface charge, reactive sites, hydrophilicity, and catalytic properties (Zhang et al., 2017). Engineered NPs are extremely small in size with unique physicochemical properties, which is not attributed to their bulk materials. The smaller sized particles penetrate effectively through the skin and blood-brain barrier for effective drug delivery (Gao, 2016). Polymeric micelles with diameters less than 50 nm were found to have the best efficacy for pancreatic tumors (Sun et al., 2012). However, it has been reported that although micelles with larger diameter (100–160 nm) survived better in the blood and accumulated in the tumor microenvironment, they had a poor drug release profile (Wang et al., 2015). Therefore, an optimal size of the nanocarriers needs to be used that finds a balance between survival in circulation and ability to penetrate the diseased tissues.

The primary interest on NP characteristics should be the composition, nature of the nanomaterial, and stability, which can be determined by the formulation status, which is either in liquid or solid form for a desirable outcome. The dissolution kinetics is the key factor in determining the efficacy of nanoformulation compared with their bulk materials. As the dissolution rate is directly proportional to the surface area, nanomaterials are more likely to dissolve in rapid rate than large-sized particles. Henceforth, nanocharacteristics assessment is the crucial step in determining the NP stability and for desirable outcome (Gatoo et al., 2014).

Route of Administration

The mode of nanodrug delivery also has a profound effect on the overall efficacy of the nanoformulations. Although most anticancer drugs have been designed for systemic administration through the intravenous route, they are expected to survive in the blood, home onto the tumor site, and dump their cargo inside the tumor cells. Besides using chemical ligands for targeting, active physical targeting using magnetic focusing has also been used. In this approach, SPIONs are often coated with organic materials to improve their colloidal stability (Lepeltier et al., 2015). It has been demonstrated to be more efficient in delivery of antitumor drugs (Lepeltier et al., 2015). Nanocarriers have also been designed to protect the therapeutic drug and penetrate the epithelial barriers of the gut, skin, and the lungs (Mitragotri et al., 2014), opening up other possible routes of administration. Indeed, efficient drug delivery using oral route has been reported by taking advantage of the formulation to accumulate in the inflamed mucosal tissues of the intestinal system (Mitragotri et al., 2014); however, microsized particles were used in the study. Transfollicular delivery of vaccines has also been demonstrated (Mittal et al., 2015); however, pulmonary delivery offers significant challenges (Paranjpe and Müller-Goymann, 2014).

Nanoparticle Toxicity

While acquiring unique properties because of their small size, there are very real possibilities that NPs

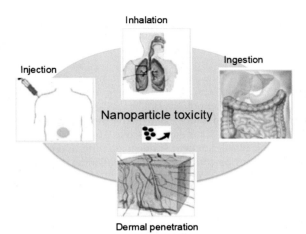

FIG. 8.5 Toxicity of Nanoparticles Depends on the Route of Administration. Nanoparticles may be administered by injection, oral, pulmonary, and dermal routes; the route of administration determines the extent to which specific cells, tissues, and organs may endure potential toxic effects.

may also become potential health hazards. However, the risk assessment of the nanomaterials has not been undertaken; their potential effect on the environment and human health is almost totally unknown. Moreover, no standardized methods are available for assessing consumer risk following use of nanomaterial-based products. The National Nanotechnology Initiative launched in 1991 by the United States realized the importance of nanotoxicity and published the first NNI Strategy for Nanotechnology-Related Environmental, Health and Safety (EHS) Research in 2008 (Schmidt, 2009) and was subsequently updated in 2011 (Roco, 2011). This becomes more important as the advent of nanomedicine may involve more direct exposure of NPs to humans. Individuals may be exposed to NPs via various routes; the route of administration (or exposure) of NPs ultimately determines the part of tissue or organ that may be more susceptible to NP-induced toxicity (Fig. 8.5).

CONCLUSION AND FUTURE PROSPECTS

These are exciting times for nanomedicine, as almost every day newer NPs are being created to address the specific challenges posed by our human body. Nevertheless, there are a number of challenges that have to be adequately addressed before the tremendous potential of nanotechnology for biomedical applications is fully realized. Although the unique properties of NPs are being exploited for their beneficial effects, they also need to be investigated for their potential harmful side effects. Although nanomedicine offers hope for improving human health, thorough in vitro and in vivo toxicological studies need to be undertaken to ensure safety.

In diagnostics, the real challenge is to develop robust and affordable point-of-care systems that are able to detect human diseases early with high sensitivity and specificity. Till date, most of the focus has been on cancer diagnosis and treatment; however, the bigger challenge is to extend the knowledge generated and technologies developed to address the infectious diseases that are the biggest killers. Ideally, in vitro diagnostic platforms for communicable diseases should be developed, which are able to identify multiple serotypes of infectious agents; furthermore, diagnostic platforms should be capable of identifying multiple pathogens causing different forms of different infectious diseases endemic in a particular geographical location or within a particular population. These diagnostic platforms should be robust and cheap with high sensitivity for early detection. Perhaps more importantly, they should be easy to use, and the readouts should be rapid and easy to interpret for a nonskilled or minimally skilled public health worker. This will allow their integration in public health programs to benefit a larger proportion of the population, especially those who are financially weak, live in substandard conditions, and are at highest risk of being exposed to multiple infectious agents. Next, safe, affordable drugs and drug delivery systems need to be developed. Although the current focus has been on using single drugs, this may not be beneficial for a large proportion of the target population. For example, tumors are inherently heterogenous; hence a single drug may not target all the tumor cells, leaving the unaffected tumor cells to proliferate, leading to relapse. A potential solution could be to use multiple drugs targeting different aspects of tumor-causing mechanisms or to specifically target cancer stem cells for long-term benefits. In a similar vein, most patients harbor multiple infectious agents at the same time, making their effective treatment more challenging. For tissue engineering, significant challenges include (1) availability of really robust nanomaterials that may be used as scaffolds, (2) cellular organization in a three-dimensional geometry that is unique for individual tissues and organs, and (3) delivery of a different combination of factors at different concentrations and in different ratios at different times.

In summary, nanotechnology holds great promise to provide diverse multimodal diagnostic platforms and therapeutic applications to radically change the scenario of human health. Efforts are currently under way to translate these scientific advancements into the clinic and should bring a new paradigm of biomedicine.

ACKNOWLEDGMENTS

The authors wish to thank the whole-hearted support and encouragement from Amity University, especially Dr. Ashok K Chauhan, Founder President, Amity Group of Institutions.

REFERENCES

Ag Seleci, D., Seleci, M., Walter, J.G., Stahl, F., Scheper, T., 2016. Niosomes as nanoparticular drug carriers: fundamentals and recent applications. Journal of Nanomaterials 2016. https://doi.org/10.1155/2016/7372306.

Ahmad, Z., Shah, A., Siddiq, M., Kraatz, H.-B., 2014. Polymeric micelles as drug delivery vehicles. RSC Advances 4, 17028—17038.

Aili, D., Mager, M., Roche, D., Stevens, M.M., 2011. Hybrid nanoparticle-liposome detection of phospholipase activity. Nano Letters 11, 1401—1405.

Akbarzadeh, A., Rezaei-Sadabady, R., Davaran, S., Joo, S.W., Zarghami, N., Hanifehpour, Y., et al., 2013. Liposome: classification, preparation, and applications. Nanoscale Res Lett 8, 1—8.

Ali, Z., Abbasi, A.Z., Zhang, F., Arosio, P., Lascialfari, A., Casula, M.F., et al., 2011. Multifunctional nanoparticles for dual imaging. Analytical Chemistry 83, 2877—2882.

Allen, T.M., Cullis, P.R., 2013. Liposomal drug delivery systems: from concept to clinical applications. Advanced Drug Delivery Reviews 65, 36—48.

An, F.-F., Zhang, X.-H., 2017. Strategies for preparing albumin-based nanoparticles for multifunctional bioimaging and drug delivery. Theranostics 7, 3667—3689.

Arruebo, M., Valladares, M., González-Fernández, Á., 2009. Antibody-conjugated nanoparticles for biomedical applications. Journal of Nanomaterials 2009. https://doi.org/10.1155/2009/439389.

Arvizo, R.R., Bhattacharyya, S., Kudgus, R.A., Giri, K., Bhattacharya, R., Mukherjee, P., 2012. Intrinsic therapeutic applications of noble metal nanoparticles: past, present and future. Chemical Society Reviews. https://doi.org/10.1039/c2cs15355f.

Asmatulu, E., 2016. Recent developments on nanomaterials and nanosafety for engineering applications. In: CAMX 2016 — Composites and Advanced Materials Expo.

Asti, A., Gioglio, L., 2014. Natural and synthetic biodegradable polymers: different scaffolds for cell expansion and tissue formation. The International Journal of Artificial Organs 37, 187—205.

Atala, A., Kasper, F.K., Mikos, A.G., 2012. Engineering complex tissues. Science Translational Medicine 4, 160rv12-160rv12.

Azimi, B., Nourpanah, P., Rabiee, M., Arbab, S., 2014. Producing gelatin nanoparticles as delivery system for bovine
serum albumin. Iranian Biomedical Journal 18, 34–40.

Babu, A., Templeton, A.K., Munshi, A., Ramesh, R., 2014.
Nanodrug delivery systems: a promising technology for
detection, diagnosis, and treatment of cancer. AAPS
PharmSciTech 15, 709–721.

Bajaj, A., Miranda, O.R., Kim, I.-B., Phillips, R.L., Jerry, D.J.,
Bunz, U.H.F., et al., 2009. Detection and differentiation
of normal, cancerous, and metastatic cells using
nanoparticle-polymer sensor arrays. Proceedings of the National Academy of Sciences of the United States of America
106, 10912–10916.

Bakry, R., Vallant, R.M., Najam-ul-Haq, M., Rainer, M.,
Szabo, Z., Huck, C.W., et al., 2007. Medicinal applications
of fullerenes. International Journal of Nanomedicine 2,
639–649.

Banerjee, A., Bandopadhyay, R., 2016. Use of dextran nanoparticle: a paradigm shift in bacterial exopolysaccharide based
biomedical applications. International Journal of Biological Macromolecules 87, 295–301.

Bangham, A.D., Standish, M.M., Watkins, J.C., 1965. Diffusion
of univalent ions across the lamellae of swollen
phospholipids. Journal of Molecular Biology 13,
238–252. IN26-IN27.

Bao, C., Conde, J., Polo, E., del Pino, P., Moros, M.,
Baptista, P.V., et al., 2014. A promising road with challenges: where are gold nanoparticles in translational
research? Nanomedicine 9, 2353–2370.

Barenholz, Y., 2012. Doxil® — the first FDA-approved nanodrug: lessons learned. Journal of Controlled Release 160,
117–134.

Binnig, G., Rohrer, H., 1982. Scanning tunneling microscopy.
Helvetica Physica Acta 55, 726–735.

Binnig, G., Quate, C.F., Gerber, C., 1986. Atomic force
microscope. Physical Review Letters 56, 930.

Bosca, A., Martina, M., Py, C., 2014. Planar patch clamp for
neuronal networks—considerations and future
perspectives. Methods in Molecular Biology 1183, 93–113.

Bronze-Uhle, E.S., Costa, B.C., Ximenes, V.F., Lisboa-
Filho, P.N., 2017. Synthetic nanoparticles of bovine serum
albumin with entrapped salicylic acid. Nanotechnology,
Science and Applications 10, 11–21.

Buhleier, E., Wehner, W., Vögtle, F., 1978. 'Cascade'- and
'Nonskid-Chain-like' syntheses of molecular cavity
topologies. Synthesis 1978, 155–158.

Bulbake, U., Doppalapudi, S., Kommineni, N., Khan, W., 2017.
Liposomal formulations in clinical use: an updated review.
Pharmaceutics 9, 12.

Chanphai, P., Bekale, L., Sanyakamdhorn, S., Agudelo, D.,
Bérubé, G., Thomas, T.J., et al., 2017. PAMAM dendrimers
in drug delivery: loading efficacy and polymer
morphology. Canadian Journal of Chemistry 95, 891–896.

Chen, H.-C., Chen, Y.-T., Tsai, R.-Y., Chen, M.-C., Chen, S.-L.,
Xiao, M.-C., et al., 2015. A sensitive and selective magnetic
graphene composite-modified polycrystalline-silicon nanowire field-effect transistor for bladder cancer diagnosis. Biosensors and Bioelectronics 66, 198–207.

Cheng, Z., Al Zaki, A., Hui, J.Z., Muzykantov, V.R., Tsourkas, A.,
2012. Multifunctional nanoparticles: cost versus benefit of
adding targeting and imaging capabilities. Science 338,
903–910.

Cheng, K., Yang, M., Zhang, R., Qin, C., Su, X., Cheng, Z., 2014.
Hybrid nanotrimers for dual T_1 and T_2-weighted magnetic
resonance imaging. ACS Nano 8, 9884–9896.

Coelho, L.C.B.B., Silva, P.M.D.S., Lima, V.L.D.M., Pontual, E.V.,
Paiva, P.M.G., Napoleão, T.H., et al., 2017. Lectins, inter-
connecting proteins with biotechnological/pharmacolog-
ical and therapeutic applications. Evidence-based
Complementary and Alternative Medicine. https://
doi.org/10.1155/2017/1594074.

Colombeau, L., Acherar, S., Baros, F., Arnoux, P., Gazzali, A.M.,
Zaghdoudi, K., et al., 2016. Inorganic nanoparticles for
photodynamic therapy. In: Topics in Current Chemistry,
pp. 113–134.

Conde, J., Dias, J.T., GrazÃ, V., Moros, M., Baptista, P.V., de la
Fuente, J.M., 2014. Revisiting 30 years of bio-
functionalization and surface chemistry of inorganic nanoparticles for nanomedicine. Frontiers in Chemistry 2, 48.

Corr, S.A., 2014. Metal oxide nanoparticles. SPR Nanoscience.
https://doi.org/10.1039/9781849737623-00204.

Cruz-Acuña, R., Quirós, M., Huang, S., Siuda, D., Spence, J.R.,
Nusrat, A., et al., 2018. PEG-4MAL hydrogels for human
organoid generation, culture, and in vivo delivery. Nature
Protocols 13, 2102–2119.

Danie Kingsley, J., Ranjan, S., Dasgupta, N., Saha, P., 2013.
Nanotechnology for tissue engineering: need, techniques
and applications. Journal of Pharmacy Research 7,
200–204.

De Jong, W.H., Borm, P.J., 2008. Drug delivery and nanoparticles: applications and hazards. International Journal of
Nanomedicine 3, 133–149.

De Witte, T.-M., Fratila-Apachitei, L.E., Zadpoor, A.A.,
Peppas, N.A., 2018. Bone tissue engineering via growth factor delivery: from scaffolds to complex matrices. Regenerative Biomaterials 5, 197–211.

DeFrates, K., Moore, R., Borgesi, J., Lin, G., Mulderig, T.,
Beachley, V., et al., 2018. Protein-based fiber materials in
medicine: a review. Nanomaterials 8, 457.

Dineshkumar, B., Krishnakumar, K., Bhatt, A.R., Paul, D.,
Cherian, J., John, a, et al., 2015. Single-walled and multi-
walled carbon nanotubes based drug delivery system: cancer therapy: a review. Indian Journal of Cancer 52,
262–264.

Ding, D., Li, K., Liu, B., Tang, B.Z., 2013. Bioprobes based on
AIE fluorogens. Accounts of Chemical Research 46,
2441–2453.

Dong, Y., Liu, B., Yuan, Y., 2018. AIEgen based drug delivery
systems for cancer therapy. Journal of Controlled Release
290, 129–137.

Drexler, K.E., 1981. Molecular engineering: an approach to the
development of general capabilities for molecular
manipulation. Proceedings of the National Academy of Sciences 78, 5275–5278.

Du, F., Ming, Y., Zeng, F., Yu, C., Wu, S., 2013. A low cytotoxic
and ratiometric fluorescent nanosensor based on carbon-

dots for intracellular pH sensing and mapping. Nanotechnology 24, 365101.

Du, J.-Z., Li, H.-J., Wang, J., 2018. Tumor-Acidity-cleavable maleic acid amide (TACMAA): a powerful tool for designing smart nanoparticles to overcome delivery barriers in cancer nanomedicine. Accounts of Chemical Research 51, 2848–2856.

Dutta, R.C., Dutta, A.K., 2009. Cell-interactive 3D-scaffold; advances and applications. Biotechnology Advances 27, 334–339.

Elahi, N., Kamali, M., Baghersad, M.H., 2018. Recent biomedical applications of gold nanoparticles: a review. Talanta 184, 537–556.

Elghanian, R., Storhoff, J.J., Mucic, R.C., Letsinger, R.L., Mirkin, C.A., 1997. Selective colorimetric detection of polynucleotides based on the distance-dependent optical properties of gold nanoparticles. Science 277, 1078–1081.

Elsabahy, M., Wooley, K.L., 2012. Design of polymeric nanoparticles for biomedical delivery applications. Chemical Society Reviews 41, 2545.

Endo, T., Kerman, K., Nagatani, N., Hiepa, H.M., Kim, D.-K., Yonezawa, Y., et al., 2006. Multiple label-free detection of antigen-antibody reaction using localized surface plasmon resonance-based core-shell structured nanoparticle layer nanochip. Analytical Chemistry 78, 6465–6475.

Fadaeian, G., Shojaosadati, S.A., Kouchakzadeh, H., Shokri, F., Soleimani, M., 2015. Targeted delivery of 5-fluorouracil with monoclonal antibody modified bovine serum albumin nanoparticles. Iranian Journal of Pharmaceutical Research 14, 395–405.

Fermo, P., Padeletti, G., 2012. The use of nano-particles to produce iridescent metallic effects on ancient ceramic objects. Journal of Nanoscience and Nanotechnology 12, 8764–8769.

Fernandes, M.P., Venkatesh, S., Sudarshan, B.G., 2015. Early detection of lung cancer using nano-nose — a review. The Open Biomedical Engineering Journal 9, 228–233.

Ferracini, R., Martínez Herreros, I., Russo, A., Casalini, T., Rossi, F., Perale, G., 2018. Scaffolds as structural tools for bone-targeted drug delivery. Pharmaceutics 10, 122.

Freidus, L.G., Pradeep, P., Kumar, P., Choonara, Y.E., Pillay, V., 2018. Alternative fluorophores designed for advanced molecular imaging. Drug Discovery Today 23, 115–133.

Gabizon, A., Tzemach, D., Gorin, J., Mak, L., Amitay, Y., Shmeeda, H., et al., 2010. Improved therapeutic activity of folate-targeted liposomal doxorubicin in folate receptor-expressing tumor models. Cancer Chemotherapy and Pharmacology 66, 43–52.

Gao, M., Tang, B.Z., 2017. Fluorescent sensors based on aggregation-induced emission: recent advances and perspectives. ACS Sensors 2, 1382–1399.

Gao, H., 2016. Progress and perspectives on targeting nanoparticles for brain drug delivery. Acta Pharmaceutica Sinica B. https://doi.org/10.1016/j.apsb.2016.05.013.

Gatoo, M.A., Naseem, S., Arfat, M.Y., Mahmood Dar, A., Qasim, K., Zubair, S., 2014. Physicochemical properties of nanomaterials: implication in associated toxic manifestations. BioMed Research International 2014. https://doi.org/10.1155/2014/498420.

Ghadi, A., Mahjoub, S., Tabandeh, F., Talebnia, F., 2014. Synthesis and optimization of chitosan nanoparticles: potential applications in nanomedicine and biomedical engineering. Caspian Journal of Internal Medicine 5, 156–161.

Giner-Casares, J.J., Henriksen-Lacey, M., Coronado-Puchau, M., Liz-Marzán, L.M., 2016. Inorganic nanoparticles for biomedicine: where materials scientists meet medical research. Materials Today 19, 19–28.

Gu, X., Ding, F., Williams, D.F., 2014. Neural tissue engineering options for peripheral nerve regeneration. Biomaterials 35, 6143–6156.

Gu, X., Kwok, R.T.K., Lam, J.W.Y., Tang, B.Z., 2017. AIEgens for biological process monitoring and disease theranostics. Biomaterials 146, 115–135.

Guerra, A., Cano, P., Rabionet, M., Puig, T., Ciurana, J., 2018. 3D-Printed PCL/PLA composite stents: towards a new solution to cardiovascular problems. Materials 11, 1679.

Guerrini, L., Rodriguez-Loureiro, I., Correa-Duarte, M.A., Lee, Y.H., Ling, X.Y., García de Abajo, F.J., et al., 2014. Chemical speciation of heavy metals by surface-enhanced Raman scattering spectroscopy: identification and quantification of inorganic- and methyl-mercury in water. Nanoscale 6, 8368–8375.

Guerrini, L., Arenal, R., Mannini, B., Chiti, F., Pini, R., Matteini, P., et al., 2015. SERS detection of amyloid oligomers on metallorganic-decorated plasmonic beads. ACS Applied Materials and Interfaces 7, 9420–9428.

Guo, B., Ma, P.X., 2014. Synthetic biodegradable functional polymers for tissue engineering: a brief review. Science China Chemistry 57, 490–500.

Guo, X., Cheng, Y., Zhao, X., Luo, Y., Chen, J., Yuan, W.-E., 2018. Advances in redox-responsive drug delivery systems of tumor microenvironment. Journal of Nanobiotechnology 16, 74.

Haythornthwaite, A., Stoelzle, S., Hasler, A., Kiss, A., Mosbacher, J., George, M., et al., 2012. Characterizing human ion channels in induced pluripotent stem cell-derived neurons. Journal of Biomolecular Screening 17, 1264–1272.

Hosseini, V., Kollmannsberger, P., Ahadian, S., Ostrovidov, S., Kaji, H., Vogel, V., et al., 2014. Fiber-assisted molding (FAM) of surfaces with tunable curvature to guide cell alignment and complex tissue architecture. Small 10, 4851–4857.

Hu, P., Chen, L., Kang, X., Chen, S., 2016. Surface functionalization of metal nanoparticles by conjugated metal—ligand interfacial bonds: impacts on intraparticle charge transfer. Accounts of Chemical Research 49, 2251–2260.

Huang, Q., Wang, Y., Chen, X., Wang, Y., Li, Z., Du, S., et al., 2018. Nanotechnology-based strategies for early cancer diagnosis using circulating tumor cells as a liquid biopsy. Nanotheranostics 2, 21–41.

Hunault, M.O.J.Y., Loisel, C., Bauchau, F., Lemasson, Q., Pacheco, C., Pichon, L., et al., 2017. Nondestructive redox quantification reveals glassmaking of rare French Gothic stained glasses. Analytical Chemistry 89, 6277—6284.

Iijima, S., Ichihashi, T., 1993. Single-shell carbon nanotubes of 1-nm diameter. Nature 363, 603—605.

Iravani, S., 2014. Bacteria in nanoparticle synthesis: current status and future prospects. International Scholarly Research Notices. https://doi.org/10.1155/2014/359316.

James, M.L., Gambhir, S.S., 2012. A molecular imaging primer: modalities, imaging agents, and applications. Physiological Reviews 92, 897—965.

Jiang, Y., Wang, Z., Dai, Z., 2016. Preparation of silicon-carbon-based Dots@Dopamine and its application in intracellular Ag(+) detection and cell imaging. ACS Applied Materials and Interfaces 8, 3644—3650.

Junk, A., Riess, F., 2006. From an idea to a vision: there's plenty of room at the bottom. American Journal of Physics 74, 825—830.

Kalepu, S., Nekkanti, V., 2015. Insoluble drug delivery strategies: review of recent advances and business prospects. Acta Pharmaceutica Sinica B. 5, 442—453.

Kedar, U., Phutane, P., Shidhaye, S., Kadam, V., 2010. Advances in polymeric micelles for drug delivery and tumor targeting. Nanomedicine: Nanotechnology, Biology and Medicine 6, 714—729.

Khawaja, A.Z., Cassidy, D.B., Al Shakarchi, J., McGrogan, D.G., Inston, N.G., Jones, R.G., 2015. Revisiting the risks of MRI with Gadolinium based contrast agents—review of literature and guidelines. Insights Imaging 6, 553—558.

Kim, W., Jung, J., 2016. Polymer brush: a promising grafting approach to scaffolds for tissue engineering. BMB Rep 49, 655—661.

Kim, S., Lee, S., Kim, K., 2018. Bone tissue engineering strategies in Co-delivery of bone morphogenetic protein-2 and biochemical signaling factors. Advances in Experimental Medicine & Biology 1078, 233—244.

Kneipp, K., Kneipp, H., Kneipp, J., 2006. Surface-enhanced Raman scattering in local optical fields of silver and gold nanoaggregates-from single-molecule Raman spectroscopy to ultrasensitive probing in live cells. Accounts of Chemical Research 39, 443—450.

Kojima, C., Turkbey, B., Ogawa, M., Bernardo, M., Regino, C.A.S., Bryant, L.H., et al., 2011. Dendrimer-based MRI contrast agents: the effects of PEGylation on relaxivity and pharmacokinetics. Nanomedicine: Nanotechnology, Biology and Medicine 7, 1001—1008.

Koller, M., 2018. Biodegradable and biocompatible polyhydroxy-alkanoates (PHA): auspicious microbial macromolecules for pharmaceutical and therapeutic applications. Molecules 23, 362.

Kono, K., Torikoshi, Y., Mitsutomi, M., Itoh, T., Emi, N., Yanagie, H., et al., 2001. Novel gene delivery systems: complexes of fusigenic polymer-modified liposomes and lipoplexes. Gene Therapy 8, 5.

Koo, Y.-E.L., Cao, Y., Kopelman, R., Koo, S.M., Brasuel, M., Philbert, M.A., 2004. Real-time measurements of dissolved oxygen inside live cells by organically modified silicate fluorescent nanosensors. Analytical Chemistry 76, 2498—2505.

Krasia-Christoforou, T., Georgiou, T.K., 2013. Polymeric theranostics: using polymer-based systems for simultaneous imaging and therapy. Journal of Materials Chemistry B 1, 3002.

Krenkel, M., Markus, A., Bartels, M., Dullin, C., Alves, F., Salditt, T., 2015. Phase-contrast zoom tomography reveals precise locations of macrophages in mouse lungs. Scientific Reports 5, 9973.

Krukemeyer, M.G., Krenn, V., Huebner, F., Wagner, W., Resch, R., 2015. History and possible uses of nanomedicine based on nanoparticles and nanotechnological progress. Journal of Nanomedicine & Nanotechnology 06. https://doi.org/10.4172/2157-7439.1000336.

Kulthe, S.S., Choudhari, Y.M., Inamdar, N.N., Mourya, V., 2012. Polymeric micelles: authoritative aspects for drug delivery. Designed Monomers and Polymers 15, 465—521.

Kumari, A., Yadav, S.K., Yadav, S.C., 2010. Biodegradable polymeric nanoparticles based drug delivery systems. Colloids and Surfaces B: Biointerfaces 75, 1—18.

Kumbar, S., Laurencin, C., Deng, M., 2014. Natural and Synthetic Biomedical Polymers. https://doi.org/10.1016/C2011-0-07330-1.

Kuna, J.J., Voïtchovsky, K., Singh, C., Jiang, H., Mwenifumbo, S., Ghorai, P.K., et al., 2009. The effect of nanometre-scale structure on interfacial energy. Nature Materials 8, 837—842.

Lamanna, G., Battigelli, A., Ménard-Moyon, C., Bianco, A., 2012. Multifunctionalized carbon nanotubes as advanced multimodal nanomaterials for biomedical applications. Nanotechnology Reviews 1, 17—29.

Langer, R., Vacanti, J.P., 1993. Tissue engineering. Science 260, 920—926.

Lee, S.H., Chung, B.H., Park, T.G., Nam, Y.S., Mok, H., 2012. Small-interfering RNA (siRNA)-Based functional micro- and nanostructures for efficient and selective gene silencing. Accounts of Chemical Research 45, 1014—1025.

Leonardi, A.A., Lo Faro, M.J., Petralia, S., Fazio, B., Musumeci, P., Conoci, S., et al., 2018. Ultrasensitive label- and PCR-free genome detection based on cooperative hybridization of silicon nanowires optical biosensors. ACS Sensors 3, 1690—1697.

Lepeltier, E.A., Nuhn, L., Lehr, C.-M., Zentel, R., 2015. Not just for tumor targeting: unmet medical needs and opportunities for nanomedicine. Nanomedicine 10, 3147—3166.

Leuvering, J.H., Thal, P.J., Van der Waart, M., Schuurs, A.H., 1981. A sol particle agglutination assay for human chorionic gonadotrophin. Journal of Immunological Methods 45, 183—194.

Li, D.-W., Qin, L.-X., Li, Y., Nia, R.P., Long, Y.-T., Chen, H.-Y., 2011. CdSe/ZnS quantum dot-Cytochrome c bioconjugates for selective intracellular O_2^- sensing. Chemical Communications 47, 8539—8541.

Li, Y., Li, N., Pan, W., Yu, Z., Yang, L., Tang, B., 2017. Hollow mesoporous silica nanoparticles with tunable structures for controlled drug delivery. ACS Applied Materials and Interfaces 9, 2123—2129.

Li, S., Poche, J.N., Liu, Y., Scherr, T., McCann, J., Forghani, A., et al., 2018. Hybrid synthetic-biological hydrogel system for adipose tissue regeneration. Macromolecular Bioscience 1800122.

Liu, Q., Liu, Y., Wu, F., Cao, X., Li, Z., Alharbi, M., et al., 2018. Highly sensitive and wearable In_2O_3 nanoribbon transistor biosensors with integrated on-chip gate for glucose monitoring in body fluids. ACS Nano 12, 1170–1178.

Long, N.V., Yang, Y., Teranishi, T., Thi, C.M., Cao, Y., Nogami, M., 2015. Biomedical applications of advanced multifunctional magnetic nanoparticles. Journal of Nanoscience and Nanotechnology 15, 10091–10107.

Luby, B.M., Charron, D.M., MacLaughlin, C.M., Zheng, G., 2017. Activatable fluorescence: from small molecule to nanoparticle. Advanced Drug Delivery Reviews 113, 97–121.

Luckachan, G.E., Pillai, C.K.S., 2011. Biodegradable polymers-a review on recent trends and emerging perspectives. Journal of Polymers and the Environment 19, 637–676.

Madaan, K., Kumar, S., Poonia, N., Lather, V., Pandita, D., 2014. Dendrimers in drug delivery and targeting: drug-dendrimer interactions and toxicity issues. Journal of Pharmacy and BioAllied Sciences 6, 139–150.

Maherani, B., Arab-Tehrany, E., Mozafari, M.R., Gaiani, C., Linder, M., 2011. Liposomes: a review of manufacturing techniques and targeting strategies. Current Nanoscience 7, 436–452.

Maiellaro, K., Taylor, W.R., 2007. The role of the adventitia in vascular inflammation. Cardiovascular Research 75, 640–648.

Maitz, M.F., 2015. Applications of synthetic polymers in clinical medicine. Biosurface and Biotribology 1, 161–176.

Makarov, V.V., Love, A.J., Sinitsyna, O.V., Makarova, S.S., Yaminsky, I.V., Taliansky, M.E., et al., 2014. 'Green' nanotechnologies: synthesis of metal nanoparticles using plants. Acta Naturae. https://doi.org/10.1039/c1gc15386b.

Mano, J.F., Silva, G.A., Azevedo, H.S., Malafaya, P.B., Sousa, R.A., Silva, S.S., et al., 2007. Natural origin biodegradable systems in tissue engineering and regenerative medicine: present status and some moving trends. Journal of The Royal Society Interface 4, 999–1030.

Mappes, T., Jahr, N., Csaki, A., Vogler, N., Popp, J., Fritzsche, W., 2012. The invention of immersion ultramicroscopy in 1912-The birth of nanotechnology? Angewandte Chemie International Edition 51, 11208–11212.

Marin, E., Briceño, M.I., Caballero-George, C., 2013. Critical evaluation of biodegradable polymers used in nanodrugs. International Journal of Nanomedicine 8, 3071–3091.

Medintz, I.L., Stewart, M.H., Trammell, S.A., Susumu, K., Delehanty, J.B., Mei, B.C., et al., 2010. Quantum-dot/dopamine bioconjugates function as redox coupled assemblies for in vitro and intracellular pH sensing. Nature Materials 9, 676–684.

Milligan, C.J., Li, J., Sukumar, P., Majeed, Y., Dallas, M.L., English, A., et al., 2009. Robotic multiwell planar patch-clamp for native and primary mammalian cells. Nature Protocols 4, 244–255.

Mitragotri, S., Burke, P.A., Langer, R., 2014. Overcoming the challenges in administering biopharmaceuticals: formulation and delivery strategies. Nature Reviews Drug Discovery 13, 655–672.

Mittal, A., Schulze, K., Ebensen, T., Weissmann, S., Hansen, S., Guzmán, C.A., et al., 2015. Inverse micellar sugar glass (IMSG) nanoparticles for transfollicular vaccination. Journal of Controlled Release 206, 140–152.

Mohajeri, M., Behnam, B., Sahebkar, A., 2018. Biomedical applications of carbon nanomaterials: drug and gene delivery potentials. Journal of Cellular Physiology 234, 298–319.

Mohamed, S., Parayath, N.N., Taurin, S., Greish, K., 2014. Polymeric nano-micelles: versatile platform for targeted delivery in cancer. Therapeutic Delivery 5, 1101–1121.

Mohammadi, S., Larsson, E., Alves, F., Dal Monego, S., Biffi, S., Garrovo, C., et al., 2014. Quantitative evaluation of a single-distance phase-retrieval method applied on in-line phase-contrast images of a mouse lung. Journal of Synchrotron Radiation 21, 784–789.

Montalti, M., Cantelli, A., Battistelli, G., 2015. Nanodiamonds and silicon quantum dots: ultrastable and biocompatible luminescent nanoprobes for long-term bioimaging. Chemical Society Reviews 44, 4853–4921.

Movassaghian, S., Merkel, O.M., Torchilin, V.P., 2015. Applications of polymer micelles for imaging and drug delivery. Wiley Interdiscip Rev Nanomedicine Nanobiotechnology 7, 691–707.

Muller, E.W., Panitz, J.A., McLane, S.B., 1968. The atom-probe field ion microscope. Review of Scientific Instruments 39, 83–86.

Nakanishi, J., Takarada, T., Yamaguchi, K., Maeda, M., 2008. Recent advances in cell micropatterning techniques for bioanalytical and biomedical sciences. Analytical Sciences 24, 67–72.

Nirosa, T., Roy, A., 2013. Carbon nanotubes – a nanotechnolgy based drug delivery system and its recent trends: a review. International Journal of PharmTech Research 5, 1765–1768.

Niu, R., Zhao, P., Wang, H., Yu, M., Cao, S., Zhang, F., et al., 2011. Preparation, characterization, and antitumor activity of paclitaxel-loaded folic acid modified and TAT peptide conjugated PEGylated polymeric liposomes. Journal of Drug Targeting 19, 373–381.

Oberlin, A., Endo, M., Koyama, T., 1976. Filamentous growth of carbon through benzene decomposition. Journal of Crystal Growth 32, 335–349.

Oblatt-Montal, M., Reddy, G.L., Iwamoto, T., Tomich, J.M., Montal, M., 1994. Identification of an ion channel-forming motif in the primary structure of CFTR, the cystic fibrosis chloride channel. Proceedings of the National Academy of Sciences of the United States of America 91, 1495–1499.

Onoue, S., Yamada, S., Chan, H.-K., 2014. Nanodrugs: pharmacokinetics and safety. International Journal of Nanomedicine 9, 1025–1037.

Pan, W., Chen, W., Jiang, X., 2010. Microfluidic western blot. Analytical Chemistry 82, 3974–3976.

Paranjpe, M., Müller-Goymann, C., 2014. Nanoparticle-mediated pulmonary drug delivery: a review. International Journal of Molecular Sciences 15, 5852–5873.

Park, E.J., Brasuel, M., Behrend, C., Philbert, M.A., Kopelman, R., 2003. Ratiometric optical PEBBLE nanosensors for real-time magnesium ion concentrations inside viable cells. Analytical Chemistry 75, 3784–3791.

Patra, J.K., Baek, K.H., 2014. Green nanobiotechnology: factors affecting synthesis and characterization techniques. Journal of Nanomaterials 2014. https://doi.org/10.1155/2014/417305.

Pelaz, B., Alexiou, C., Alvarez-Puebla, R.A., Alves, F., Andrews, A.M., Ashraf, S., et al., 2017. Diverse applications of nanomedicine. ACS Nano 11, 2313–2381.

Pellá, M.C.G., Lima-Tenório, M.K., Tenório-Neto, E.T., Guilherme, M.R., Muniz, E.C., Rubira, A.F., 2018. Chitosan-based hydrogels: from preparation to biomedical applications. Carbohydrate Polymers 196, 233–245.

Pina, S., Oliveira, J.M., Reis, R.L., 2015. Natural-based nanocomposites for bone tissue engineering and regenerative medicine: a review. Advanced Materials 27, 1143–1169.

Polo, E., del Pino, P., Pelaz, B., Grazu, V., de la Fuente, J.M., 2013. Plasmonic-driven thermal sensing: ultralow detection of cancer markers. Chemical Communications 49, 3676–3678.

Polyak, A., Ross, T.L., 2017. Nanoparticles for SPECT and PET imaging: towards personalized medicine and theranostics. Current Medicinal Chemistry. https://doi.org/10.2174/092986732466170830095553.

Ramos, A.P., Cruz, M.A.E., Tovani, C.B., Ciancaglini, P., 2017. Biomedical applications of nanotechnology. Biophysical Reviews. https://doi.org/10.1007/s12551-016-0246-2.

Ramos-Perez, V., Cifuentes, A., Coronas, N., de Pablo, A., Borrós, S., 2013. Modification of carbon nanotubes for gene delivery vectors. Methods in Molecular Biology 1025, 261–268.

Rana, S., Le, N.D.B., Mout, R., Saha, K., Tonga, G.Y., Bain, R.E.S., et al., 2015. A multichannel nanosensor for instantaneous readout of cancer drug mechanisms. Nature Nanotechnology 10, 65–69.

Rao, J.P., Geckeler, K.E., 2011. Polymer nanoparticles: preparation techniques and size-control parameters. Progress in Polymer Science 36, 887–913.

Raza, A., Hayat, U., Rasheed, T., Bilal, M., Iqbal, H.M.N., 2018. Redox-responsive nano-carriers as tumor-targeted drug delivery systems. European Journal of Medicinal Chemistry 157, 705–715.

Reddy, R., Reddy, N., 2018. Biomimetic approaches for tissue engineering. Journal of Biomaterials Science, Polymer Edition 29, 1667–1685.

Reneker, D.H., Yarin, A.L., 2008. Electrospinning jets and polymer nanofibers. Polymer 49, 2387–2425.

Renth, A.N., Detamore, M.S., 2012. Leveraging "raw materials" as building blocks and bioactive signals in regenerative medicine. Tissue Engineering Part B Reviews 18, 341–362.

Rivera Gil, P., Hühn, D., del Mercato, L.L., Sasse, D., Parak, W.J., 2010. Nanopharmacy: inorganic nanoscale devices as vectors and active compounds. Pharmacological Research 62, 115–125.

Rivera Gil, P., Vazquez-Vazquez, C., Giannini, V., Callao, M.P., Parak, W.J., Correa-Duarte, M.A., et al., 2013. Plasmonic nanoprobes for real-time optical monitoring of nitric oxide inside living cells. Angewandte Chemie International Edition in English 52, 13694–13698.

Rivron, N.C., Liu, J., Rouwkema, J., De Boer, J., Van Blitterswijk, C.A., 2008. Engineering vascularised tissues in vitro. European Cells and Materials 15, 27–40.

Roco, M.C., 2011. The long view of nanotechnology development: The National Nanotechnology Initiative at 10 years. Journal of Nanoparticle Research 13, 427–445.

Rosenman, G., Beker, P., Koren, I., Yevnin, M., Bank-Srour, B., Mishina, E., et al., 2011. Bioinspired peptide nanotubes: deposition technology, basic physics and nanotechnology applications. Journal of Peptide Science 17, 75–87.

Rubilar, O., Rai, M., Tortella, G., Diez, M.C., Seabra, A.B., Durán, N., 2013. Biogenic nanoparticles: copper, copper oxides, copper sulphides, complex copper nanostructures and their applications. Biotechnology Letters 35, 1365–1375.

Ruska, E., 1987. The development of the electron microscope and of electron microscopy. Bioscience Reports 7, 607–629.

Sanles-Sobrido, M., Rodríguez-Lorenzo, L., Lorenzo-Abalde, S., González-Fernández, A., Correa-Duarte, M.A., Alvarez-Puebla, R.A., et al., 2009. Label-free SERS detection of relevant bioanalytes on silver-coated carbon nanotubes: The case of cocaine. Nanoscale 1, 153–158.

Sattler, K.D., 2011. Handbook of Nanophysics. 3, Nanoparticles and Quantum Dots. CRC Press.

Schmidt, C.W., 2009. Nanotechnology-related environment, health, and safety research. Environmental Health Perspectives 117, A158–A161.

Schwendener, R.A., 2014. Liposomes as vaccine delivery systems: a review of the recent advances. Therapeutic Advances in Vaccines 2, 159–182.

Sercombe, L., Veerati, T., Moheimani, F., Wu, S.Y., Sood, A.K., Hua, S., 2015. Advances and challenges of liposome assisted drug delivery. Frontiers in Pharmacology 6, 286.

Shao, N., Su, Y., Hu, J., Zhang, J., Zhang, H., Cheng, Y., 2011. Comparison of generation 3 polyamidoamine dendrimer and generation 4 polypropylenimine dendrimer on drug loading, complex structure, release behavior, and cytotoxicity. International Journal of Nanomedicine 6, 3361–3372.

Shcharbin, D.G., Klajnert, B., Bryszewska, M., 2009. Dendrimers in gene transfection. Biochemistry 74, 1070–1079.

Shin, T.H., Choi, J.S., Yun, S., Kim, I.S., Song, H.T., Kim, Y., et al., 2014. T(1) and T(2) dual-mode MRI contrast agent for enhancing accuracy by engineered nanomaterials. ACS Nano 8, 3393–3401.

Shinde, S.B., Fernandes, C.B., Patravale, V.B., 2012. Recent trends in in-vitro nanodiagnostics for detection of pathogens. Journal of Controlled Release 159, 164–180.

Siddiqui, N., Asawa, S., Birru, B., Baadhe, R., Rao, S., 2018. PCL-based composite scaffold matrices for tissue engineering applications. Molecular Biotechnology 60, 506–532.

Singh, R., Lillard, J.W., 2009. Nanoparticle-based targeted drug delivery. Experimental and Molecular Pathology 86, 215–223.

Singh, N., Jenkins, G.J.S.S., Asadi, R., Doak, S.H., 2010. Potential toxicity of superparamagnetic iron oxide nanoparticles (SPION). Nano Reviews 1, 1—15.

Siu, K.S., Zheng, X., Liu, Y., Zhang, Y., Zhang, X., Chen, D., et al., 2014. Single-walled carbon nanotubes noncovalently functionalized with lipid modified polyethylenimine for sirna delivery in vitro and in vivo. Bioconjugate Chemistry 25, 1744—1751.

Stirland, D.L., Nichols, J.W., Miura, S., Bae, Y.H., 2013. Mind the gap: a survey of how cancer drug carriers are susceptible to the gap between research and practice. Journal of Controlled Release 172, 1045—1064.

Stuart, M.A.C., Huck, W.T.S., Genzer, J., Müller, M., Ober, C., Stamm, M., et al., 2010. Emerging applications of stimuli-responsive polymer materials. Nature Materials 9, 101.

Suarato, G., Bertorelli, R., Athanassiou, A., 2018. Borrowing from nature: biopolymers and biocomposites as smart wound care materials. Front Bioeng Biotechnol 6, 137.

Sumner, J.P., Aylott, J.W., Monson, E., Kopelman, R., 2002. A fluorescent PEBBLE nanosensor for intracellular free zinc. The Analyst 127, 11—16.

Sun, Q., Radosz, M., Shen, Y., 2012. Challenges in design of translational nanocarriers. Journal of Controlled Release 164, 156—169.

Sun, Q., Sun, X., Ma, X., Zhou, Z., Jin, E., Zhang, B., et al., 2014. Integration of nanoassembly functions for an effective delivery cascade for cancer drugs. Advanced Materials 26, 7615—7621.

Sun, T., Li, Y., Huang, Y., Zhang, Z., Yang, W., Du, Z., et al., 2016. Targeting glioma stem cells enhances anti-tumor effect of boron neutron capture therapy. Oncotarget 7, 43095—43108.

Sun, H., Dong, Y., Feijen, J., Zhong, Z., 2018. Peptide-decorated polymeric nanomedicines for precision cancer therapy. Journal of Controlled Release 290, 11—27.

Sutradhar, K.B., Amin, M.L., 2014. Nanotechnology in cancer drug delivery and selective targeting. ISRN Nanotechnol. https://doi.org/10.1155/2014/939378.

Szefler, B., 2018. Nanotechnology, from quantum mechanical calculations up to drug delivery. International Journal of Nanomedicine 13, 6143—6176.

Taniguchi, N., 1974. On the Basic Concept of Nanotechnology. Proc Intl Conf Prod Eng Tokyo, Part II, Japan Soc Precis Eng, pp. 18—23.

Tong, R., Gabrielson, N.P., Fan, T.M., Cheng, J., 2012. Polymeric nanomedicines based on poly(lactide) and poly(lactide-co-glycolide). Current Opinion in Solid State & Materials Science 16, 323—332.

Tsai, C.-H., Wang, P.-Y., Lin, I.-C., Huang, H., Liu, G.-S., Tseng, C.-L., 2018. Ocular drug delivery: role of degradable polymeric nanocarriers for ophthalmic application. International Journal of Molecular Sciences 19, 2830.

Ulbrich, K., Holá, K., Šubr, V., Bakandritsos, A., Tuček, J., Zbořil, R., 2016. Targeted drug delivery with polymers and magnetic nanoparticles: covalent and noncovalent approaches, release control, and clinical studies. Chemistry Review 116, 5338—5431.

Valentini, P., Fiammengo, R., Sabella, S., Gariboldi, M., Maiorano, G., Cingolani, R., et al., 2013. Gold-nanoparticle-based colorimetric discrimination of cancer-related point mutations with picomolar sensitivity. ACS Nano 7, 5530—5538.

Vallabani, N.V.S., Singh, S., 2018. Recent advances and future prospects of iron oxide nanoparticles in biomedicine and diagnostics. 3 Biotech 8, 279.

Vance, M.E., Kuiken, T., Vejerano, E.P., McGinnis, S.P., Hochella, M.F., Hull, D.R., 2015. Nanotechnology in the real world: redeveloping the nanomaterial consumer products inventory. Beilstein Journal of Nanotechnology 6, 1769—1780.

Vishinkin, R., Haick, H., 2015. Nanoscale sensor technologies for disease detection via volatolomics. Small 11, 6142—6164.

Wakaskar, R.R., 2017. General overview of lipid—polymer hybrid nanoparticles, dendrimers, micelles, liposomes, spongosomes and cubosomes. Journal of Drug Targeting 1—8.

Walmsley, G.G., McArdle, A., Tevlin, R., Momeni, A., Atashroo, D., Hu, M.S., et al., 2015. Nanotechnology in bone tissue engineering. Nanomedicine: Nanotechnology, Biology and Medicine 11, 1253—1263.

Wang, J., Mao, W., Lock, L.L., Tang, J., Sui, M., Sun, W., et al., 2015. The role of micelle size in tumor accumulation, penetration, and treatment. ACS Nano 9, 7195—7206.

Wang, J., Liu, R., Liu, B., 2016. Cadmium-containing quantum dots: current perspectives on their application as nanomedicine and toxicity concerns. Mini Reviews in Medicinal Chemistry 16, 905—916.

Weissenböck, A., Wirth, M., Gabor, F., 2004. WGA-grafted PLGA-nanospheres: preparation and association with Caco-2 single cells. Journal of Controlled Release 99, 383—392.

Weissig, V., Pettinger, T.K., Murdock, N., 2014. Nanopharmaceuticals (part 1): products on the market. International Journal of Nanomedicine 9, 4357—4373.

Weng, X., Wang, M., Ge, J., Yu, S., Liu, B., Zhong, J., et al., 2009. Carbon nanotubes as a protein toxin transporter for selective HER2-positive breast cancer cell destruction. Molecular BioSystems 5, 1224.

Wu, S., Li, Z., Han, J., Han, S., 2011. Dual colored mesoporous silica nanoparticles with pH activable rhodamine-lactam for ratiometric sensing of lysosomal acidity. Chemical Communications 47, 11276—11278.

Xu, H., Aylott, J.W., Kopelman, R., 2002. Fluorescent nano-PEBBLE sensors designed for intracellular glucose imaging. The Analyst 127, 1471—1477.

Xu, W., Kattel, K., Park, J.Y., Chang, Y., Kim, T.J., Lee, G.H., 2012. Paramagnetic nanoparticle T1 and T2 MRI contrast agents. Physical Chemistry Chemical Physics 14, 12687.

Xu, Y., Wang, X., Zhang, W.L., Lv, F., Guo, S., 2017. Recent progress in two-dimensional inorganic quantum dots. Chemical Society Reviews. https://doi.org/10.1039/c7cs00500h.

Yang, W.-W., Pierstorff, E., 2012. Reservoir-based polymer drug delivery systems. Journal of Laboratory Automation 17, 50–58.

Yang, D., Yang, F., Hu, J., Long, J., Wang, C., Fu, D., et al., 2009. Hydrophilic multi-walled carbon nanotubes decorated with magnetite nanoparticles as lymphatic targeted drug delivery vehicles. Chemical Communications 4447–4449.

Ye, H., Zhang, K., Kai, D., Li, Z., Loh, X.J., 2018. Polyester elastomers for soft tissue engineering. Chemical Society Reviews 47, 4545–4580.

Yu, Z., Yu, M., Zhang, Z., Hong, G., Xiong, Q., 2014. Bovine serum albumin nanoparticles as controlled release carrier for local drug delivery to the inner ear. Nanoscale Research Letters 9, 343.

Yue, K., Trujillo-de Santiago, G., Alvarez, M.M., Tamayol, A., Annabi, N., Khademhosseini, A., 2015. Synthesis, properties, and biomedical applications of gelatin methacryloyl (GelMA) hydrogels. Biomaterials. https://doi.org/10.1016/j.biomaterials.2015.08.045.

Zhang, W., Zhang, Z., Zhang, Y., 2011. The application of carbon nanotubes in target drug delivery systems for cancer therapies. Nanoscale Research Letters 6, 1–22.

Zhang, S., Sun, H.-J., Hughes, A.D., Moussodia, R.-O., Bertin, A., Chen, Y., et al., 2014a. Self-assembly of amphiphilic Janus dendrimers into uniform onion-like dendrimersomes with predictable size and number of bilayers. Proceedings of the National Academy of Sciences 111, 9058–9063.

Zhang, Y., Zhang, L., Sun, J., Liu, Y., Ma, X., Cui, S., et al., 2014b. Point-of-care multiplexed assays of nucleic acids using microcapillary-based loop-mediated isothermal amplification. Analytical Chemistry 86, 7057–7062.

Zhang, Y., Chen, R., Xu, L., Ning, Y., Xie, S., Zhang, G.-J., 2015a. Silicon nanowire biosensor for highly sensitive and multiplexed detection of oral squamous cell carcinoma biomarkers in saliva. Analytical Sciences 31, 73–78.

Zhang, C., Li, C., Liu, Y., Zhang, J., Bao, C., Liang, S., et al., 2015b. Gold nanoclusters-based nanoprobes for simultaneous fluorescence imaging and targeted photodynamic therapy with superior penetration and retention behavior in tumors. Advanced Functional Materials 25, 1314–1325.

Zhang, J., Tang, H., Liu, Z., Chen, B., 2017. Effects of major parameters of nanoparticles on their physical and chemical properties and recent application of nanodrug delivery system in targeted chemotherapy. International Journal of Nanomedicine. https://doi.org/10.2147/IJN.S148359.

Zhou, S.A., Brahme, A., 2008. Development of phase-contrast X-ray imaging techniques and potential medical applications. Physica Medica 24, 129–148.

Zhu, X., Chen, S., Luo, Q., Ye, C., Liu, M., Zhou, X., 2015. Body temperature sensitive micelles for MRI enhancement. Chemical Communications 51, 9085–9088.

Zylberberg, C., Matosevic, S., 2016. Pharmaceutical liposomal drug delivery: a review of new delivery systems and a look at the regulatory landscape. Drug Delivery 23, 3319–3329.

CHAPTER 9

Cancer Cytosensing Approaches in Miniaturized Settings Based on Advanced Nanomaterials and Biosensors

BUDDHADEV PUROHIT • ASHUTOSH KUMAR • KULDEEP MAHATO •
SHARMILI ROY • PRANJAL CHANDRA, MSC, MTECH, PHD

INTRODUCTION

Among all the fatal diseases humankind has ever encountered, cancer has long been a major health concern across the globe. As per the latest annual cancer report by the World Health Organization, there are more than 200 different types of cancers reported, namely, pulmonary, breast, gastrointestinal. These have counted 8.8 million deaths and additionally 14 million new cases annually (Stewart and Wild, 2017; Siegel et al., 2018). Fundamentally, cancer is a disorder caused because of the imbalances in cell cycle homeostasis, where the checkpoints related to cell division failed to operate synchronously. The cell cycle checkpoints are the regulator for controlled proliferation of the cells, which is a normal phenomenon in living beings. There are several factors, which influences the origin and progression of such abnormality, which are either hereditary or sporadic mutations. Apart from these, environmental factors also play an important role in different stages of cancer (Mahato et al., 2017a). The occurrence of such changes or mutations in some important genes, that is, tumor suppressor gene, protooncogene, and genes involved in DNA repair system, which are a part of cell cycle regulation that leads to cancer development by disrupting the cellular homeostasis.

Conventional approaches to the cancer detections use the imaging tests followed by confirmatory biopsy techniques, where high-end sophisticated instrument, namely, X-ray, ultrasound-based imaging, mammography, nuclear magnetic resonance, computed tomography, positron emission tomography—based scanning

method are commonly used. Furthermore, several extensive lab protocols based on the genomics and proteomics principles such as polymerase chain reaction, mass spectroscopy, and electrophoresis are followed to detect a particular gene sequence, cancerous protein to identify the cancer type (Smith et al., 2002). Although these methods are highly improvised over time for better diagnostic results, these methods still require a considerable amount of time, dedicated laboratory space, particular environmental condition to conduct the test, high costs, and skilled human resources. Hence, the outcome of these tests becomes a function of multiple factors that can lead to errors in results. Not only this, the development process of cancer often takes a very long time, that is, from a single base mutation to the metastatic movement of neoplastic cells such that the appearance of susceptible symptoms often takes place at very late stage of disease. Thus, the early diagnosis is of foremost demand, which offers a better recovery from the diseases and decrease in the mortality rate (Etzioni et al., 2003).

Biosensors are rapidly gaining pace for the detection of different diseases. Biosensor is an analytical devise consisting of (1) a biological molecule acting as recognition interface, (2) a transducer for quantifiable signal generation, (3) an amplifier for signal amplification and background noise cancellation, and (4) a digital platform to show the result (Mahato and Chandra, 2019a). Several reports demonstrated development of biosensors as a cost-effective, reliable, and highly efficient method for disease detection. The biosensors follow certain principles for the detection

Nanotechnology in Modern Animal Biotechnology. https://doi.org/10.1016/B978-0-12-818823-1.00009-0

methodologies, and based on these principles, biosensors are classified into different subgroups, that is, optical, piezoelectric, and electrochemical, of which the electrochemical biosensors hold more exciting promises for better diagnostic methodologies. Recent developments in the nanotechnology lead to miniaturization of the analytical devices causing a revolution in the biosensor development. Miniaturized sensors made the point-of-care (POC) diagnostic application a reality by producing result in real time for clinical as well as daily health monitoring (Chandra, 2015a,b; Mahato et al., 2017a; Baranwal and Chandra, 2018). Biosensors use biomarkers (a molecule of biological origin that refers to a certain disease or condition) as the recognition molecules for cancer detection (Chandra et al., 2013a). The biomolecule can be a genetic biomarker, cancer protein antigen, metabolic biomarkers, and whole live cells detected for the recognition of a particular cancer type (Pallela et al., 2016; Chandra et al., 2015). The traditional lab-based detection methodologies use a wide range of molecules of biological origin, viz., proteins, genes, miRNA, glycoproteins, which require extensive extraction procedures generally compromised by the sensing performance. In addition, extensive efforts have been put to replace such lab-based techniques by using various biosensing strategies that commonly target biological macromolecule-based biomarkers. In practical perspective, the extraction of such macromolecule-based biomarkers is not only tedious but also time-consuming, which limits the usage of such evident device in emergency as well as hospital setting. To overcome such limitations, the cell-based detections are universally accepted because of their simplistic methods, viz., no sample preparation. For instance, CTC has found a prevalent attention in the clinical biosensor development, as it confirms the important transition phase of cancer advancement. Cell-based biosensor is also termed as "cytosensors," which is a type of biosensor that detects a whole cell following the interaction with cell surface biomarkers. The detection principle of cancer cytosensor is based on the fact that each cancer cell type has a signature cell surface expression pattern that is different from other cancer cell type as well as the normal noncancer cells. The proteins, glycoproteins, and carbohydrate moieties on the cell surface are common target molecules for cancer cytosensors, which are detected by various biosensing methods, viz., optical, electrochemical (amperometry, conductimetry, potentiometry, electrochemical impedance spectroscopy (EIS), electrochemiluminescence, and photoelectrochemiluminescence) methods (Akhtar et al., 2018; Bansal

et al., 2017; Chandra et al., 2013a; Zhu et al., 2014, Chandra et al., 2015). There is a complete array of biomarkers expressed on the cell surface of cancerous cells already known, for developing a sensor based on this property (Chatterje and Zetter, 2005). The cytosensors require a minimal protocol for sample preparation for the detection, removing several laboratory-based methods, which increases its efficiency (Tepeli et al., 2015). Limited work has been done to exploit the true potential of cytosensors for cancer cell detection as compared with other biosensors. Therefore, in this report, we have comprehensively described the importance of various cell surface biomarkers, utilization of nanomaterials for signal enhancement, and the recent trends in cancer detection by cytosensor technology that we think would contribute to present literature for better understanding for the scientific community on whole cancer cell diagnosis.

CANCER BIOMARKERS USED IN CYTOSENSOR

Cancer progression follows a multistep pathway of unrestricted proliferation by self-sufficient growth signal, evasion of growth suppressor, cell migration from one site to another (metastasis), formation of new blood vessels (angiogenesis), and avoiding the cell death (Hanahan and Weinburg, 2011). Therefore, the cells try to adopt to the microenvironment for its survival, growth, and evasion to other parts of the body, which leads to certain modification in the cell metabolism, functioning, and architecture. Some of these changes, that is, expression of certain genes, proteins, glycoproteins, are very specific for a particular cancer cell type or types and are of great source of information from a detection point of view (Chandra et al., 2013a). The modifications expressed on the outer cell surface can be detected by the cytosensors where the proteins, glycoproteins, and carbohydrates act as the target molecules such as mucin 1 (MUC-1) protein in the cell surface (Mahato et al., 2017a). There are many types of classification of the cancer biomarkers available in the literature (Sawyers., 2008). The target marker molecules such as proteins are recognized and captured by antibody, aptamers, and nanoenzymes (Chandra, 2013a). Glycosylation, one such modification of proteins in normal cell, plays an important role in the cell-cell communication, cell division and growth, and immune system (Mahato et al., 2018a). Nevertheless, protein glycosylation in cancer is of great importance in disease progression, helping the malignant cells to surpass the immunological activity. This makes glycoprotein an important biomarker for cancer detection (Jelinek and Kolusheva, 2004). An extensively studied glycoprotein is carcinoembryonic antigen (CEA), an important marker for cancer detection often found in higher concentration in cancer patient's referrers to cancer development but not specific for a particular cancer type. Another glycoprotein marker, the 1MUC-1 protein is a cysteine-rich protein with a 20-amino-acid extracellular domain that acts as a recognition molecule for several cancer cell lines. A number of studies on this glycoprotein are conducted to detect several cancer cell types (Cheng et al., 2009). Sialic acid (SA)—associated glycoproteins have also been used in several cytosensor studies reported with several cancer types (Cao et al., 2015; Liu et al., 2010). This exposes its nine-carbon backbone on the cell surface, which expresses increasingly with tumor advancement imparting a net negative charge on the cell surface helping in cell-cell repulsion and eventually leads to the metastasis (Qian et al., 2012). Folic acid receptor overexpressed in several cancer cell types has been used in several cytosensors (Dervisevic et al., 2017b). With the advancement of nanotechnology, the cancer cell detection methodologies using the cancer biomarkers also got a boost to achieve a very small detection limit.

NANOMATERIALS USED IN CYTOSENSOR

Nanomaterials with their unique properties enhanced the range of functioning of a biosensor. Researchers across the world are working on exploiting the exceptional properties of the nanoscale materials, including nanotubes, nanosheets, metal nanoparticles (MNs), and conducting polymers, because of their competence to aid in the efficient electron transfer from bioreceptors to sensing electrode (Kumar et al., 2018a; Won et al., 2013; Noh et al., 2012; Chandra et al., 2011). The higher surface to volume ratio enables the particles to capture more biological analyte, more surface for attachment of recognizing molecules, and better conductivity for signal enhancement (Mahato et al., 2016). The nanoparticles behave differently in comparison with their constituting individual atoms, and again they change the property when present in close conjugation with a different complementary material forming a composite (Prasad et al., 2016b). All these properties made nanoparticle an extremely exciting molecule to develop biosensor with enhanced efficiency (Yadav et al., 2013). Nanoscale detection platform by utilizing nanoparticles miniaturized the detection instrumentation, removing dedicated lab space, extremely sophisticated detection environment, and trained human resources (Mahato et al., 2017b). This made the development of POC detection probes, which proved the global effect on cancer detection methodologies (Mahato et al., 2018b; Prasad et al., 2016a; Chandra, 2013b, 2016). Nanoparticles that are based on carbon provide a new possibility in development of modern novel biosensors because of their excellent mechanical characteristics, outstanding biocompatibility, and improved electrical conductivity. Recently, graphene, a two-dimensional (2D) carbon-based nanomaterial, is extensively used as the matrix material in designing of novel cytosensors. Graphene shows an alluring feature of biocompatibility, readily availability of sites to dock analyte or analyte recognition elements, high surface area of detection, and most importantly very low cost of synthesis. Carbon nanotubes are another set of material used in cytosensor construction, which shows efficient electron transfer potential between the electrode and the reaction site. MNs of noble metals with interesting optical, electromagnetic, and chemical size-dependent properties have been comprehensively used not only for their significant technical contemplation but also for their several industrial applications. Gold, silver, and platinum nanoparticles are another set of the materials extensively used because of their relative chemical inertness, biocompatibility, electrical conductivity, and optical property (Choudhary et al., 2016; Baranwal et al., 2018a). Nanoparticles, particularly gold (AuNPs) and silver (AgNPs), have fascinated the consideration of researchers in the scope of cell sensing (Koh et al., 2011; Shim, 2013; Chandra et al., 2010). It has been found that the application of AuNPs in the design of biosensing platform enhanced the analytical performances such as sensitivity and LOD, because of the enhanced electron transfer in between bioreceptor site and electrode. As compared with other MNs, AuNPs have considerably very low cytotoxic effects, because of which they have a greater possibility to be utilized in fabrication of cytosensor (Connor et al., 2005; Baranwal et al., 2018c; Mahato et al., 2019).

Metal oxide nanomaterials (MONs) have freshly raised into noteworthy materials, which provide an effective surface for immobilization of biorecognition elements also having predicted alignment as well as virtuous conformation and astonishing biological activity ensuring the enriched biosensing features (Baranwal et al., 2016). d and f block MONs such as Zn, Zr, Ce, Hf, Gd, Sn, Mn, and Fe have been established to be nontoxic and compatible with biological counterparts and showed catalytic properties, as well as these can be fabricated into fascinating shapes at nanolevel (Solanki et al., 2011; Deka et al., 2018; Kumar et al.,

2018b; Saxena et al., 2018). These MONs presented upgraded direct electron transfer kinetics and robust adsorption ability, also providing suitable microenvironments for the adherence of bioreceptors and consequently improving the charge transfer and better biosensing characteristics. Numerous immobilizing platforms have been fabricated with help of various MONs, which have exceptional electrical and optical properties because of unique surface to volume ratio, superior work function of interacting surface, great surface response activity, astonishing catalytic competence, and solid adsorption ability (Mandal et al., 2018).

Polymers particularly conducting in nature are utilized as optimistic aspirants as active materials for various works including development of modern cytosensors. These polymers show astonishing beneficial properties including good energy density, great electronic property, and good stability (Zhu et al., 2012b). With enormous surface areas, outstanding conductivity, and limitless flexibility, conducting polymers have captivated widespread attention in design and development of modern cytosensors. Polymers such as polyaniline, polypyrrole (PPy), and poly(ethylenimine) or their composites with other nanomaterials are ideal constituents for assembling a novel cytosensor, as they can deliver the stacking and stability to bioreceptors. Chitosan, another polymer, holds great promises in the development of cytosensor (Baranwal et al., 2018b; Bhatnagar et al., 2018). In this chapter, we have provided some insights of application of these nanomaterials as signal-enhancing materials of modern cancer cytosensors.

CYTOSENSORS FOR CANCER DETECTION

The currently used methodologies of detecting whole cell cancer cells are described in the following section emphasizing the biomarkers involved and strategies and nanomaterials utilized for enhanced signal generation. The recent development in cytosensor technology has been comprehensively described with the most common type of cancer occurring in the world population taken into consideration.

Lung Cancer

Lung cancer is the most common among all the cancer types and causes the highest number of deaths globally. Lung cancer can be grouped into two main domains, that is, small-cell lung carcinoma (SCLC) and non–small-cell lung carcinoma (NSCLC). SCLC is further classified into small-cell, mixed cell/large-cell carcinoma, and combined small-cell carcinoma, whereas NSCLC is classified into squamous cell carcinoma, adenocarcinoma, bronchoalveolar carcinoma, and large-cell carcinoma. NSCLC is a very aggressive disease accounting for more than 85% of incidence and causes a 5-year survival rate of less than 20% (Sharma et al., 2007). Considering the lung is the center for blood oxygenation, the chances of metastasis are also very high and this leads to the fact that lung cancer can be highly invasive. So a quick detection method for lung cancer is of high clinical significance (Sung and Cho, 2008). Routine lab protocols follow imaging and molecular biology techniques, but detection of the early stages of the tumor growth is not satisfactory.

A549 cell is the extensively characterized NSCLC cell line used in biosensor studies related to lung cancer. Lu et al. (2015) used m-aminophenol-based resin immobilized with AuNP for the detection of cancer cell line A549. The resin conjugated to the AuNP was a suitable immobilization platform, possesses higher surface-enhanced Raman scattering, suitable to bind immunoglobulin, and facilitates the electron transfer for enhanced signal generation. In this work, authors have immobilized Au-resin onto the glassy carbon electrode (GCE) surface followed with EGFR antibody to specifically bind to the A549 lung cancer cells. The biosensor achieved a linear range with a detection of $0-10^6$ cells/mL with a limit of detection of 5 cells/mL. Similarly, in another study, Mir et al. (2015) developed an electrochemical detection method for the same A549 human NSCLC cells selectively as shown in Fig.9.1. The MUC-1 aptamer immobilized on the self-assembly of 4-([2,2′:5′,2″- terthiophen]-3′-yl) benzoic acid (TTBA) on AuNPs and conjugated with hydrazine. The detection was done by chronoamperometry method. The range of detection is 15 to 1×10^6 cells/mL, and the lower limit of detection is of 8 cells/mL. Kara et al. (2016) developed an EIS method to detect the NSCLC cell detection based on the aptamer on screen-printed electrode. They detected A549 cell line through the aptamer and used hepatocarcinoma and HeLa cell lines as a negative control for selectivity of the sensor. The range of the lower limit of detection was 163.7 cells/mL.

Ma et al. reported development of a cytosensor for the detection of SA, an overexpressed marker glycoprotein on the A549 human lung cancer cells. Fig.9.2 shows that a lectin molecule *Sambucus nigra* agglutinin (SNA) was used to specifically recognize the cell and capture the A549 cell, based on its molecular recognition with SA (Ma et al., 2017). Hollow hornlike PPy and chitosan-AuNP was developed as the sensing platform. The cytosensor achieved a very low detection limit of 2 cells/mL. In another study, Zhang et al. (2018) reported a cytosensor based on the dual catalytic hairpin

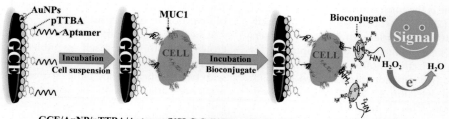

GCE/AuNP/pTTBA/Aptamer/NSLC Cell/ Bioconjugate(Aptamer-AuNP-Hydrazine)

FIG. 9.1 The schematic detection of the cancer cell by specific aptamer for MUC-1 on the cell surface and subsequent hydrogen peroxide formation for signal generation. (Reproduced with permission from Mir, T. A., Yoon, J.-H., Gurudatt, N., Won, M.-S. Shim, Y.-B. 2015. Ultrasensitive cytosensing based on an aptamer modified nanobiosensor with a bioconjugate: detection of human non-small-cell lung cancer cells. Biosensors and Bioelectronics 74, 594–600.)

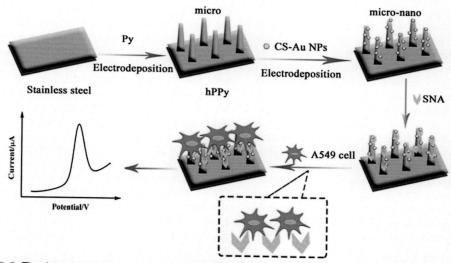

FIG. 9.2 The detection of A549 cells by SNA on chitosan-AuNP targeting the SA on the cancer cell surface. (Reproduced with permission from Ma, S., Hu, S., Wang, Q., Liu, Y., Zhao, G., Zhang, Q., Mao, C., Zhao, B. 2017. Evaluation of sialic acid based on electrochemical cytosensor with 3D micro/nanostructured sensing interface. Analytical Methods 9, 6171–6176.)

assembly method, a highly selective isothermal method of hairpin assembly devoid of any enzymes. A549 lung cancer cell line is used as model system with overexpressed MUC-1 as the recognition element. The developed cytosensor had a detection limit of 30 cells/mL and a linear range of detection of $50–10^6$ cells/mL. A lower limit of detection is one of the most important features of a cytosensor, which enables a sensor to detect a cancer cell in the earliest stage of disease progression when the concentration of the mutated neoplastic cell is very low. A cytosensor developed to detect the lung cancer cells in the early stage can reduce the global cancer burden significantly.

Breast Cancer

Breast cancer is the most predominant form of cancer in women, exceeding the occurrence of lung cancer. Currently, mammography is treated as the gold standard of preliminary diagnosis for early detection of any suspected tumor growth followed by other tests. Based on genetic and protein markers, several biosensors are developed for early detection of breast cancer, which is available in literature (Mittal et al., 2017; Zhu et al., 2012a). Among all the biomarkers, BRCA1, BRCA2, and HER 2 are treated as the most important biomarkers associated with breast cancer (Chandra et al., 2013b)

MCF-7 cell line is the most studied cell line of breast cancer. Li et al. (2010) described a method to specifically identify MCF-7 cell line with both the markers MUC-1 and CEA simultaneously by an aptamer for MUC-1 on a gold electrode followed by an anti-CEA cadmium sulfide (CdS) nanoparticle labeled anti-CEA with a minimum detection 3.3×10^2 cells/mL by voltammetric method. Zhu et al. (2013) developed an aptamer-cell-aptamer sandwich system for the detection of the cancer cell line MCF-7. The MCF7 cells were captured on the gold electrode by an MUC-1 aptamer. A secondary MUC-1 aptamer conjugated with horseradish peroxidase (HRP) was used to amplify the signal by reduction of H_2O_2. This double detection method used HRP-catalyzed and thionine-mediated H_2O_2 reduction to provide enhanced sensitivity as well as selectivity. The range of detection was $1-10^7$ cells/mL and a limit of detection of 100 cells/mL. In another work, Chen et al. (2014), using the same Au electrode and MUC-1 aptamer simultaneously, detected MUC-1 and folic acid receptor, which is highly expressed in metastatic breast cancer. MCF-7 cells are captured on an MUC-1 aptamer, and monodispersed nanoparticle conjugated with folic acid binds to folic acid receptor, leading to better signal on SPR with a detection limit of 500 cells/mL. Zhao et al. (2013) constructed a simple design of Au electrode functionalized with folate attached to DNA. Folate receptors are overexpressed in the cancer cells, and using folate as a recognition molecule, the cancer cells can be detected and the bound cell protects the DNA from nuclease degradation. To double check, exonuclease 1 was used, as it degrades the DNA unbound to the MCF-7 cell as described in Fig. 9.3.

The range of detection is 10^2-10^6 cells/mL with a limit of detection of 67 cells/mL.

Chen et al. in another report, used calix[4]arene crown ether to form a self-assembled monolayer on the gold surface and a selectively developed peptide with two distinct peptide parts (spacer and recognition sequences) attached to it. The first sequence binds to the calix[4]arene Au electrode, whereas the far-end sequence binds specifically to the MCF-7 cell. The measurement was done by SPR, and limit of detection was 197 cells/mL. Sheng et al. (2015) developed a method for the detection of breast cancer cell line MCF7 in an Au electrochemical platform, which enhanced signal amplification strategy by biotin-streptavidin-conjugated rolling circle amplification as shown in Fig. 9.4. The aptamer bound to the overexpressed surface protein brings its conformational changes and amplifies the signal. MUC-1 and CEA are the most common markers found in invasive breast cancer. An electrochemical biosensor developed by Wang et al. (2017) captured the MCF-7 cells on a gold electrode modified with polydA aptamer, as polydA has a greater binding affinity with the GE. They further treated it with another aptamer MUC-1/Au/nanoparticle/GO for DPV. The limit of detection is 8 cells/mL, and the range of detection is $10-10^5$ cells.

Li et al. (2017) constructed a glutathione-modified Au electrode for quantitatively measuring the cancer cell concentration based on the reactive oxygen species (ROS). The cancer cells upon induction generate larger amount of ROS as compared with the normal cell. Here, the glutathione-modified Au NP binds to the human breast cancer HepG2 cell and gets dimerized by

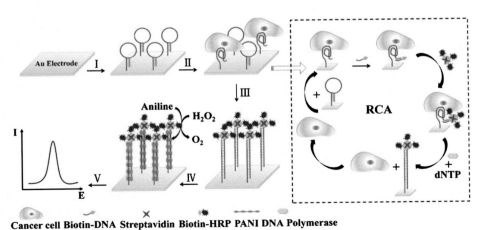

Cancer cell Biotin-DNA Streptavidin Biotin-HRP PANI DNA Polymerase

FIG. 9.3 Outline of the signal generation based on the exonuclease activity. (Reproduced with permission from Zhao, J., Zhu, L., Guo, C., Gao, T., Zhu, X., Li, G. 2013. A new electrochemical method for the detection of cancer cells based on small molecule-linked DNA. Biosensors and Bioelectronics 49, 329–333.)

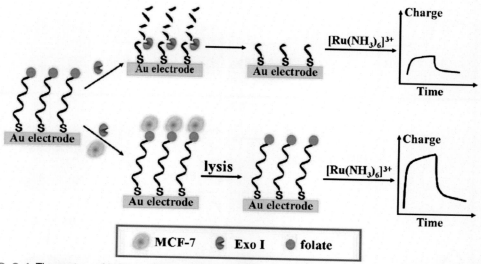

FIG. 9.4 The capture of MCF-7 cell and signal generation by the cytosensor. (Reproduced with permission from Sheng, Q., Cheng, N., Bai, W., Zheng, J. 2015. Ultrasensitive electrochemical detection of breast cancer cells based on DNA-rolling-circle-amplification-directed enzyme-catalyzed polymerization. Chemical Communications 51, 2114–2117.)

the action of ROS, and the nanoparticle aggregation leads to ROS signal amplification. The limit of detection is 25 cells/mL. Tian et al. (2017) developed a cytosensor with GCE as sensing platform. The GCE was modified rGO/AuNP for enhanced signal amplification. MUC-1 aptamer was used to specifically recognize and capture the MCF-7 cells. CuO coenzyme was used for its peroxidase activity, and CuO-S1 probe was made to recognize the cell. The cytosensor achieved a detection range from 50 to 7×10^3 cells/mL and detection limit of 27 cells/mL. In another study, Li et al. used a magnetic GCE/reduced graphene oxide/molybdenum disulfide (rGO/MoS$_2$) to effectively recognize and isolate the MCF-7 cell (Tian et al., 2018). Fe$_3$O$_4$ NPs enriched with MCF-7 cell-specific aptamer are used to separate the cells. The Fe$_3$O$_4$ NPs are adsorbed on the MCF-7 cell and bound to MGCE by application of a strong magnetic field. The rGO/MoS$_2$ and Fe$_3$O$_4$ act as enzymes synergistically to catalyze the reduction of 3,3,5,5-tetramethylbenzidine (TMB)-hydrogen peroxide (H$_2$O$_2$) system. The linear range of detection found to be 15–45 cells/mL and limit of detection 6 cells/mL.

Tang et al. (2018) reported a cytosensor for detection of MCF-7 cells by using Ag nanoflower and acetylene black because of their high surface area and biocompatibility (Tang et al., 2018). Ag nanoflower has high HRP activity catalyzing H$_2$O$_2$ reduction that enhances the signal amplification many fold. Ab- and AuNPs-modified gold electrodes act as the sensing platform. Protein G was used on it as it captures antibody on a proper orientation that expose more binding sites to bind the MCF-7 cells. The nanomaterials used in the platform increased the surface area for reaction by 5fold, and the cytosensor shows increased sensitivity, as the Ab1 recognizes and captures the cell of interest, whereas the Ab2 amplifies the signal by reducing H$_2$O$_2$ as shown in Fig. 9.5. The cytosensor achieved a linear range detection of $20-1 \times 10^6$ cells/mL and the detection limit of 3 cells/mL. Zanghelini et al. (2017) developed a POC system to distinguish between the MCF-7 and an invasive cancer cell line (Zanghelini et al., 2017). They used titanium dioxide (TiO$_2$) butterfly membrane coated with gold as the detection platform, as TiO$_2$ is known for its minimal protein degradation activity. Lectin molecules such as wheat germ agglutinin and concanavalin A (ConA) have strong affinity towards N-acetyl glucosamine (GlcNAc) and mannose moieties found on N-glycans, used to detect the cancerous cell from the normal fibroblast cell. By the electrochemical impedance method, the proposed cytosensor demonstrated limits of detection as low as 10 cells/mL and a linear range from 10 to 1.0×10^6 cells/mL. Hua et al. (2013) developed a cytosensor based on the same overexpressed MUC-1 on MCF-7 by using two

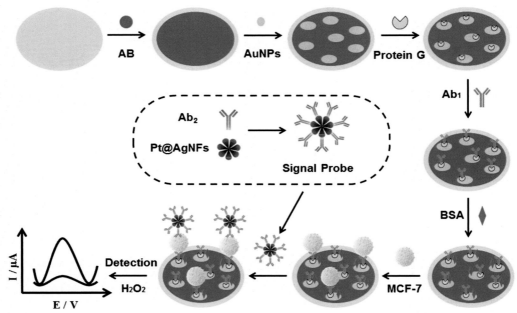

FIG. 9.5 The use of Ag nanoflower in MCF-7 cell detection. (Reproduced with permission from Tang, S., Shen, H., Hao, Y., Huang, Z., Tao, Y., Peng, Y., Guo, Y., Xie, G., Feng, W. 2018. A novel cytosensor based on Pt@ Ag nanoflowers and AuNPs/Acetylene black for ultrasensitive and highly specific detection of Circulating Tumor Cells. Biosensors and Bioelectronics.)

different aptamers. MB covalently conjugated 1MUC-1 aptamer (Apt1) recognizes the cell, whereas the nucleolin aptamer AS1411Apt2-modified QDs/SiO$_2$ nanoparticle is used as label. Both fluorescence and electrochemical signals of the system were measured, and the limit of detection of the system was found to be 201 and 85 cells/mL by photoluminescence and square-wave voltammetric method, respectively, showing higher efficiency of the electrochemical approach. Su et al. (2014) in another experiment developed an aptamer-based electrochemiluminescent method based on MUC-1 aptamer. A three-dimensional macroporous AuNPs@graphene with enhanced surface with AuNP bound to ConA for capturing the cell was developed, and another aptamer conjugated with carbon quantum dots was used to develop chemiluminescent signal. The range of detection was 500$-$2 × 10^7 cells/mL, and limit of detection was found to be 230 cells/mL. MUC-1 is the predominantly targeted marker for developing a cytosensor to detect breast cancer and with the improved detection limit of recently developed methods; cytosensor can play a significant role in early breast cancer detection.

Gastointestinal Cancer

Gastrointestinal cancer is the cancer that infects the organs in digestive system, mostly related with diet, alcoholism, hereditary, and some microorganisms. The symptoms are very nonspecific, leading to its delayed diagnosis. Gastrointestinal cancer causes a massive global cancer burden with a high mortality rate. Cytosensors developed for the detection of gastrointestinal cancer are discussed with references to some common cancer types.

Stomach cancer, also known as gastric cancer, goes unsuspected for a longer period because of its unspecific symptoms, delaying its diagnosis and further treatment. One of the causes of stomach cancer is *Helicobacter pylori* (Uemura et al., 2001), a gram-negative bacterium involved in the disease progression and that changes the microenvironment in the stomach. Several reports of biosensor of the stomach cancer are developed based on the detection of the causative microorganism and its metabolic products (Zilberman et al., 2015). Dervisevic et al. (2017a) developed a method to detect the human Caucasian gastric adenocarcinoma (AGS) cancer cell line by boronic acid$-$functionalized pencil graphite electrode (PGE). The detection principle was based on the interaction between boronic acid and cell surface carbohydrates (SA). By EIS, the linear range of detection was 1×10^1 to 1×10^6 cells/mL, and the lower limit of detection was 10 cells/mL. The same group conducted another experiment to compare the effectiveness of boronic acid versus folic acid to recognize the AGS cancer cell line (Dervisevic et al., 2017b). In this experiment, a normal human kidney cell line HEK-239 was used for selectivity testing of the cytosensor. The gold electrode surface was modified by cysteamine and followed with covalent interaction with ferrocene-cored polyamidoamide (Fc-PAMAM) dendritic structures with its amine-terminated branch. Then boronic acid and folic acid were functionalized on two separate platforms with a coupling agent to compare the sensitivity. The linear range of detection was equal for both the platform with 100 to 1×10^6 cells/mL, but the lower limit of detection for the sensing platform using folic acid performed better with 20 cells/mL as compared with 28 cells/mL for boronic acid. This shows high efficiency of folic acid in the detection of cancer cells. Tabrizi et al. (2017) developed a protocol for detection of the adenocarcinoma gastric cell (AGS) by an SPE-AuNP-aptamer platform in flow injection microcarrier device. The SPE-AuNP-aptamer acts as the primary detecting platform to capture the AGS cell, whereas the secondary aptamer-Au@Ag was used for signal generation in presence of H_2O_2. Au-Ag was used because of its biocompatibility, small size, and electrocatalytic activity to H_2O_2. The nanoparticle, as a redox probe, is better than the peroxidase enzyme for signal generation because of its stability, no critical immobilization process, cost-effectiveness, ease in preparation, and higher sensitivity. The linear range of detection is $10-5 \times 10^5$ cells/mL, and the limit of detection was 6 cells/mL achieved by amperometric method.

Liver cancer, another type of gastric cancer, also known as hepatic cancer, is largely associated with hepatitis b, hepatitis c, and alcoholism (Danaei et al., 2005; Perz et al., 2006). Among all, the hepatocellular carcinoma (HCC) or malignant hepatoma contributes more than two-thirds of all cases of liver cancer (Sanna et al., 2016). Sun et al. (2018) reported human liver hepatocellular carcinoma (HepG2) detection method using dendritic nanostructures. A DNA nanohedron-based TLS11a aptamers specific for HepG2 was immobilized on a disposable screen-printed gold electrode (SPGE) working as sensor platform. Two-hybrid PdPt nanocage was developed with complementary cDNA (cDNA1 and cDNA2) with HRP and hemin to develop two nanoprobes, respectively. When immobilized on SPGE, the nanoprobes form a self-assembled dendritic structure (DS). HepG2 cell, when present in the system,

competed with the DS to bind to the aptamer, causing a dip in the signal where the limit of detection was found to be 5 cells/mL. The cell can be isolated after the detection by simply breaking the Au-S bond that made the cells available for further studies. Jiang et al. (2018) also reported the detection using the aptamer and nanocage formation as shown in Fig. 9.6. Here the aptamer was immobilized on the gold nanoparticle deposited on the ITO. Two-ferrocene cDNA immobilized on the PtNPs with complementary sequence acts as the nanoprobes bind to the aptamer form nanocage. The cell competed to bind the aptamer and shows a detection range of $50-1 \times 10^6$ cells/mL with limit of detection of 15 cells/mL.

Sun et al. (2015) developed a reusable cytosensor where they captured the HpeG2 cell on the thiolated TLS11a aptamer-fabricated Au electrode and used G-quadruplex/hemin/aptamer-AuNPs-HRP nanoprobe to recognize the cell and amplify the signal. The range of detection was $100-10^7$ cells/mL with the lower limit of detection of 30 cells/mL. Potential of 0.9–1.7 V was used to regenerate the bare gold electrode by breaking the Au-S bond with the aptamer for further use. Sun et al. developed another sensor with GCE and (Fe_3O_4/MnO_2/Au@Pd) as probe for the detection of the cell. Fe_3O_4 and MnO_2 are used for their larger surface area, longer stability, and catalytic reduction activity in presence of H_2O_2 (Sun et al., 2016). The sensor achieved a wide detection range of 1×10^2 to 1×10^7 cells/mL and a detection limit of 15 cells/mL.

Colorectal cancer, the third most cause of cancer death annually, largely develops in the distal region of the large intestine, that is, colon; rectum is associated with the hereditary, infection, and diet (Siegel et al., 2018; Rabeneck et al., 2003). Colorectal cancer follows a specific pattern for disease progression, known as adenoma-carcinoma sequence (Fearon and Vogelstein, 1990). The adenoma on the gut lining accumulates genetic and epigenetic changes for over a decade and gets converted in to the carcinoma; this is where the detection of the developing cancer can be done. Colonoscopy is currently the most reliable method for detection of the colorectal cancer, but with the limitation of high cost and not suitable for huge population screening. Some detection methods based on genetic and molecular markers were reported earlier (Chung et al., 2018; Lam et al., 2016; Imperiale et al., 2014; Ren et al., 2015). Cao et al. (2015) developed a BSA-Ag nanoflower with porus architecture for the detection of the human colon cancer cell line DLD-1 on a modified GCE. The BSA-Ag nanoflower was conjugated with a lectin molecule (SNA) to detect SA, which is overexpressed on the DLD-1 cell. The porous architecture

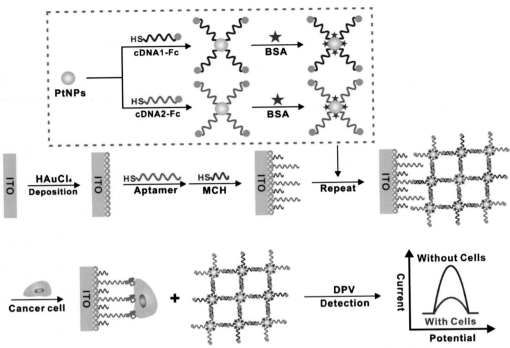

FIG. 9.6 Formation of the nanocage by PtNP and cDNA for the detection of HepG2 cells. (Image reproduced with permission from Jiang, Y., Sun, D., Liang, Z., Chen, L., Zhang, Y., Chen, Z. 2018. Label-free and competitive aptamer cytosensor based on layer-by-layer assembly of DNA-platinum nanoparticles for ultrasensitive determination of tumor cells. Sensors and Actuators B: Chemical 262, 35–43.)

contributes to the excellent biocompatibility and long retention of cell because of its increased surface area. The sensor achieved a range of detection 1.35×10^2 to 1.35×10^7 cells/mL and detection limit of 40 cells/mL. In another experiment, Yaman et al. (2017) described the self-assembly of a diphenyl alaninamide–based peptide nanoparticle (PNPs) for detection of DLD-1 colon cancer cell line. PNP was immobilized on the PGE, and the attachment of the cells was measured by EIS. The range of detection is 2×10^2 to 2.0×10^5 cells/mL, and the limit of detection is 100 cells/mL.

With the changing diet habit and the modern lifestyle, the occurrence of colorectal cancer is expected to increase rapidly in near future (Arnold et al., 2017). The survival colorectal cancer mostly depends on the stage of diagnosis (Gatta et al., 2000), which makes the cytosensor an important tool in cancer diagnostics, especially in the developing countries where the risk of colorectal cancer is increasing very fast (Stevens et al., 2012).

Prostate Cancer

Prostate cancer occurs mostly in men aged more than 50 years. Transrectal ultrasound-guided prostate biopsy (TRUS-Bx) is the traditionally followed method to detect the occurrence of prostate cancer. Prostate cancer is commonly associated with the high level of prostate-specific cancer (PSA) (Schoots et al., 2015). Reports from previous studies show PSA concentration can be incorporated to existing diagnostic method for efficient early detection of prostate (Catalona et al., 1994). Yadegari et al. (2017) reported a cytosensor for the detection of prostate metastatic cancer cell Du-145 through an electrochemical method applying a dual recognition and enhanced signal amplification procedure. AntiCD-166 monoclonal antibody immobilized on gold electrode was used to recognize and hold the target cells. A graphene (G)/gold nanoparticle (GNP)/HRP-conjugated trastuzumab antibody was utilized further to capture and recognize the cell. Both of the antibodies are developed for specifically binding to the prostate cancer cell overexpressing cluster of differentiation

166 (CD166) and human epidermal growth factor receptor 2 (HER2) on cell surface. Con A was used to enhance the cell capture capacity and stability of the system. The cytosensor shows a lower detection limit of 20 cells/mL and a linear range of detection of $10^2 - 10^6$ cells/mL.

Ovarian Cancer

Uterine cancer, the second most common type of cancer in woman, is associated with hormonal abnormality, hereditary, and a virus named human papilloma virus (Lux et al., 2006; Walboomers et al., 1999). Though papanicolaou test (Pap test) is recommended for the detection of cervical cancer; it is not error free (Kinde et al., 2013). Among all the cell lines used in modern research, HeLa cell line, a uterine cancer cell line, is a treasure for cancer research owing to exceptional characteristics.

Jiang et al. (2014) developed a method of recognizing the cervical cancer HeLa cell line based on the electrochemiluminescence of CdTe nanoparticles-GSH-modified ITO platform. Using folate and folate receptor, they shown have a significant decrease in signal when folate-overexpressing HeLa cell line is present in the sample that blocks the transfer of electron from oxygen to CdTe/GSH on ITO electrodes. The sensor achieved a linear range of detection of 3.5×10^3 to 3.5×10^5 cells/mL with a limit of detection of 3.5×10^3 cells/mL. Ni et al. (2016) also developed a probe for detection of HeLa cells based on the same folate receptor expressed on the cell membrane. Folate and ferrocene are conjugated on the 3' and 5' of the ssDNA subjected to exonuclease activity on the detecting platform of ITO. Folate molecule got bound with the folic acid receptor on the HeLa cell surface and shield the ssDNA from exonuclease activity. The similar negative charges of ssDNA and ITO surface developed an electrostatic repulsion force, keeping them far from each other, which resulted in an increased distance between the ITO surface and the ferrocene molecule attached on one side of the ssDNA. The increased distance between the redox molecule ferrocene and the sensor surface led to a significant decrease in the electrochemical signal. The LOD was 50 cells/mL and the linear range of detection was 10−104 cells/mL. Wu et al. (2013) reported a cytosensor for selective detection of the HeLa cell by carbon nanodots@Ag nanocomposite that is highly sensitive. GCE grafted with amino-functionalized graphene (by PAH) led to more efficient electron transfer, functionalized with c-dot@Ag nanocomposite and cysteine for binding of folic acid on the surface. Folic acid is used for the detection of the folic acid receptor on the cell surface of the HeLa cells, which when present decrease the electrochemical current by blocking the c-dot@Ag nanocomposite. The LOD of the sensor was found to be 10 cells/mL, and the linear range of detection was $10−1 \times 10^5$ cells/mL. Li et al. (2012) developed a rectifier-based detection platform of HeLa cell where capturing through folic acid receptor decreases the current by blocking electron transfer. The sensor achieved a detection limit of 10 cells/mL and is able to detect even in presence of other cells.

CONCLUSION

A number of cancer cytosensors have been developed targeting the biomarkers of various origin, viz., lung, breast, and gastrointestinal with a modified expression profile that helps in recognizing the cell with no further isolation of genetic material or any other extensive treatment, making it cost-effective and time-saving. Furthermore, the use of nanomaterials has paved various advanced cytosensing platforms, leading to a gradual increase in analytical performances over the years. In recent days, various microfluidic-based strategies have been used to the biosensing modules to introduce stable and guided flow of the sample for detection of the analytes. The current trend of cytosensing follows a dual detection procedure which brings more precision to the detection system. In this chapter, we have summarized the various detection methods and their advancement over time to detect the cancer cells at a very early stage of disease progression.

ACKNOWLEDGMENTS

Dr. Pranjal Chandra thanks DST Ramanujan Fellowship (SB/S2/RJN-042/2015) for financial assistance. Research fellowship to BP, KM, and AK by IIT Guwahati is also acknowledged. SR acknowledges the National Post Doctorate Fellowship Grant (PDF/2017/002924) supported by DST-SERB from the Government of India.

REFERENCES

Akhtar, M.H., Hussain, K.K., Gurudatt, N.G., Chandra, P., Shim, Y.B., 2018. Ultrasensitive dual probe immunosensor for the monitoring of nicotine induced-brain derived neurotrophic factor released from cancer cells. Biosensors and Bioelectronics 116, 108−115.

Arnold, M., Sierra, M.S., Laversanne, M., Soerjomataram, I., Jemal, A., Bray, F., 2017. Global patterns and trends in colorectal cancer incidence and mortality. Gut 66, 683−691.

Bansal, S., Jyoti, A., Mahato, K., Chandra, P., Prakash, R., 2017. Highly sensitive in vitro biosensor for enterotoxigenic *Escherichia coli* detection based on ssDNA anchored on

PtNPs-chitosan nanocomposite. Electroanalysis 29, 2665–2671.

Baranwal, A., Chandra, P., 2018. Clinical implications and electrochemical biosensing of monoamine neurotransmitters in body fluids, in vitro, in vivo, and ex vivo models. Biosensors and Bioelectronics 121, 137–152.

Baranwal, A., Kumar, A., Priyadharshini, A., Oggu, G.S., Bhatnagar, I., Srivastava, A., Chandra, P., 2018a. Chitosan: an undisputed bio-fabrication material for tissue engineering and bio-sensing applications. International Journal of Biological Macromolecules 110, 110–123.

Baranwal, A., Mahato, K., Srivastava, A., Maurya, P.K., Chandra, P., 2016. Phytofabricated metallic nanoparticles and their clinical applications. RSC Advances 6, 105996–106010.

Baranwal, A., Srivastava, A., Chandra, P., 2018b. A systematic study on phyto-synthesized silver nanoparticles and their antimicrobial mode of action. In: Advances in Microbial Biotechnology. Apple Academic Press, pp. 553–572.

Baranwal, A., Srivastava, A., Kumar, P., Bajpai, V.K., Maurya, P.K., Chandra, P., 2018c. Prospects of nanostructure materials and their composites as antimicrobial agents. Frontiers in Microbiology 9, 422.

Bhatnagar, I., Mahato, K., Ealla, K.K.R., Asthana, A., Chandra, P., 2018. Chitosan stabilized gold nanoparticle mediated self-assembled gliP nanobiosensor for diagnosis of invasive Aspergillosis. International Journal of Biological Macromolecules 110, 449–456.

Cao, H., Yang, D.-P., Ye, D., Zhang, X., Fang, X., Zhang, S., Liu, B., Kong, J., 2015. Protein-inorganic hybrid nanoflowers as ultrasensitive electrochemical cytosensing Interfaces for evaluation of cell surface sialic acid. Biosensors and Bioelectronics 68, 329–335.

Catalona, W.J., Richie, J.P., Ahmann, F.R., M'liss, A.H., Scardino, P.T., Flanigan, R.C., Dekernion, J.B., Ratliff, T.L., Kavoussi, L.R., Dalkin, B.L., 1994. Comparison of digital rectal examination and serum prostate specific antigen in the early detection of prostate cancer: results of a multicenter clinical trial of 6,630 men. The Journal of Urology 151, 1283–1290.

Chandra, P., 2013a. Advances in clinical diagnosis through electrochemical aptamer sensors. Journal of Bioanalysis & Biomedicine 5, 4442–4449.

Chandra, P., 2013b. Miniaturized multiplex electrochemical biosensor in clinical bioanalysis. Journal of Bioanalysis & Biomedicine 5, e122.

Chandra, P., 2015a. Advance diagnosis of drug resistance in cancer: towards point-of-care electronic nanodevice. Journal of Analytical & Bioanalytical Techniques 6, e120.

Chandra, P., 2015b. Electrochemical nanobiosensors for cancer diagnosis. Journal of Analytical & Bioanalytical Techniques 6.

Chandra, P., Das, D., Abdelwahab, A.A., 2010. Gold nanoparticles in molecular diagnostics and therapeutics. Digest Journal of Nanomaterials & Biostructures (DJNB) 5.

Chandra, P., Noh, H.-B., Pallela, R., Shim, Y.-B., 2015. Ultrasensitive detection of drug resistant cancer cells in biological matrixes using an amperometric nanobiosensor. Biosensors and Bioelectronics 70, 418–425.

Chandra, P., Noh, H.-B., Shim, Y.-B., 2013a. Cancer cell detection based on the interaction between an anticancer drug and cell membrane components. Chemical Communications 49, 1900–1902.

Chandra, P., 2016. Nanobiosensors for Personalized and Onsite Biomedical Diagnosis. The Institution of Engineering and Technology.

Chandra, P., Suman, P., Mukherjee, M., Kumar, P., 2013b. HER2 protein biomarker based sensor systems for breast cancer diagnosis. Journal of Molecular Biomarkers & Diagnosis 5, e119.

Chandra, P., Zaidi, S.A., Noh, H.B., Shim, Y.B., 2011. Separation and simultaneous detection of anticancer drugs in a microfluidic device with an amperometric biosensor. Biosensors and Bioelectronics 28, 326–332.

Chatterjee, S.K., Zetter, B.R., 2005. Cancer biomarkers: knowing the present and predicting the future. Future Oncology 1 (1), 37–50.

Chen, H., Hou, Y., Ye, Z., Wang, H., Koh, K., Shen, Z., Shu, Y., 2014. Label-free surface plasmon resonance cytosensor for breast cancer cell detection based on nano-conjugation of monodisperse magnetic nanoparticle and folic acid. Sensors and Actuators B: Chemical 201, 433–438.

Cheng, A.K., Su, H., Wang, Y.A., Yu, H.-Z., 2009. Aptamer-based detection of epithelial tumor marker mucin 1 with quantum dot-based fluorescence readout. Analytical Chemistry 81, 6130–6139.

Choudhary, M., Yadav, P., Singh, A., Kaur, S., Ramirez-Vick, J., Chandra, P., Arora, K., Singh, S.P., 2016. CD 59 targeted ultrasensitive electrochemical immunosensor for fast and noninvasive diagnosis of oral cancer. Electroanalysis 28, 2565–2574.

Chung, S., Chandra, P., Koo, J.P., Shim, Y.-B., 2018. Development of a bifunctional nanobiosensor for screening and detection of chemokine ligand in colorectal cancer cell line. Biosensors and Bioelectronics 100, 396–403.

Connor, E.E., Mwamuka, J., Gole, A., Murphy, C.J., Wyatt, M.D., 2005. Gold nanoparticles are taken up by human cells but do not cause acute cytotoxicity. Small 1, 325–327.

Danaei, G., Vander Hoorn, S., Lopez, A.D., Murray, C.J., Ezzati, M., Group, C.R.A.C., 2005. Causes of cancer in the world: comparative risk assessment of nine behavioural and environmental risk factors. The Lancet 366, 1784–1793.

Deka, S., Saxena, V., Hasan, A., Chandra, P., Pandey, L.M., 2018. Synthesis, characterization and in vitro analysis of $\alpha\text{-}Fe_2O_3\text{-}GdFeO_3$ biphasic materials as therapeutic agent for magnetic hyperthermia applications. Materials Science and Engineering: C 92, 932–941.

Dervisevic, M., Senel, M., Sagir, T., Isik, S., 2017a. Highly sensitive detection of cancer cells with an electrochemical cytosensor based on boronic acid functional polythiophene. Biosensors and Bioelectronics 90, 6–12.

Dervisevic, M., Şenel, M., Sagir, T., Isik, S., 2017b. Boronic acid vs. Folic acid: a comparison of the bio-recognition performances by impedimetric cytosensors based on ferrocene cored dendrimer. Biosensors and Bioelectronics 91, 680–686.

Etzioni, R., Urban, N., Ramsey, S., Mcintosh, M., Schwartz, S., Reid, B., Radich, J., Anderson, G., Hartwell, L., 2003. Early detection: the case for early detection. Nature Reviews Cancer 3, 243.

Fearon, E.R., Vogelstein, B., 1990. A genetic model for colorectal tumorigenesis. Cell 61, 759–767.

Gatta, G., Capocaccia, R., Sant, M., Bell, C., Coebergh, J., Damhuis, R., Faivre, J., Martinez-Garcia, C., Pawlega, J., DE Leon, M.P., 2000. Understanding variations in survival for colorectal cancer in Europe: a EUROCARE high resolution study. Gut 47, 533–538.

Hanahan, D., Weinberg, R.A., 2011. Hallmarks of cancer: the next generation. Cell 144, 646–674.

Hua, X., Zhou, Z., Yuan, L., Liu, S., 2013. Selective collection and detection of MCF-7 breast cancer cells using aptamer-functionalized magnetic beads and quantum dots based nano-bio-probes. Analytica Chimica Acta 788, 135–140.

Imperiale, T.F., Ransohoff, D.F., Itzkowitz, S.H., Levin, T.R., Lavin, P., Lidgard, G.P., Ahlquist, D.A., Berger, B.M., 2014. Multitarget stool DNA testing for colorectal-cancer screening. New England Journal of Medicine 370, 1287–1297.

Jelinek, R., Kolusheva, S., 2004. Carbohydrate biosensors. Chemical Reviews 104, 5987–6016.

Jiang, H., Wang, X., 2014. Label-free detection of folate receptor (+) cells by molecular recognition mediated electrochemiluminescence of CdTe nanoparticles. Analytical Chemistry 86, 6872–6878.

Jiang, Y., Sun, D., Liang, Z., Chen, L., Zhang, Y., Chen, Z., 2018. Label-free and competitive aptamer cytosensor based on layer-by-layer assembly of DNA-platinum nanoparticles for ultrasensitive determination of tumor cells. Sensors and Actuators B: Chemical 262, 35–43.

Kara, P., Erzurumlu, Y., Kirmizibayrak, P.B., Ozsoz, M., 2016. Electrochemical aptasensor design for label free cytosensing of human non-small cell lung cancer. Journal of Electroanalytical Chemistry 775, 337–341.

Kinde, I., Bettegowda, C., Wang, Y., Wu, J., Agrawal, N., Shih, I.-M., Kurman, R., Dao, F., Levine, D.A., Giuntoli, R., 2013. Evaluation of DNA from the Papanicolaou test to detect ovarian and endometrial cancers. Science Translational Medicine 5, 167ra4-167ra4.

Koh, W.C.A., Chandra, P., Kim, D.-M., Shim, Y.-B., 2011. Electropolymerized self-assembled layer on gold nanoparticles: detection of inducible nitric oxide synthase in neuronal cell culture. Analytical Chemistry 83, 6177–6183.

Kumar, P., Patra, J.K., Chandra, P., 2018a. Advances in Microbial Biotechnology: Current Trends and Future Prospects. CRC Press.

Kumar, A., Sharma, S., Pandey, L.M., Chandra, P., 2018b. Nanoengineered material based biosensing electrodes for enzymatic biofuel cells applications. Materials Science for Energy Technologies 1 (1), 38–48.

Lam, K., Pan, K., Linnekamp, J.F., Medema, J.P., Kandimalla, R., 2016. DNA methylation based biomarkers in colorectal cancer: a systematic review. Biochimica et Biophysica Acta (BBA) – Reviews on Cancer 1866, 106–120.

Li, A., Liu, H., Ouyang, P., Yang, P.-H., Cai, H.-H., Cai, J., 2017. A sensitive probe for detecting intracellular reactive oxygen species via glutathione-mediated nanoaggregates to enhance Resonance Rayleigh scattering signals. Sensors and Actuators B: Chemical 246, 190–196.

Li, H., Li, D., Liu, J., Qin, Y., Ren, J., Xu, S., Liu, Y., Mayer, D., Wang, E., 2012. Electrochemical current rectifier as a highly sensitive and selective cytosensor for cancer cell detection. Chemical Communications 48, 2594–2596.

Li, T., Fan, Q., Liu, T., Zhu, X., Zhao, J., Li, G., 2010. Detection of breast cancer cells specially and accurately by an electrochemical method. Biosensors and Bioelectronics 25, 2686–2689.

Liu, A., Peng, S., Soo, J.C., Kuang, M., Chen, P., Duan, H., 2010. Quantum dots with phenylboronic acid tags for specific labeling of sialic acids on living cells. Analytical Chemistry 83, 1124–1130.

Lu, W., Wang, H.-Y., Wang, M., Wang, Y., Tao, L., Qian, W., 2015. Au nanoparticle decorated resin microspheres: synthesis and application in electrochemical cytosensors for sensitive and selective detection of lung cancer A549 cells. RSC Advances 5, 24615–24624.

Lux, M.P., Fasching, P.A., Beckmann, M.W., 2006. Hereditary breast and ovarian cancer: review and future perspectives. Journal of Molecular Medicine 84, 16–28.

Ma, S., Hu, S., Wang, Q., Liu, Y., Zhao, G., Zhang, Q., Mao, C., Zhao, B., 2017. Evaluation of sialic acid based on electrochemical cytosensor with 3D micro/nanostructured sensing interface. Analytical Methods 9, 6171–6176.

Mahato, K., Baranwal, A., Srivastava, A., Maurya, P.K., Chandra, P., 2016. Smart materials for biosensing applications. In: Techno-Societal 2016, International Conference on Advanced Technologies for Societal Applications. Springer, pp. 421–431.

Mahato, K., Chandra, P., 2019a. Paper based miniaturized immunosensor for naked eye ALP detection based on digital image colorimetry integrated with smartphone. Biosensors and Bioelectronics 128, 9–16.

Mahato, K., Kumar, A., Maurya, P.K., Chandra, P., 2017a. Shifting paradigm of cancer diagnoses in clinically relevant samples based on miniaturized electrochemical nanobiosensors and microfluidic devices. Biosensors and Bioelectronics 100, 411–428.

Mahato, K., Kumar, S., Srivastava, A., Maurya, P.K., Singh, R., Chandra, P., 2018a. Electrochemical immunosensors: fundamentals and applications in clinical diagnostics. In: Handbook of Immunoassay Technologies. Elsevier.

Mahato, K., Maurya, P.K., Chandra, P., 2018b. Fundamentals and commercial aspects of nanobiosensors in point-of-care clinical diagnostics. 3 Biotech 8, 149.

Mahato, K., Nagpal, S., Shah, M.A., Srivastava, A., Maurya, P.K., Roy, S., Jaiswal, A., Singh, R., Chandra, P., 2019. Gold nanoparticle surface engineering strategies and their applications in biomedicine and diagnostics. 3 Biotech 9, 57.

Mahato, K., Srivastava, A., Chandra, P., 2017b. Paper based diagnostics for personalized health care: emerging technologies and commercial aspects. Biosensors and Bioelectronics 96, 246–259.

Mandal, R., Baranwal, A., Srivastava, A., Chandra, P., 2018. Evolving trends in bio/chemical sensors fabrication incorporating bimetallic nanoparticles. Biosensors and Bioelectronics 117, 546–561.

Mir, T.A., Yoon, J.-H., Gurudatt, N., Won, M.-S., Shim, Y.-B., 2015. Ultrasensitive cytosensing based on an aptamer modified nanobiosensor with a bioconjugate: detection of human non-small-cell lung cancer cells. Biosensors and Bioelectronics 74, 594–600.

Mittal, S., Kaur, H., Gautam, N., Mantha, A.K., 2017. Biosensors for breast cancer diagnosis: a review of bioreceptors, biotransducers and signal amplification strategies. Biosensors and Bioelectronics 88, 217–231.

Noh, H.B., Chandra, P., Moon, J.O., Shim, Y.B., 2012. In vivo detection of glutathione disulfide and oxidative stress monitoring using a biosensor. Biomaterials 33, 2600–2607.

Ni, J., Wang, Q., Yang, W., Zhao, M., Zhang, Y., Guo, L., Qiu, B., Lin, Z., Yang, H.-H., 2016. Immobilization free electrochemical biosensor for folate receptor in cancer cells based on terminal protection. Biosensors and Bioelectronics 86, 496–501.

Pallela, R., Chandra, P., Noh, H.-B., Shim, Y.-B., 2016. An amperometric nanobiosensor using a biocompatible conjugate for early detection of metastatic cancer cells in biological fluid. Biosensors and Bioelectronics 85, 883–890.

Perz, J.F., Armstrong, G.L., Farrington, L.A., Hutin, Y.J., Bell, B.P., 2006. The contributions of hepatitis B virus and hepatitis C virus infections to cirrhosis and primary liver cancer worldwide. Journal of Hepatology 45, 529–538.

Prasad, A., Mahato, K., Chandra, P., Srivastava, A., Joshi, S.N., Maurya, P.K., 2016a. Bioinspired composite materials: applications in diagnostics and therapeutics. Journal of Molecular and Engineering Materials 4, 1640004.

Prasad, A., Mahato, K., Maurya, P., Chandra, P., 2016b. Biomaterials for biosensing applications. Journal of Analytical & Bioanalytical Techniques 7, e124.

Qian, R., Ding, L., Bao, L., He, S., Ju, H., 2012. In situ electrochemical assay of cell surface sialic acids featuring highly efficient chemoselective recognition and a dual-functionalized nanohorn probe. Chemical Communications 48, 3848–3850.

Rabeneck, L., Davila, J.A., EL-Serag, H.B., 2003. Is there a true "shift" to the right colon in the incidence of colorectal cancer? American Journal of Gastroenterology 98, 1400.

Ren, A., Dong, Y., Tsoi, H., Yu, J., 2015. Detection of miRNA as non-invasive biomarkers of colorectal cancer. International Journal of Molecular Sciences 16, 2810–2823.

Sanna, C., Rosso, C., Marietti, M., Bugianesi, E., 2016. Non-alcoholic fatty liver disease and extra-hepatic cancers. International Journal of Molecular Sciences 17, 717.

Saxena, V., Chandra, P., Pandey, L.M., 2018. Design and characterization of novel Al-doped ZnO nanoassembly as an effective nanoantibiotic. Applied Nanoscience 8 (8), 1925–1941.

Sawyers, C.L., 2008. The cancer biomarker problem. Nature 452, 548.

Schoots, I.G., Roobol, M.J., Nieboer, D., Bangma, C.H., Steyerberg, E.W., Hunink, M.M., 2015. Magnetic resonance imaging–targeted biopsy may enhance the diagnostic accuracy of significant prostate cancer detection compared to standard transrectal ultrasound-guided biopsy: a systematic review and meta-analysis. European Urology 68, 438–450.

Sharma, S.V., Bell, D.W., Settleman, J., Haber, D.A., 2007. Epidermal growth factor receptor mutations in lung cancer. Nature Reviews Cancer 7, 169.

Sheng, Q., Cheng, N., Bai, W., Zheng, J., 2015. Ultrasensitive electrochemical detection of breast cancer cells based on DNA-rolling-circle-amplification-directed enzyme-catalyzed polymerization. Chemical Communications 51, 2114–2117.

Shim, Y.B., 2013. Gold nanoparticles and nanocomposites in clinical diagnostics using electrochemical methods. Journal of Nanoparticles 2013.

Siegel, R.L., Miller, K.D., Jemal, A., 2018. Cancer statistics, 2018. CA: A Cancer Journal for Clinicians 68, 7–30.

Smith, R.A., Cokkinides, V., von Eschenbach, A.C., Levin, B., Cohen, C., Runowicz, C.D., Sener, S., Saslow, D., Eyre, H.J., 2002. American Cancer Society guidelines for the early detection of cancer. CA: A Cancer Journal for Clinicians 52, 8–22.

Solanki, P.R., Kaushik, A., Agrawal, V.V., Malhotra, B.D., 2011. Nanostructured metal oxide-based biosensors. NPG Asia Materials 3, 17.

Stevens, G.A., Singh, G.M., Lu, Y., Danaei, G., Lin, J.K., Finucane, M.M., Bahalim, A.N., Mcintire, R.K., Gutierrez, H.R., Cowan, M., 2012. National, regional, and global trends in adult overweight and obesity prevalences. Population Health Metrics 10, 22.

Stewart, B., Wild, C.P., 2017. World Cancer Report 2014. Health.

Su, M., Liu, H., Ge, L., Wang, Y., Ge, S., Yu, J., Yan, M., 2014. Aptamer-Based electrochemiluminescent detection of MCF-7 cancer cells based on carbon quantum dots coated mesoporous silica nanoparticles. Electrochimica Acta 146, 262–269.

Sun, D., Lu, J., Chen, D., Jiang, Y., Wang, Z., Qin, W., Yu, Y., Chen, Z., Zhang, Y., 2018. Label-free electrochemical detection of HepG2 tumor cells with a self-assembled DNA nanostructure-based aptasensor. Sensors and Actuators B: Chemical 268, 359–367.

Sun, D., Lu, J., Chen, Z., Yu, Y., Mo, M., 2015. A repeatable assembling and disassembling electrochemical aptamer cytosensor for ultrasensitive and highly selective detection of human liver cancer cells. Analytica Chimica Acta 885, 166–173.

Sun, D., Lu, J., Zhong, Y., Yu, Y., Wang, Y., Zhang, B., Chen, Z., 2016. Sensitive electrochemical aptamer cytosensor for highly specific detection of cancer cells based on the hybrid nanoelectrocatalysts and enzyme for signal amplification. Biosensors and Bioelectronics 75, 301–307.

Sung, H.-J., Cho, J.-Y., 2008. Biomarkers for the lung cancer diagnosis and their advances in proteomics. BMB reports 41, 615–625.

Tabrizi, M.A., Shamsipur, M., Saber, R., Sarkar, S., Sherkatkhameneh, N., 2017. Flow injection amperometric sandwich-type electrochemical aptasensor for the determination of adenocarcinoma gastric cancer cell using aptamer-Au@ Ag nanoparticles as labeled aptamer. Electrochimica Acta 246, 1147–1154.

Tang, S., Shen, H., Hao, Y., Huang, Z., Tao, Y., Peng, Y., Guo, Y., Xie, G., Feng, W., 2018. A novel cytosensor based on Pt@ Ag nanoflowers and AuNPs/Acetylene black for ultrasensitive and highly specific detection of circulating tumor cells. Biosensors and Bioelectronics 104, 72–78.

Tepeli, Y., Demir, B., Timur, S., Anik, U., 2015. An electrochemical cytosensor based on a PAMAM modified glassy carbon paste electrode. RSC Advances 5, 53973–53978.

Tian, L., Qi, J., Qian, K., Oderinde, O., Cai, Y., Yao, C., Song, W., Wang, Y., 2018. An ultrasensitive electrochemical cytosensor based on the magnetic field assisted binanozymes synergistic catalysis of Fe$_3$O$_4$ nanozyme and reduced graphene oxide/molybdenum disulfide nanozyme. Sensors and Actuators B: Chemical 260, 676–684.

Tian, L., Qi, J., Qian, K., Oderinde, O., Liu, Q., Yao, C., Song, W., Wang, Y., 2017. Copper (II) oxide nanozyme based electrochemical cytosensor for high sensitive detection of circulating tumor cells in breast cancer. Journal of Electroanalytical Chemistry 812, 1–9.

Uemura, N., Okamoto, S., Yamamoto, S., Matsumura, N., Yamaguchi, S., Yamakido, M., Taniyama, K., Sasaki, N., Schlemper, R.J., 2001. *Helicobacter pylori* infection and the development of gastric cancer. New England Journal of Medicine 345, 784–789.

Walboomers, J.M., Jacobs, M.V., Manos, M.M., Bosch, F.X., Kummer, J.A., Shah, K.V., Snijders, P.J., Peto, J., Meijer, C.J., Muñoz, N., 1999. Human papillomavirus is a necessary cause of invasive cervical cancer worldwide. The Journal of Pathology 189, 12–19.

Wang, K., He, M.-Q., Zhai, F.-H., He, R.-H., Yu, Y.-L., 2017. A novel electrochemical biosensor based on polyadenine modified aptamer for label-free and ultrasensitive detection of human breast cancer cells. Talanta 166, 87–92.

Won, S.Y., Chandra, P., Hee, T.S., Shim, Y.B., 2013. Simultaneous detection of antibacterial sulfonamides in a microfluidic device with amperometry. Biosensors and Bioelectronics 39, 204–209.

Wu, L., Wang, J., Ren, J., Li, W., Qu, X., 2013. Highly sensitive electrochemiluminescent cytosensing using carbon nanodot@ Ag hybrid material and graphene for dual signal amplification. Chemical Communications 49, 5675–5677.

Yadav, S.K., Chandra, P., Goyal, R.N., Shim, Y.B., 2013. A review on determination of steroids in biological samples exploiting nanobio-electroanalytical methods. Analytica Chimica Acta 762, 14–24.

Yadegari, A., Omidi, M., Yazdian, F., Zali, H., Tayebi, L., 2017. An electrochemical cytosensor for ultrasensitive detection of cancer cells using modified graphene—gold nanostructures. RSC Advances 7, 2365–2372.

Yaman, Y.T., Akbal, Ö., Bolat, G., Bozdogan, B., Denkbas, E.B., Abaci, S., 2017. Peptide nanoparticles (pnps) modified disposable platform for sensitive electrochemical cytosensing of dld-1 cancer cells. Biosensors and Bioelectronics 104, 50–57.1.

Zanghelini, F., Frías, I.A., Rêgo, M.J., Pitta, M.G., Sacilloti, M., Oliveira, M.D., Andrade, C.A., 2017. Biosensing breast cancer cells based on a three-dimensional TIO$_2$ nanomembrane transducer. Biosensors and Bioelectronics 92, 313–320.

Zhang, Y., Luo, S., Situ, B., Chai, Z., Li, B., Liu, J., Zheng, L., 2018. A novel electrochemical cytosensor for selective and highly sensitive detection of cancer cells using binding-induced dual catalytic hairpin assembly. Biosensors and Bioelectronics 102, 568–573.

Zhao, J., Zhu, L., Guo, C., Gao, T., Zhu, X., Li, G., 2013. A new electrochemical method for the detection of cancer cells based on small molecule-linked DNA. Biosensors and Bioelectronics 49, 329–333.

Zhu, C., Yang, G., Li, H., Du, D., Lin, Y., 2014. Electrochemical sensors and biosensors based on nanomaterials and nanostructures. Analytical Chemistry 87, 230–249.

Zhu, X., Yang, J., Liu, M., Wu, Y., Shen, Z., Li, G., 2013. Sensitive detection of human breast cancer cells based on aptamer–cell–aptamer sandwich architecture. Analytica Chimica Acta 764, 59–63.

Zhu, Y., Chandra, P., Shim, Y.-B., 2012a. Ultrasensitive and selective electrochemical diagnosis of breast cancer based on a hydrazine—Au nanoparticle—aptamer bioconjugate. Analytical Chemistry 85, 1058–1064.

Zhu, Y., Chandra, P., Song, K.-M., Ban, C., Shim, Y.-B., 2012b. Label-free detection of kanamycin based on the aptamer-functionalized conducting polymer/gold nanocomposite. Biosensors and Bioelectronics 36, 29–34.

Zilberman, Y., Sonkusale, S.R., 2015. Microfluidic optoelectronic sensor for salivary diagnostics of stomach cancer. Biosensors and Bioelectronics 67, 465–471.

CHAPTER 10

Nanotherapeutics: A Novel and Powerful Approach in Modern Healthcare System

ASHUTOSH KUMAR • SHARMILI ROY • ANANYA SRIVASTAVA •
MASTAN MUKRAM NAIKWADE • BUDDHADEV PUROHIT • KULDEEP MAHATO •
V.G.M. NAIDU • PRANJAL CHANDRA, MSC, MTECH, PHD

INTRODUCTION

Nanotherapeutics is recently originated from applications of nanotechnology and has a wide range of utilities in the medical field (Noh et al., 2012; Mahato et al., 2016). Nanotechnology is formed with nanolevel of molecular structure with atomic modifications of objects having size range 1—100 nm. One nanometer equals to 1,000,000 part of 1 mm, and the actual meaning of the nano is "dwarf" (Bharali and Mousa, 2010; Ventola, 2012). Nanomedicine involves a wide range of various aspects together with diverse biological tools, whereas nanobiosensors are mostly concerned toward point-of-care diagnostic devices (Chandra et al., 2015; Mahato et al., 2017). Nanomedicines are also being used as nanorobots with the concepts of making as molecular nanotechnology. In general, arrangement, structures, and conjugates of biological macromolecules with nanostructures provide synergistic effects in medical diagnostic applications. The most interesting concern about nanomaterials is their size that shows similarity with biological macromolecules and hence suitable for in vitro as well as in vivo use. Therefore, by considering compatibility of nanoformulations with biological objects, numerous analytical tools, kits for diagnosis, and drug delivery vehicles are developed till date. Unlike conventional medication system, drugs attached onto the nanoparticles (NPs) act more specifically with less side effects. Nanotherapeutics is opening new doors to improve the efficacy and safety aspects of conventional therapeutic system (Wagner et al., 2006; Freitas, 2005). Because of collaboration of national as well as international agencies, excellent research has been going on for development of novel systems for drug delivery, biosensing, gene therapy, and advanced imaging technologies.

In this chapter, we have documented the current contexts including the role of nanotherapeutics with intention to support scientists and research workers regarding nanodelivery system such as emulsions, gels, solid lipid particles, dendrimeric structures, capsules, and sponges at nanolevel. Applications of nanotechnology with respect to nanotherapeutics in different aspects such as biosensor system (Chandra, 2013), blood purification, diagnosis, and imaging technology are explained. It is an attempt to give an overlook to healthcare professionals about growth and development in this field with amalgamation of nanotechnology.

NANO FORMULATIONS

Around 40% of drugs presently have few limitations such as low solubility, less stability, and extremely poor performance with minute biological counterparts (Baumgartner et al., 2014). These problems lead to new direction of researchers to explore advancement of nanotechnology in the dome nanoformulations. Conventionally, there are two stages of cell targeting methods such as passive targeting and active targeting, where it showed significant difference in cell behavior between normal tissue and effected tissue, respectively. Compared with conventional process, using nanoformulations for cancer therapy is undoubtedly a big step as nanotherapeutics arena. Marketed nanoformulations of capsule show good chemical stability with general properties of reproducibility and biocompatibility

Nanotechnology in Modern Animal Biotechnology. https://doi.org/10.1016/B978-0-12-818823-1.00010-7

(Fig. 10.1). These formulations are getting researcher's attraction because of their coatings that protect the drugs and other substances. Several processes have been developed as nanoformulations for better drug delivery system, which we have mentioned briefly in the following sections.

Nanogel

Nanogels are defined as tiny, spherical, semipermeable polymer networks with internal spaces that are capable of encapsulating the molecules of various sizes, and they are found to be highly soluble drug delivery systems (Vinogradov and Senanayake, 2013). The drug can be added immediately in nanogel, as it gets swell up in water, ultimately providing a scope of materialization of hard, compact nanocapsule as a drug-carrying vehicle by reduction in the volume of solvent. As it is having high biocompatibility, wetness, and good structural features, it possesses distinctive application in the carrier-based drug delivery system. Gels after formation give large superficial space to bioconjugate and interior structure to incorporate biomolecules. There are various methods of preparation of nanogels such as photolithography, continuous micromolding, and reformation of polymers. The unique feature of bioconjugates combined with nanogels and ligands can identify specific receptors that are present on the infected cells. It has good self-degradability feature, which aids in the drug release and also supports easy clear-up of remaining part of nanogel formulation (Oh et al., 2008). Recently, Peter

et al. exhibited the information about core/shell nanogel preparations with *N*-isopropyl methacrylamide for controlled release as chemotherapeutics. This study presented about new technique of successful drug delivery path for hydrophobic drug release such as doxorubicin (DOX). Applying these nanogels onto DOX showed improvement on drug delivery system, and it helped to separate hydrophobic materials into shell gel. However, there was a limitation present in this system, which was the slow release of drug. We anticipate that modification of this method with various NPs can improve the release rate in future (Jaiswal et al., 2015).

Nanoemulsion

The term "nanoemulsion" scientifically means fine oil dispersion/water, balanced by an artificial thin film of surfactant within the range of 20–600 nm. These films are formed as droplets onto interfacial films. Owing to small size, nanoemulsions are transparent (Fig. 10.2A). These are heterogeneous mixtures of droplets of oil in water resulting into formation of nanodroplets with small size distribution. It can formulate three types of nanoemulsion formulations such as (1) oil into water where oil presents in dispersed phase and water presents as continuous phase, (2) water in oil where water is in diffused phase and oil is in continuous phase, and (3) bicontinuous nanoemulsion. The good feature of these kinds of formulations can mask the unpleasant taste of oily liquids that also provide stability from oxidation and hydrolysis. These formulations preferred

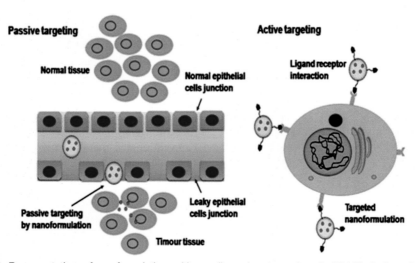

FIG. 10.1 Representations of nanoformulations with new dimensions toward medical field for better enhancement of cancer detection. (Reproduced with permission from Sahoo, A. K., Verma, A. and Pant, P. 2017. Nanoformulations for cancer therapy. Nanotechnology Applied to Pharmaceutical Technology, 157–181.)

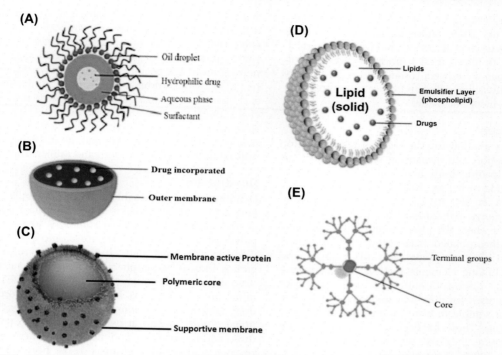

FIG. 10.2 Different nanoformulations for drug delivery. **(A)** Solid lipid nanoparticle, **(B)** nanosponge, **(C)** nanocapsule, **(D)** nanoemulsion, and **(E)** dendimer. (Reproduced with permission with Lin, C.-H., Chen, C.-H., Lin, Z-C. and Fang, J-Y. 2017. Recent advances in oral delivery of drugs and bioactive natural products using solid lipid nanoparticles as the carriers. Journal of Food and Drug Analysis 25, 219–234; Hu, C. J., Fang, R. H., Copp, J., Luk, B. T. and Zhang, L. 2013. A biomimetic nanosponge that absorbs pore-forming toxins. Nature Nanotechnology 8, 336.)

over other conventional formulations, as they provide better biocompatibility. Nowadays, nanoemulsions are widely used and explored to target various anticancer drugs and photosensitizers so these are very useful for diagnosis of biological materials and biochemical mediators (Kothamasu et al., 2012). In 2017, Patel et al. invented about novel folate-targeted theranostic nanoemulsion for imaging anticancer activity toward ovarian cancers in in vitro and in vivo applications. To enhance the therapeutic effect, they induced folate receptors as nanoemulsion form to reduce the toxicity. Folate-targeted gadolinium is used as a theranostic agent that helps to improve interaction between cell and cell surface ligand density. Consequently, with this research, it was observed that folate-targeted theranostic nanoemulsion reduced the tumor growth against ovarian cancer model (Patel et al., 2018).

Nanocapsules

Nanocapsules consist of liquid, semisolid, or solid core where the drug can be encapsulated, which is enclosed by a polymer membrane prepared by natural or artificial (synthetic) means. It has been observed that the size of nanocapsules falls into the range from 10 to 100 nm. Nanocapsules are having lipid core that is generally prepared by the technique of precipitation (Fig. 10.2B). Nanocapsules are currently in research demand for biosensing because of their protective coating properties that aid to have oxidation easily for electrochemical studies. Besides, these NPs are largely in use as drug carriers. Most of them are used in numerous diseases such as cancer, HIV, Alzheimer's, and many crucial syndromes (Peters et al., 2018). The characteristics like this can be analyzed by various techniques such as diffraction X-ray, photoelectron spectroscopy, electron microscopic system, and transmission electronic microscope—based techniques (Kothamasu et al., 2012). Nanocapsules give a wide range of applications in medical field such as agrochemicals, biomedical engineering, sanitizing agents, cosmetics, and in sewage treatment (Patel et al., 2018). In addition, these nanocapsules can be explored as potent drug for cancer management (Peters et al., 2018), radiotherapy (Bouclier et al., 2008), autoremedial, contamination treatment (Deutsch et al., 1986), and useful for agricultural studies. In future, these nanocapsulation technologies will open a fresh era toward successful delivery of biologically active drugs and other substances with respect to their specific tissue.

Nanosponges

Nanosponges are drawing the focus of researchers for drug formulation and dispensing system, as they are capable of loading both hydrophilic and lipophilic moieties inside in it (Fig. 10.2C) (Trotta and Cavalli, 2009; Trotta et al., 2012). Nanosponges are tiny, nontoxic, spongy colloid-like architecture that has various voids where the pharmaceutically active moieties can be incorporated. Beta-cyclodextrins are generally utilized for the preparation of nanosponges. Some cross-linking elements such as hexamethylene diisocyanate, carbonyldiimidazole, and carbonate diphenyl are used for their preparation and synthesis. They are not soluble in water in any kind of carbon-based solvents. These nanosponges are sterilized by their own nature; they have pH range of 2–11 and are stable upto 300°C (Selvamuthukumar et al., 2012; Ahmed et al., 2013).

Solid Lipid Nanoparticles

NPs of lipid made along with hard matricular structures are called as solid lipid nanoparticles (SLNs). These formulations of SLNs are generally prepared in such a way that water will be in continuous phase and oil is in dispersed phase by using a lipid solid form (Fig. 10.2D) (Müller et al., 2000). These SLNs have a wide range of advantages such as high biocompatibility, easy to scale up, no requirement of the use of organic solvents, good bioavailability, and controlled as well as effective release of drug (Jores et al., 2003). Along with these merits, it has some limitations too such as drug expulsion phenomenon, less drug incorporation capacity, and crystalline nature of lipid that makes these NPs not as much of good for therapeutic drug delivery applications (Fang et al., 2013). However, this can be checked with the novel synthesis procedure having different bioconjugates for fabrication of SLNs. Recently, a drug known as 1-cyclopropyl-6-fluoro-4-oxo-7-piperazin-1-ylquinoline-3-carboxylic acid has been used to prepare SLNs by ultrasonic melting-emulsification, which has shown higher antibacterial

activities (Shazly, 2017). Ciprofloxacin SLNs prepared using octadecanoic acid have shown boosted bursting effect, which results into effective release of drug. Hence, these kinds of particles have good stability at normal room temperature up to 120 days. SLNs can be utilized with various routes of drug administration such as dermal (Dingler et al., 1999), ocular (Cavalli et al., 2002), and rectal (Sznitowska et al., 2001), tested thoroughly in vivo and in vitro evaluation. Nowadays, nanopearl and nanobase formulations are being widely utilized in marketplaces (Souto and Müller, 2008).

Dendrimers

Dendrimers are 3D, highly branched, globular, and polymeric nanostructures. Important characteristics of these dendrimers are water solubility, low index of polydispersity, adjustable structures, and presence of voids in the inner part, and numerous functional groups located at peripheral part discriminate them from other nanoformulation system (Fig. 10.2E). Terminal functional groups, in general, give platform for drug targeting and conjugation. Additionally, functional groups present at peripheral part also provide those ideal properties and flexibility. From various research studies on dendrimers, it has been observed that this material is used for delivery of therapeutic substances such as polyamidoamine (PAMAM). Their production is done by reacting amine group with methyl acrylate to form new two branches that have esterified end dendrimer. PAMAM dendrimer shows no immunogenic properties with water solubility and has an amine functional group at terminal end that can be changed according to the need of effective targeting of drug. In addition, the solubility natures of dendrimers are explored for their delivery through ocular, by mouth, transdermal, and respiratory means. Structure alteration can resolve issues related to toxicities. Recent studies suggested that peptide with essential amino acids at terminal of dendrimers showed significant increase in transdermal entry of 2-(3-benzoylphenyl) propanoic acid (Patri et al., 2002). The outcomes from another study showed that combination of dendrimeric peptide aided skin permeation of ketoprofen just in 30 min of ultrasound exposure acting as synergistic application. For the first time, this study exposed that the synthesized peptide dendrimers have increased the ratio of transdermal permeation of ketoprofen and enhancement of displayed synthesized peptides upto 3.25 times in comparison with passive diffusion (Manikkath et al., 2017).

APPLICATIONS OF NANOTHERAPEUTICS SYSTEM

Nanomaterials for Implantation

From the past few years, there has been continuous need of reformative medication and substitution aimed at organ injury or failure. Nanoplanned implantations have been utilized to control the discharge of therapeutic drugs and management of pathological conditions. Drug modification has been performed at the scaling level of nano- to microgram level. Recently, titanium nanomaterial—based implantations got design to control the release of drug. Titanium tubes are now being used for hormone delivery and medication in a blend to assure quick regaining of bone loss, which shows enhanced osteoblastic capacity of attachment (Gulati et al., 2016).

Nanomaterials for Nonviral Gene and Protein Delivery in Cancer Therapy

NPs have various roles in diagnostic parameters of disease such as checking the efficiency of pharmaceutically active moiety, apoptotic behavior of cancerous cell, and gene therapy (Fig. 10.3) (Wang et al., 2012; Pérez-Herrero and Fernandez-Medarde, 2015; Tu and Eric, 2015; Ghaz-Jahanian et al., 2015; Morille et al., 2008). The effectiveness of gene therapy basically deals with development of safe, efficient, effective vehicle system that will give strong evidence of delivering foreign material into specific targeted cells (Kai et al., 2015). NPs have high surface area, and hence it is possible to functionalize surface effectively with different groups. This functionalized part has been used to detect and bind cancerous cells (Nie et al., 2007). The selection

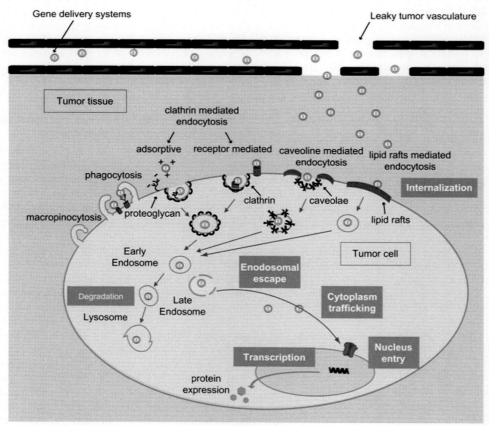

FIG. 10.3 Schematic representations of different factors of a gene delivery system to penetrate into a tumor cell. (Reproduced with the permission from Morille, M., Passirani, C., Vonarbourg, A., Clavreul, A. and Benoit, J. 2008. Progress in developing cationic vectors for non-viral systemic gene therapy against cancer. Biomaterials 29, 3477–3496.)

of specific moiety is important to ensure effective and efficient delivery of drug. The vectors of nonviral consists of polymeric DNA (protamine)-liposomal complex (Li et al., 1998), polyethylenemine (Bragonzi et al., 2000), and liposome (El-Aneed, 2004). Review says that these NPs are capable of forming complex with gene and could deliver various drugs to its target (Wang et al., 2015). The formulation prepared by these NPs should protect the sensor targets for biosensor device preparation and encapsulate nuclear components for positive genetic therapy. From the past decade, considerable research has been going on to develop the nonviral delivery system for gene therapy (Ge et al., 2016). One of the recent studies on nanocarrier for the delivery of plasmid to cancer cells by Misra et al. reported for the first time about a cationic cholesterol-based nanocarrier (plasmid; p53-EGFP-C3) delivery to cancer cells, where a plasmid was used for cell detection as well as for gene therapy. This gene therapy considered a huge advantage when it is capable of treatment toward cancer cells. In this case, the regulation of virus transfer of particular gene has been shown with wild-type p53 that transduced into tumor cell. Unfortunately, this system showed low effectiveness due to lack of transduced cell production, which will affect the tumor cell for cancer treatment (Misra et al., 2014). Nowadays, charge reversal polymer and nanoformulated gels have a great potential for delivery of drug for cancer treatment. They have capacity to carry substances such as DNA and can directly bring at the site of nucleus and show enhanced uptake by cell. Zhang et al. (2014) showed the localization at nucleus by such positively charged polymeric material. Recently, various charge reversal substances were developed as to deliver drug effectively and efficiently in the treatment of cancer. Combination of medication can be used for better prognosis and results in drug delivery system (Mujokoro et al., 2016).

Usage of Nanomaterial in the Photodynamic Therapy

Photodynamic therapy (PDT) is known as nontoxic dye treatment, where these photosensitive dyes are used for treatment for skin cancer. For example, photosensitizers (PS) and eosin dye are used as nontoxic drugs with combination of light along with various nanomaterials for tumor cell treatment. Furthermore, PDT has been used as nontoxic drug with safe visible red light, which leads to an extra step toward new developments of clinical analysis in gene therapy (Castano et al., 2004). NPs-based PDT is generally used for the treatment of infections such as acne, sun-burned skins, and skin complications. Depending on which body part is affected, these nanomaterials are injected on definite skin part or stream of blood and stimulated with exposure of external light. These particles absorb a particular wavelength of light to repair the tissue system. On reaching the excited stage, these particles interact with oxygen that ultimately results in formation of high-powered free radicals. PDT also kills those blood vessels that generally provide blood to cancerous tissue. In another study, reported by Tian et al. (2011), using graphene oxide (GO) with PDT has been used as cancer therapy. They have conjugated GO with six-armed amine group of PEG using π-π stacking theory (Fig. 10.4).

They also showed positive outcomes with the combinations of mild light-triggered heated PDT and GO effectively (Tian et al., 2011). This therapy is becoming famous because of its less toxicity and less side effects as compared with other complementary long-term therapies. Therapies like this have unique capability, as it is a noninvasive as compared with surgical procedures. Generally, it is considered as target-specific and cost-effective as compared with normal neoplastic therapies. NPs of gold have shown some specific therapeutic functions by targeting specific effected tissues, organs, and cancer cells (Loo et al., 2004; Chidambaram et al., 2011; Conde et al., 2013).

Applications of Nanoparticles in Medical Imaging for Diagnostic Approach

The recent advances in biological imaging system became a powerful means for diagnostic purpose and evolutionary approach for various pathological conditions. It gives us freedom to detect and count the cells involved in various kinds of diseases (Han and Zhou, 2016; Lee et al., 2016). In recent times, drastic improvement is done in the imaging system. The NPs conjugated with radiolabeling give effective results as compared with normal systems of imaging. NPs have very specific nature to target desired tissue for image-based diagnosis and therapy is necessary to avoid the nonspecific cell binding. NPs tak part in various techniques of imaging such as magnetic resonance imaging and ultrasonography; they are capable of giving better results than normal imaging systems (Leng et al., 2015). One of the important applications of NPs is in quantum dots (QDs), which have quantum energy to emit the light. They are having very good applications in the field of tumor removal approaches with better precision. QD was used in magnetic resonance imaging, and it was found that these are good in displaying tumor affected sites (Su et al., 2017). In another study, it was found that selenide cadmium showed better illumination effect when exposed to UV light. The only limitation of this material is toxicity of the chemical

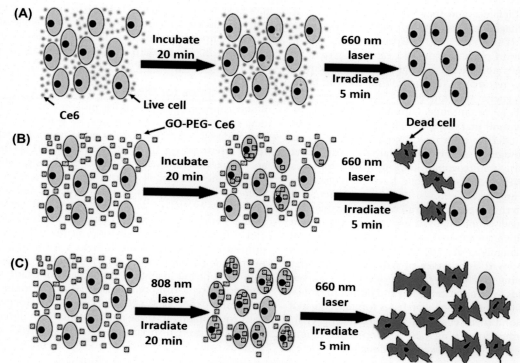

FIG. 10.4 Schematic representations of PDT-enhanced therapy for mutated cell treatment. **(A)** Cells were treated with GO-PEG-Ce6, **(B)** 20 min of incubation at dark place by the 660 nm laser for control analysis, **(C)** To induce the effect of PDT, GO-PEG-Ce6 was incubated and exposed at 808 nm laser. (Reproduced with permission from Tian, B., Wang, C., Zhang, S., Feng, L. and Liu, Z. 2011. Photothermally enhanced photodynamic therapy delivered by nano-graphene oxide. ACS Nano 5, 7000–7009.)

compound from which the QDs are prepared (Zhu et al., 2017; Kominkova et al., 2017; Manshian et al., 2017). The particles such as poly(lactic-co-glycolic acid) have been considered as a good polymer because of their biocompatibility and biodegradability. In one study, it was found that the 5-ethylamino-9-diethylaminobenzo[a]phenothiazinium (EtNBS) capsulated with PLGA that showed decreased trends in side effect. In this study, they have used NPs encapsulation to conquer the toxicity effects and few drawbacks. Surprisingly, they found that with EtNBS and poly(lactic-co-glycolic acid) as encapsulating compounds, the toxicity drastically reduced. Hence, these kinds of approaches help to reduce the toxicity for the treatment of cancer cells (Hung et al., 2016; Klein et al., 2012).

Role of Nanoparticles in Biosensors

Biosensor is an electrical tool that is able to analyze or target with combination of transducer and bioconjugation unit (Baranwal et al., 2018; Mahato et al., 2018; Mahato et al., 2018; Mandal et al., 2018). Biosensor devices are made up of transducer and biorecognition element that is sensitive toward biological elements as an analytical target such as organelles, antibodies, enzymes, and nucleic acid (Holzinger et al., 2014). In the field of medical diagnostics, the biosensor devices are widely utilized for quick study of different diseases and various microbial recognitions. Modern material science has been evolved and enhanced novel biosensors with the aid of conducting NPs such as AuNPs, silver nanoparticles (AgNPs), and platinum nanoparticles (PtNPs) that are established as proper matrices (Martin et al., 2014). These are suitable for good sensitivity associated with biosensor devices (Chen et al., 2011). However, another type of NPs combined with magnetic properties (size less than 100 nm) are having immense effect on biosensor because of their highly charged surface region and lower precipitation rate (Wang et al., 2010). These magnetic NPs are effective for tissue diffusion (Arruebo et al., 2007) and biological therapies and they have potential effect of medical diagnosis and drug delivery therapies (Fig. 10.5) (Duan et al., 2015; Upendra et al., 2016). The circulation phenomenon and capillary system of magnetic NPs are fruitful for designing better biosensor.

Recently, researchers reported about the combination of magnetic and AuNPs to detect Ebola virus with the help of immune chromatographic assays. They have found that using this combination together was much effective than typical conventional process owing to usage of sensitive colloidal gold strip (Saha et al., 2016). Simultaneously, the enzymatic properties of these NPs mentioned above have been used for different analytical studies for advancement of biosensor technologies to gain an outstanding quantification of real sample analysis and to determine their high sensitivity compared with typical conventional process (Ganguly et al., 2017). Recently, few research groups observed importance of nanoport technology for designing analytical sensing device to convert nucleic acids to an electrical signal and further the signal matched with conventional databases for quick on-spot identification (Zheng et al., 2005; Amatatongchai et al., 2017). Furthermore, using these developments, various kinds of sensor chips have been designed to detect

biomarkers such as cancer, HIV, Alzheimer's, and Parkinson's diseases (Zhang et al., 2017). Nanobiosensor-based electrochemical process has a huge aspect in the development of modern diagnostic devices (Peng et al., 2015; Wang et al., 2013).

Nanoparticles for Blood Purification

Blood consists of various cells, proteins, and cellular materials, which we can separate by using magnetic powered cell separation system as conventional technique (Herrmann et al., 2013; Kang et al., 2014). There are some experiments performed previously, which showed the removal of various harmful components from the body fluid such as bacteria, toxins, and some pretentious matter by taking help of circuit network similar to the dialysis system (Lunardi et al., 2009; Berry and Curtis, 2003). The dialysis principle is based on diffusion and ultrafiltration through a semipermeable membrane (Lunardi et al., 2009). Therefore, in recent studies, applying magnetic NPs–based purification by targeting specific compound made easier step for blood purification (Lee et al., 2013) (Fig. 10.6). Alongside, covalently linking to these nanoparticles with different antibodies, protein-specific matter and synthetic compounds over their external surfaces gave enhancement

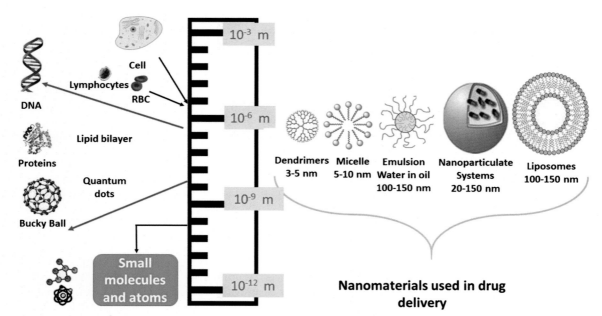

FIG. 10.5 Schematic representation of magnetic nanoparticles for drug delivery applications as controlled release of drugs from nanostructured as magnetic functional materials. (Reproduced with permission from Upendra, L., Minakshi, P., Brar, B., Guray, M., Ikbal, Ranjan, K., Bansal, N., Khurana, S. K. and Manimegalai, J. 2016. Nanodiagnostics: a new frontier for veterinary and medical sciences. Journal of Experimental Biology and Agricultural Sciences 4, 307–320.)

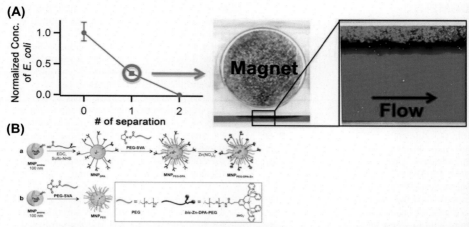

FIG. 10.6 Schematic representation of microfluidic bacterial blood separation by magnetic nanoparticles (MNPs). **(A)** Magnetic microfluidic devices applied to remove MNPs with ligands that bind to gram-positive/negative. **(B)** Formation of MNPs with the help of polyethylene glycol. (Reproduced with permission from Schumacher, C.M., Herrmann, I. K., Bubenhofer, S. B., Gschwind, S., Hirt, A-M., Beck-Schimmer, B., Gunther, D. and Stark, A. J. 2013. Quantitative recovery of magnetic nanoparticles from flowing blood: trace analysis and the role of magnetization. Advanced Functional Materials 23, 4888–4896.)

of blood purification with microfluidic system (Schumacher et al., 2013). The magnetically active NPs linked with different functional groups have capability to bind with a specific object present in blood circulation and various other fluids of body. After that, this fluid subjected to an external magnetic field resulted into aggregation of NPs close to the pole of magnet; by this method, various components from the blood have been separated (Yung et al., 2009; Faunce, 2007).

SAFETY AND HAZARDS OF NANOTHERAPEUTICS

This most challenging part needs to be taken in consideration about the after-use effect of nanotherapeutics such as their accumulation, classification, analysis, and depiction of data related to safety regarding application of clinical significance (Ryman-Rasmussen et al., 2006). Additionally, addition of NPs shows certain health hazards as they can penetrate the biological stoppages present in the human body and reach to various systems that are biologically interlinked (Maynard, 2006; Xia et al., 2006). According to studies, NPs can cause fatal hazard to cell membrane, genetic material, and various organelles, as they can readily form free radicals (Penn et al., 2005). Hazardous materials can absorb on cell surface or can enter into the cellular microstructure or can trigger some

immunogenic responses by having interaction with the receptors available on superfacial side of cell (Vallhov et al., 2006; Gaspar and Duncan, 2009). Several attempts have been explored in this regard, and improvement in near future is expected (Ehmann et al., 2013).

CONCLUSION AND FUTURE PROSPECTIVE

In the medical field, utility of nanomedicines already has many breakthroughs and shows progress toward becoming a vital aspect of healthcare system in future. The point is concluded from ongoing work of research patents and publication accepted in such an area: nano-drug delivery system. This chapter describes about the advantages/disadvantages over the other conventional drug delivery systems. Nanotherapeutics such as gels, emulsions, SLNs, sponges, and capsule have been given in detailed format. The utilities of nanomedicines in the field of neoplastic condition have been discussed. The conversion of scientific work into products of market has remained a challenge, by taking into consideration the speed of evolution in the development of nanomedicine system. We can have wide varieties of effective medicine in near future associated with NPs. Nanotechnology will become a very powerful tool for quick diagnosis with improved selectivity and specificity. In future, we can have more precise disease diagnosis for its early stage control.

ACKNOWLEDGMENTS

Dr. Pranjal Chandra thanks DST Ramanujan Fellowship (SB/S2/RJN−042/2015) for financial assistance. Dr. Ananya Srivastava thanks Prof. (Dr.) U.S.N Murty, Director, NIPER—Guwahati for his help and encouragement.

REFERENCES

Ahmed, R.Z., Patil, G., Zaheer, A., 2013. Nanosponges—a completely new nano-horizon: pharmaceutical applications and recent advances. Drug Development and Industrial Pharmacy 39, 1263−1272.

Amatatongchai, M., Sroysee, W., Chairam, S., Nacapricha, D., 2017. Amperometric flow injection analysis of glucose using immobilized glucose oxidase on nano-composite carbon nanotubes-platinum nanoparticles carbon paste electrode. Talanta 166, 420−427.

Arruebo, M., Fernandez-Pacheco, R., Ibarra, M.R., Santamaria, J., 2007. Magnetic nanoparticles for drug deliver. Nano Today 2, 22−32.

Baranwal, A., Srivastava, A., Kumar, P., Bajpai, V.K., Maurya, P.K., Chandra, P., 2018. Prospects of nanostructure materials and their composites as antimicrobial agents. Frontiers in Microbiology 9, 422.

Baumgartner, R., Eitzlmayr, A., Matsko, N., Tetyczka, C., Khinast, J., Roblegg, E., 2014. Nano-extrusion: a promising tool for continous manufacturing of solid nanoformulations. International Journal of phermaceutics 477, 1−11.

Berry, C.C., Curtis, A.S.G., 2003. Functionalisation of magnetic nanoparticles for applications in biomedicine. Journal of Physics D: Applied Physics 36, R198.

Bharali, D.J., Mousa, S.A., 2010. Emerging nanomedicines for early cancer detection and improved treatment: current perspective and future promise. Pharmacology & Therapeutics 128, 324−335.

Bouclier, C., Moine, L., Hillaireau, H., Marsaud, V., Connault, E., Opolon, P., Couvreur, P., Fatta, E., Rnoir, J.M., 2008. Physicochemical characteristics and preliminary in vivo biological evaluation of nanocapsules loaded with siRNA targeting estrogen receptor alpha. Biomacromolecules 9, 2881−2890.

Bragonzi, A., Dina, G., Villa, A., Calori, G., Biffi, A., Bordignon, C., Assael, B.M., Conese, M., 2000. Biodistribution and transgene expression with nonviral cationic vector/DNA complexes in the lungs. Gene Therapy 7, 1753.

Castano, A.P., Demidova, T.N., Hamblin, M.R., 2004. Mechanisms in photodynamic therapy: part one—photosensitizers, photochemistry and cellular localization. Photodiagnosis and Photodynamic Therapy 1, 279−293.

Cavalli, R., Gasco, M.R., Chetoni, P., Burgalassi, S., Saettone, M.F., 2002. Solid lipid nanoparticles (SLN) as ocular delivery system for tobramycin. International Journal of Pharmaceutics 238, 241−245.

Chandra, P., 2013. Miniaturized multiplex electrochemical biosensor in clinical bioanalysis. Journal of Bioanalysis and Biomedicine 5, e122.

Chandra, P., Noh, H.B., Pallela, R., Shim, Y.B., 2015. Ultrasensitive detection of drug resistant cancer cells in biological matrixes using an amperometric nanobiosensor. Biosensors and Bioelectronics 70, 418−425.

Chen, J.-P., Yang, P.-C., Ma, Y.-H., Wu, T., 2011. Characterization of chitosan magnetic nanoparticles for in situ delivery of tissue plasminogen activator. Carbohydrate Polymers 84, 364−372.

Chidambaram, M., Manavalan, R., Kathiresan, 2011. Nanotherapeutics to overcome conventional cancer chemotherapy limitations. Journal of Pharmacy & Pharmaceutical Sciences 14, 67−77.

Conde, J., Fuente, J.M.D.L., Baptista, P.V., 2013. Nanomaterials for reversion of multidrug resistance in cancer: a new hope for an old idea? Frontiers in Pharmacology 4, 134.

Deutsch, E., Libson, K., Vanderheyden, J.L., Ketring, A.R., Maxon, H.R., 1986. The chemistry of rhenium and technetium as related to the use of isotopes of these elements in therapeutic and diagnostic nuclear medicine. International Journal of radiation applications and instrumentation. Part B. Nuclear Medicine and Biology 13, 465−477.

Dingler, A., Blum, R.P., Niehus, H., Muller, R.H., Gohla, S., 1999. Solid lipid nanoparticles (SLNTM/LipopearlsTM) a pharmaceutical and cosmetic carrier for the application of vitamin E in dermal products. Journal of Microencapsulation 16, 751−767.

Duan, D., Fan, K., Zhang, D., Tan, S., Liang, M., Liang, M., Liu, Y., Zhang, J., Zhang, P., Liu, W., Qiu, X., Kobinger, G.P., Gao, G.F., Yan, X., 2015. Nanozyme-strip for rapid local diagnosis of Ebola. Biosensors and Bioelectronics 74, 134−141.

Ehmann, F., Sakai-Kato, Duncan, R., Ossa, D.H.P., Pita, R., Vidal, J.-M., Kohli, A., Tothfalusi, L., Sanh, A., Tinton, S., Robert, J.-L., Lima, B.S., Amati, M.P., 2013. Next-generation nanomedicines and nanosimilars: EU regulators' initiatives relating to the development and evaluation of nanomedicines. Nanomedicine 8, 849−856.

El-Aneed, A., 2004. An overview of current delivery systems in cancer gene therapy. Journal of Controlled Release 94, 1−14.

Fang, C.-L., Al-Suwayeh, S.A., Fang, J.Y., 2013. Nanostructured lipid carriers (NLCs) for drug delivery and targeting. Recent Patents On Nanotechnology 7, 41−55.

Faunce, T.A., 2007. Nanotherapeutics: new challenges for safety and cost-effectiveness regulation in Australia. Nanotechnology 186, 189−191.

Freitas, R.A., 2005. Nanotechnology, nanomedicine and nanosurgery. International Journal of Surgery 3, 243−246.

Ganguly, J., Saha, S., Bera, A., Ghosh, M., 2017. Exploring electro-optic effect and third-order nonlinear optical susceptibility of impurity doped quantum dots: interplay between hydrostatic pressure, temperature and noise. Optics Communications 387, 166−173.

Gaspar, R., Duncan, R., 2009. Polymeric carriers: preclinical safety and the regulatory implications for design and development of polymer therapeutics. Advanced Drug Delivery Reviews 61, 1220–1231.

Ge, Y., Ma, Y., Li, L., 2016. The application of prodrug-based nano-drug delivery strategy in cancer combination therapy. Colloids and Surfaces B: Biointerfaces 146, 482–489.

Ghaz-Jahanian, M.A., Abbaspour-Aghdam, F., Anarjan, N., Berenjian, A., Jafarizadeh-Malmiri, 2015. Application of chitosan-based nanocarriers in tumor-targeted drug delivery. Molecular Biotechnology 57 (3), 201–218.

Gulati, K., Kogawa, M., Prideaux, M., Findlay, D.M., Atkins, G.J., Losic, D., 2016. Drug-releasing nano-engineered titanium implants: therapeutic efficacy in 3D cell culture model, controlled release and stability. Materials Science and Engineering: C 69, 831–840.

Han, L., Zhou, Z., 2016. Synthesis and characterization of liposomes nano-composite-particles with hydrophobic magnetite as a MRI probe. Applied Surface Science 376, 252–260.

Herrmann, I.K., Schlegel, A., Graf, R., Schumacher, C.M., Senn, N., Hasler, M., Gschwind, S., Hirt, A., Gunther, D., Clavien, P., Starck, W.J., Beck-Schimmer, B., 2013. Nanomagnet-based removal of lead and digoxin from living rats. Nanoscale 5, 8718–8723.

Holzinger, M., Goff, A.L., Cosnier, S., 2014. Nanomaterials for biosensing applications: a review. Frontiers in Chemistry 2, 63.

Hu, C.J., Fang, R.H., Copp, J., Luk, B.T., Zhang, L., 2013. A biomimetic nanosponge that absorbs pore-forming toxins. Nature Nanotechnology 8, 336, 2013.

Zhu, X., Wu, G., Lu, N., Yuan, X., Li, B., 2017. A miniaturized electrochemical toxicity biosensor based on graphene oxide quantum dots/carboxylated carbon nanotubes for assessment of priority pollutants. Journal of Hazardous Materials 324, 272–280.

Hung, H.-I., Klein, O.J., Peterson, S.W., Rokosh, S.R., Osseiran, S., Nowell, N.H., Evams, C.L., 2016. PLGA nanoparticle encapsulation reduces toxicity while retaining the therapeutic efficacy of EtNBS-PDT in vitro. Scientific Reports 6, 33234.

Jaiswal, M., Dudhe, R., Sharma, P.K., 2015. Nanoemulsion: an advanced mode of drug delivery system. 3 Biotech 5, 123–127.

Jores, K., Mehnert, W., Mader, K., 2003. Physicochemical investigations on solid lipid nanoparticles and on oil-loaded solid lipid nanoparticles: a nuclear magnetic resonance and electron spin resonance study. Pharmaceutical Research 20, 1274–1283.

Kai, W., Qian, H., Fuming, Q., Meihua, S., 2015. Non-viral delivery systems for the application in p53 cancer gene therapy. Current Medicinal Chemistry 22, 4118–4136.

Kang, J.H., Super, M., Yung, C.W., Cooper, R.M., Domansky, K., Graveline, A.R., Mammoto, T., Berhet, J.B., Tobin, H., Cartwright, M.J., Watters, A.L., Rottman, M., Waterhouse, A., Mammoto, A., Gamini, N., Roadas, M.J., Kole, A., Jiang, A., Valentin, T.M., Diaz, A., Takahashi, K., Ingber, D.E., 2014. An extracorporeal blood-cleansing device for sepsis therapy. Nature Methodology 20, 1211–1216.

Klein, O.J., Bhayana, B., Park, Y.J., Evans, C.L., 2012. In vitro optimization of EtNBS-PDT against hypoxic tumor environments with a tiered, high-content, 3D model optical screening platform. Molecular Pharmaceutics 9, 3171–3182.

Kominkova, M., Milosavljevic, V., Vitek, P., Polanska, H., Cihalova, K., Dostalova, S., Hynstova, V., Guran, R., Kopel, P., Richtera, L., Masarik, M., Brtnickly, M., Kynicky, J., Zitka, O., Adam, V., 2017. Comparative study on toxicity of extracellularly biosynthesized and laboratory synthesized CdTe quantum dots. Journal of Biotechnology 241, 193–200.

Kothamasu, P., Kanumur, H., Ravur, N., Maddu, C., Parasuramrajam, R., Thanqavel, S., 2012. Nanocapsules: the weapons for novel drug delivery systems. BioImpacts 2, 71.

Lee, J.-L., Jeong, K.J., Hashimoto, M., Kwon, A.H., Rwei, A., Shankarappa, S.A., Tsui, J.H., Kohane, D.S., 2013. Synthetic ligand-coated magnetic nanoparticles for microfluidic bacterial separation from blood. Nano Letters 14, 1–5.

Lee, J., Gordon, A.C., Kim, H., Park, W., Cho, S., Lee, B., Larson, A.C., Rozhkova, E.A., Kim, D.-H., 2016. Targeted multimodal nano-reporters for pre-procedural MRI and intra-operative image-guidance. Biomaterials 109, 69–77.

Leng, J., Li, J., Ren, J., Deng, L., Lin, C., 2015. Star–block copolymer micellar nanocomposites with Mn, Zn-doped nano-ferrite as superparamagnetic MRI contrast agent for tumor imaging. Materials Letters 152, 185–188.

Li, S., Rizzo, M.A., Bhattacharya, S., Huang, L., 1998. Characterization of cationic lipid-protamine–DNA (LPD) complexes for intravenous gene delivery. Gene Therapy 5, 930.

Lin, C.-H., Chen, C.-H., Lin, Z.-C., Fang, J.-Y., 2017. Recent advances in oral delivery of drugs and bioactive natural products using solid lipid nanoparticles as the carriers. Journal of Food and Drug Analysis 25, 219–234.

Loo, C., Lin, A., Hirsch, L., 2004. Nanoshell-enabled photonics-based imaging and therapy of cancer. Technology in Cancer Research and Treatment 3, 33–40.

Lunardi, G., Armirotti, A., Nicodemo, M., Cavallini, L., Demonte, G., Vannozzi, M.O., Venturini, M., 2009. Comparison of temsirolimus pharmacokinetics in patients with renal cell carcinoma not receiving dialysis and those receiving hemodialysis: a case series. Clinical Therapeutics 31, 1812–1819.

Mahato, K., Baranwal, A., Srivastava, A., Maurya, P.K., Chandra, P., 2016. Smart materials for biosensing applications. In: Techno-Societal 2016, International Conference on Advanced Technologies for Societal Applications. Springer, pp. 421–431.

Mahato, K., Kumar, A., Maurya, P.K., Chandra, P., 2017. Shifting paradigm of cancer diagnoses in clinically relevant samples based on miniaturized electrochemical nanobiosensors and microfluidic devices. Biosensors and Bioelectronics 100, 411–428.

Mahato, K., Kumar, S., Srivastava, A., Maurya, P.K., Singh, R., Chandra, P., 2018. Electrochemical Immunosensors: fundamentals and applications in clinical diagnostics. Handook of Immunoassay Technologies 359–414.

Mahato, K., Maurya, P.K., Chandra, P., 2018. Fundamentals and commercial aspects of nanobiosensors in point-of-care clinical diagnostics. 3 Biotech 8, 149.

Mandal, R., Baranwal, A., Srivastava, A., Chandra, P., 2018. Evolving trends in bio/chemical sensors fabrication incorporating bimetallic nanoparticles. Biosensors and Bioelectronics 117, 546−561.

Manikkath, J., Hegde, A.R., Kalthur, G., Parekh, H.S., Mutalik, S., 2017. Influence of peptide dendrimers and sonophoresis on the transdermal delivery of ketoprofen. International Journal of Pharmaceutics 521, 110−119.

Manshian, B.B., Jimenez, J., Himmelreich, U., Soenen, J.S., 2017. Personalized medicine and follow-up of therapeutic delivery through exploitation of quantum dot toxicity. Biomaterials 127, 1−12.

Martin, M., Salazar, P., Villalonga, R., Campuzano, S., Pingarron, J.M., Gonzalez-Mora, J.L., 2014. Preparation of core-shell Fe_3O_4 @ poly (dopamine) magnetic nanoparticles for biosensor construction. Journal of Materials Chemistry B 2, 739−746.

Maynard, A.D., 2006. Nanotechnology: assessing the risks. Nano Today 1, 22−33.

Misra, S.K., Naz, S., Kondaiah, P., Bhattacharya, S., 2014. A cationic cholesterol based nanocarrier for the delivery of p53-EGFP-C3 plasmid to cancer cells. Biomaterials 35, 1334−1346.

Morille, M., Passirani, C., Vonarbourg, A., Clavreul, A., Benoit, J., 2008. Progress in developing cationic vectors for non-viral systemic gene therapy against cancer. Biomaterials 29, 3477−3496.

Mujokoro, B., Adabi, M., Sadroddiny, E., Adabi, M., Khosravani, M., 2016. Nano-structures mediated co-delivery of therapeutic agents for glioblastoma treatment: a review. Materials Science and Engineering: C 69, 1092−1102.

Müller, R.H., Mader, K., Gohla, S., 2000. Solid lipid nanoparticles (SLN) for controlled drug delivery-a review of the state of the art. European Journal of Pharmaceutics and Biopharmaceutics 50, 161−177.

Nie, S., Xing, Y., Kim, G.J., Simons, J.W., 2007. Nanotechnology applications in cancer. Nanotechnology Applications in Cancer Annual Review of Biomedical Engineering 9, 257−288.

Noh, H.-B., Lee, K.-S., Chandra, P., Won, M.-S., Shim, Y.-B., 2012. Application of a Cu−Co alloy dendrite on glucose and hydrogen peroxide sensors. Electrochemica Acta 61, 36−43.

Oh, J.K., Drumright, R., Siegwart, D.J., Matyjaszewski, K., 2008. The development of microgels/nanogels for drug delivery applications. Progress in Polymer Science 33, 448−477.

Patel, N.R., Piroyan, A., Ganta, S., Morse, A.B., Candiloro, K.M., Solon, A.L., Nack, A.H., Galati, C.A., Bora, C., Maglaty, M.A., O'brien, S.W., Litwin, S., Davis, B., Connolly, D.C., Coleman, T.P., 2018. In vitro and in vivo evaluation of a novel folate-targeted theranostic nanoemulsion of docetaxel for imaging and improved anticancer activity against ovarian cancers. Cancer Biology & Therapy 19, 554−564.

Patri, A.K., Majoros, I.S., Baker, J.R., 2002. Dendritic polymer macromolecular carriers for drug delivery. Current Opinion in Chemical Biology 6, 466−471.

Peng, H., Brimijoin, S., Hrabovska, A., Targosova, K., Krejci, E., Blake, T.A., Johnson, R.C., Masson, P., Lockridge, O., 2015. Comparison of 5 monoclonal antibodies for immunopurification of human butyrylcholinesterase on Dynabeads: KD values, binding pairs, and amino acid sequences. Chemico-Biological Interactions 240, 336−345.

Penn, A., Murphy, G., Barker, S., Henk, W., Penn, L., 2005. Combustion-derived ultrafine particles transport organic toxicants to target respiratory cells. Environmental Health Perspectives 113, 956.

Pérez-Herrero, E., Fernandez-Medarde, A., 2015. Advanced targeted therapies in cancer: drug nanocarriers, the future of chemotherapy. European Journal of Pharmaceutics and Biophermaceutics 93, 52−79.

Peters, J.T., Hutchinson, S.S., Lizana, N., Verma, I., Peppas, N.A., 2018. Synthesis and characterization of poly(N-isopropyl methacrylamide) core/shell nanogels for controlled release of chemotherapeutics. Chemical Engineering Journal 340, 58−65.

Ryman-Rasmussen, J.P., Riviere, J.E., Monterio-Riviere, N.A., 2006. Penetration of intact skin by quantum dots with diverse physicochemical properties. Toxicological Sciences 91, 159−165.

Saha, S., Ganguly, J., Pal, S., Ghosh, M., 2016. Influence of anisotropy and position-dependent effective mass on electro-optic effect of impurity doped quantum dots in presence of Gaussian white noise. Chemical Physics Letters 658, 254−258.

Sahoo, A.K., Verma, A., Pant, P., 2017. Nanoformulations for cancer therapy. Nanotechnology Applied to Pharmaceutical Technology 157−181.

Schumacher, C.M., Herrmann, I.K., Bubenhofer, S.B., Gschwind, S., Hirt, A.-M., Beck-Schimmer, B., Gunther, D., Stark, A.J., 2013. Quantitative recovery of magnetic nanoparticles from flowing blood: trace analysis and the role of magnetization. Advanced Functional Materials 23, 4888−4896.

Selvamuthukumar, S., Anandam, S., Kannan, K., Rajappan, M., 2012. Nanosponges: a novel class of drug delivery system-review. Journal of Pharmacy & Pharmaceutical Sciences 15, 103−111.

Shazly, G.A., 2017. Ciprofloxacin controlled-solid lipid nanoparticles: characterization, in vitro release, and antibacterial activity assessment. BioMed Research International 2017, 9.

Souto, E.B., Müller, R.H., 2008. Cosmetic features and applications of lipid nanoparticles (SLN®, NLC®). International Journal of Cosmetic Science 30, 157−165.

Su, X., Chan, C., Shi, J., Tsang, M.-K., Pan, Y., Cheng, C., Gerile, O., Yang, M., 2017. A graphene quantum dot@ Fe_3O_4@ SiO_2 based nanoprobe for drug delivery sensing and dual-modal fluorescence and MRI imaging in cancer cells. Biosensors and Bioelectronics 92, 489−495.

Sznitowska, M., Gajewska, M., Janicki, S., Radwanska, A., Lukowski, 2001. Bioavailability of diazepam from aqueous-organic solution, submicron emulsion and solid lipid nanoparticles after rectal administration in rabbits. European Journal of Pharmaceutics and Biopharmaceutics 522, 159–163.

Tian, B., Wang, C., Zhang, S., Feng, L., Liu, Z., 2011. Photothermally enhanced photodynamic therapy delivered by nanographene oxide. ACS Nano 5, 7000–7009.

Trotta, F., Cavalli, R., 2009. Characterization and applications of new hyper-cross-linked cyclodextrins. Composite Interfaces 16, 39–48.

Trotta, F., Zanetti, M., Cavalli, R., 2012. Cyclodextrin-based nanosponges as drug carriers. Beilstein Journal of Organic Chemistry 8, 2091–2099.

Tu, N., Eric, R., 2015. Reinvention of chemotherapy: drug conjugates and nanoparticles. Current Opinion in Oncology 27, 232–242.

Upendra, L., Minakshi, P., Brar, B., Guray, M., Ikbal, Ranjan, K., Bansal, N., Khurana, S.K., Manimegalai, J., 2016. Nanodiagnostics: a new frontier for veterinary and medical sciences. Journal of Experimental Biology and Agricultural Sciences 4, 307–320.

Vallhov, H., Qin, J., Johansson, S.M., Ahlborg, N., Muhammed, M.A., Scheynius, A., Gabrielsson, S., 2006. The importance of an endotoxin-free environment during the production of nanoparticles used in medical applications. Nano Letters 6, 1682–1686.

Ventola, C.L., 2012. The nanomedicine revolution: Part 1: emerging concepts. Pharmacology & Therapeutics 37, 512–525.

Vinogradov, S.V., Senanayake, T., 2013. Nanogel–drug conjugates: a step towards increasing the chemotherapeutic efficacy. Nanomedicine 8, 1229–1232.

Wagner, V., Dullaart, A., Bock, A.K., Zweck, A., 2006. The emerging nanomedicine landscape. Nature Biotechnology 24, 1211–1217.

Wang, F., Banerjee, D., Liu, Y., Chen, X., Liu, X., 2010. Upconversion nanoparticles in biological labeling, imaging and therapy. Analyst 135, 1839–1854.

Wang, A.Z., Langer, R., Farokhzad, O.C., 2012. Nanoparticle delivery of cancer drugs. Annual Review of Medicine 63, 185–198.

Wang, T., Zhou, Y., Lei, C., Lei, J., Yang, Z., 2013. Development of an ingenious method for determination of Dynabeads protein A based on a giant magnetoimpedance sensor. Sensors and Actuators B: Chemical 186, 727–733.

Wang, Y., Rajala, A., Rajala, V.S., 2015. Lipid nanoparticles for ocular gene delivery. Journal of Functional Biomaterials 6, 379–394.

Xia, T., Kovochich, M., Brant, J., Hotze, M., Sempf, J., Oberley, T., Sioutas, C., Yeh, J.I., Wiesner, M.R., Nel, A.E., 2006. Comparison of the abilities of ambient and manufactured nanoparticles to induce cellular toxicity according to an oxidative stress paradigm. Nano Letters 6, 1794–1807.

Yung, C.W., Fiering, J., Mueller, A.J., Ingber, D.E., 2009. Micromagnetic–microfluidic blood cleansing device. Lab on a Chip 9, 1171–1177.

Zhang, B., Wang, K., Si, J., Sui, M., Shen, Y., 2014. Charge-Reversal Polymers for Biodelivery. Wiley-VCH Verlag GmbH & Co. KGaA.

Zhang, Y., Wang, L., Yu, J., Yang, H., Pan, G., Miao, L., Song, Y., 2017. Three-dimensional macroporous carbon supported hierarchical ZnO-NiO nanosheets for electrochemical glucose sensing. Journal of Alloys and Compounds 698, 800–806.

Zheng, G., Patolsky, F., Cui, Y., Wang, W.U., Lieber, C.M., 2005. Multiplexed electrical detection of cancer markers with nanowire sensor arrays. Nature Biotechnology 23, 1294–1301.

Zhu, X., Wu, G., Lu, N., Yuan, X., Li, B., 2017. A miniaturized electrochemical toxicity biosensor based on graphene oxide quantum dots/carboxylated carbon nanotubes for assessment of priority pollutants. Journal of Hazardous Materials 324, 272–280.

Index

Note: Page numbers followed by "t" indicate tables and "f" indicate figures.

Transmission electron microscope, 113
Trimethyl chitosan, 65
Tumor necrosis factor-α (TNF-α), 105

U
Ulcerative colitis, 104
Urea biosensor, 82

W
Waveguide-based optical biosensors, 10

Y
Yellow Springs Instrument (YSI), 75

Z
Zinc oxide nanoparticles (ZnONPs), 22—23
antioxidant effect, 32
applications, 29—30
oxidative stress, 100—101
uses, 114—115

Printed in the United States
By Bookmasters